T0297840

CAMBRIDGE LIBRARY COLLECTION

Books of enduring scholarly value

History of Medicine

It is sobering to realise that as recently as the year in which On the Origin of Species was published, learned opinion was that diseases such as typhus and cholera were spread by a 'miasma', and suggestions that doctors should wash their hands before examining patients were greeted with mockery by the profession. The Cambridge Library Collection reissues milestone publications in the history of Western medicine as well as studies of other medical traditions. Its coverage ranges from Galen on anatomical procedures to Florence Nightingale's common-sense advice to nurses, and includes early research into genetics and mental health, colonial reports on tropical diseases, documents on public health and military medicine, and publications on spa culture and medicinal plants.

The Elements of Medical Chemistry

The physician and author John Ayrton Paris (1785–1856), several of whose other medical and popular works have been reissued in the Cambridge Library Collection, published the first edition of his *Pharmacologia* in 1812. It was immediately successful, and went into eight further editions until 1843. The third edition, of 1820, has been reissued in this series. This book, published in 1825, was intended as a companion volume, providing a 'grammar' of chemistry for the medical student. After an imaginary dialogue on the importance of chemistry, between a provincial physician and 'the author', to whom the former is entrusting his son for his medical education, the book moves systematically from the general application of chemistry to medicine, through topics such as gravity, crystallization and electricity, to the detail of the actions of specific elements, and tables of relevant weights and measures, providing fascinating insights into the history of medical education.

The Elements of Medical Chemistry

*Embracing Only Those Branches of Chemical Science
which are Calculated to Illustrate or Explain
the Different Objects of Medicine,
and to Furnish a Chemical Grammar
to the Author's Pharmacologia*

JOHN AYRTON PARIS

CAMBRIDGE
UNIVERSITY PRESS

University Printing House, Cambridge, CB2 8BS, United Kingdom

Cambridge University Press is part of the University of Cambridge.
It furthers the University's mission by disseminating knowledge in the pursuit of
education, learning and research at the highest international levels of excellence.

www.cambridge.org
Information on this title: www.cambridge.org/9781108069908

© in this compilation Cambridge University Press 2014

This edition first published 1825
This digitally printed version 2014

ISBN 978-1-108-06990-8 Paperback

This book reproduces the text of the original edition. The content and language reflect
the beliefs, practices and terminology of their time, and have not been updated.

Cambridge University Press wishes to make clear that the book, unless originally published
by Cambridge, is not being republished by, in association or collaboration with,
or with the endorsement or approval of, the original publisher or its successors in title.

THE ELEMENTS

OF

MEDICAL CHEMISTRY;

EMBRACING ONLY

THOSE BRANCHES OF CHEMICAL SCIENCE WHICH ARE CALCULATED
TO ILLUSTRATE OR EXPLAIN THE DIFFERENT OBJECTS
OF MEDICINE; AND TO FURNISH

A CHEMICAL GRAMMAR TO THE AUTHOR's PHARMACOLOGIA.

Illustrated by numerous Engravings on Wood.

BY

JOHN AYRTON PARIS, M.D. F.R.S. F.L.S.

Fellow of the Royal College of Physicians of London;

Honorary Member of the Board of Agriculture; Fellow of the Philosophical
Society of Cambridge; and of the Royal Medical Society of Edinburgh;
and late Senior Physician to the Westminster Hospital.

*" The objects of Science are so multiplied that it is high time to subdivide them.
Thus the numerous branches of an overgrown family in the Patriarchal ages found
it necessary to separate; and the convenience of the whole, and the strength and increase of each branch, were promoted by the separation."* PRIESTLEY.

LONDON:

PRINTED AND PUBLISHED BY W. PHILLIPS;
George Yard, Lombard Street;

SOLD ALSO BY

T. AND G. UNDERWOOD, FLEET STREET; W. AND C. TAIT, EDINBURGH;
AND HODGES AND M'ARTHUR, DUBLIN.

1825.

A DIALOGUE

*Between the Author, and a Practitioner who is about
to direct the Medical Studies of his Son.*

PRACTITIONER.

I trust, my good friend, that you will give me some
credit for the alacrity with which I have obeyed your
orders. My son has accompanied me to town, and we
are both impatient to avail ourselves of your friendly
advice.

AUTHOR.

The Medical Lectures will commence next week,
and you were prudent to get a few leisure days for the
preliminary arrangements; I need scarcely add that
my services are at your disposal, although I feel no
small degree of diffidence in offering advice to a practitioner who has been nearly thirty years in the active
exercise of the profession.

PRACTITIONER.

On the score of experience, I claim but one privi-
lege, or rather indulgence; to question a little more
freely, than might otherwise be permitted, the expediency of several of the prevailing opinions, without

a

incurring the penalty of Midas. The truth is that for the last twenty years I have been so absorbed in medical practice, that I have neither found leisure nor inclination to enquire into the improvements of medical education; and I therefore apply to you for such information as may supply my deficiency.

AUTHOR.

During the period you mention, general knowledge has made a rapid stride; and it would have been " passing strange" had not medical science participated in the advancement; but I am by no means satisfied that our system of *teaching* has been improved. It is true that amongst our Metropolitan lecturers, may be ranked some of the first philosophers of the age; but there are many competitors, some of whom, to maintain their ground, have introduced a system of " *grinding*," or " *cramming*" as it is technically called in our universities, which allures pupils, from the assistance it affords them in passing an examination at the College of Surgeons, or at Apothecaries' Hall, but is, in my humble judgment, ill calculated to impart solid information.

PRACTITIONER.

This remark is not new to me. I have often been astonished at the harlequinade dexterity with which a college license has transformed the new student into the sapient practitioner, and the humble scholar into the dogmatical lecturer.

AUTHOR.

The evils which arise from such a system are to be easily avoided. But before I proceed to offer you any advice upon this subject, you must inform me of the views you entertain respecting the future destiny of your son. Do you intend that he shall succeed you in country practice, or will you send him forth as a surgeon or physician. After the extraordinary success which has attended your townsman, I conclude that nothing short of the rank of a London Physician will bound your expectations.

PRACTITIONER.

The case, my dear Sir, to which you allude, must tend to extinguish rather than to encourage such ambition. I never dwell upon the extraordinary success of that man, but the Persian fable immediately intrudes itself upon my recollection. " A drop of water fell out of a cloud into the sea, and finding itself lost in such an immensity of fluid matter, broke out into the following reflection. ' Alas! what an insignificant creature am I in this prodigious ocean of waters; my existence is of no concern to the universe; I am reduced to a kind of nothing, and am the least of the works of God.' It so happened that an oyster which lay in the neighbourhood of this drop, chanced to gape and swallow it up in the midst of this its humble soliloquy. ' The drop,' says the fable, ' hardened in the shell, until by degrees it was ripened into a pearl, which falling into the hands of a diver, after a long

series of adventures, is at present that famous pearl
which is fixed on the top of the Persian diadem.'

AUTHOR.

From which I am, of course, to conclude that you
consider the worldly success of a physician to be alone
dependent upon accidental circumstances which no
wisdom can foresee, nor any prudence control.

PRACTITIONER.

Such, I believe, was the opinion of no less a person
than Dr. Samuel Johnson, whose judgment you will
scarcely venture to question.

AUTHOR.

I am ready to admit that " *Victory is not always
to the strong;*" and I am well acquainted with
instances in which superiority of success has been
united with inferiority of pretensions; but, as Dr.
Young has very justly remarked, whatever may be the
accidental irregularities inseparable from the operation
of moral causes, it must be admitted that every man's
chance of success in his profession will be in some
measure proportionate to his merits and his talents;—
in the lottery of physic, as in all other lotteries, the
chance of a man who holds ten tickets must be decidedly better than one who is possessed but of five.
The same argument will apply to the courteous address and behaviour of a practitioner; we have seen
the most unpolished and even indecorous manners
distinguish the most successful, but are we, on that

account, to argue that urbanity and kindness are of no avail? Dr. Radcliffe told Dr. Mead, as a great secret, that the true way to succeed in Physic was to use every body ill; but Dr. Mead used nobody ill, and succeeded better than Dr. Radcliffe.

PRACTITIONER.

Whatever opinion I may entertain upon this subject, it will not influence my determination to afford all the advantages of education to my son; " We cannot command success, but we will do more, Sempronius, we will deserve it." For the same reason, I do not think it necessary to come to any immediate decision respecting his future plan of life; for, however necessary it may be to separate the different departments of practice in a great city, no part of the elementary or preliminary studies should be neglected by a student of any description; thus prepared he can afterwards pursue any line of practice which existing circumstances may render most eligible. My son is now nineteen years of age, he has been grounded in classical knowledge, and possesses some general ideas upon the subjects of Natural Philosophy; with regard to Physic, with the exception perhaps of a little information on subjects of Pharmacy, he is entirely ignorant. Under these circumstances, how are we to proceed?

AUTHOR.

I should recommend Anatomy as the first object of his pursuit, a knowledge of this branch cannot be too early attained; enter him therefore at once with some good teacher, and after he has attended one course of lectures, let him commence dissection; at the same time, he should be admitted as a pupil at some Hospital, in order that he may become practically acquainted with the history and treatment of disease; his attendance also at lectures on the Theory and Practice of Physic will be necessary. Such a course of study will be amply sufficient for his first season. In the second year of his residence in London, he may attend to lectures on Surgery, and Midwifery; and on Chemistry, and the Materia Medica. The third year may be advantageously spent in Edinburgh, which is unquestionably one of the first medical schools in the world, greatly inferior, * however, to London, with regard to anatomy; and on that account, the first year of study will be better spent in London; and he will be thus enabled to receive a greater profit from the excellent clinical and practical lectures which are annually delivered in Edinburgh. During the Summer recesses, he will do well to cultivate a knowledge of Botany, and to enter upon a course of medical reading; he may, at the same time, amuse himself with the repetition of the various chemical experiments which he had witnessed during his attendance at lectures.

PRACTITIONER.

What is your opinion respecting the advantage of taking notes at the several lectures.

---

* In consequence of the difficulty of obtaining recent subjects. This embarrassment, however, now threatens the London schools.

AUTHOR.

No pupil should make the attempt during a first course; for, in the first instance, the subjects will be so novel, as to require the entire devotion of the student's mind; and if he attempt to reduce any part to writing, he will lose the thread of the discourse, and be unable to follow and comprehend the lecturer. When, however, he retires to his chamber, after the lecture, much advantage will arise by his making memoranda of such facts or opinions as may have appeared to him the most striking. In the second course, the practice of taking notes is of decided utility; but the use of short hand, on these occasions, is, in every way to be reprobated. It converts, says Dr. Young, the writer into a mere machine; it employs him in copying words, instead of digesting and compressing thoughts; and unless he has two or three more hours to bestow on the same subject after the lecture, which very few lectures are worth, his manuscript remains in a form almost as inconvenient for reference, as if it were written in an unknown language.

PRACTITIONER.

All these hints will be of great value to us.

AUTHOR.

I have nothing more, I believe, to add with respect to the London Lectures. If your object be to educate your son as a Physician, lose no time in admitting him at Cambridge or Oxford.

PRACTITIONER.

But these are not medical schools.

AUTHOR.

It is most extraordinary to me that this senseless cry should be so long continued. Is technical knowledge all that is required for the accomplished physician? do not the liberal pursuits, which are so successfully cultivated in those seats of learning, contribute to the elevation of the understanding, to say nothing of the gentlemanly manners and feelings which are thus acquired by intercourse with the most exalted characters of the age? Why is the rank of the regular physician so much higher in this than in any other country? because he must receive his education in the same school as that in which the nobles and statesmen of the land are instructed.

PRACTITIONER.

I thank you much for the sketch you have afforded me; I shall be guided by it, filling up the outline according to the circumstances and particular views which may present themselves. I am also, to speak candidly, not a little pleased to find that you have laid less stress upon the necessity of Chemistry, than I had anticipated.

AUTHOR.

Mistake me not.—I have certainly not assigned to it more importance than I conceived necessary; but you must distinctly understand that I consider no man capable of practising the profession of physic without a competent knowledge of it; and I am quite certain that, without its assistance, he can never discharge his duty to himself, or to his patient.

PRACTITIONER.

I have now been settled as a medical practitioner for upwards of five and twenty years; and I need scarcely observe that in my younger days Chemistry was scarcely regarded as a branch of medical study; my knowledge on this subject is necessarily, therefore, extremely imperfect; but I feel no hesitation in declaring, that in no single instance do I remember ever having felt an embarrassment at the bed side, or in the Surgery, from my deficiency; and I am strongly inclined to regard it rather in the light of an accomplishment, than in that of a necessary and indispensable qualification.

AUTHOR.

Your argument, if such it can be called, amounts to no more than this,—that you have never felt the want of that which you never knew.

PRACTITIONER.

If I have never known the advantages of Chemistry, I have, at least, experienced some of its evils; and I think you will admit, notwithstanding all your prejudices in its favour, that the Physiologist can have but few obligations to it; recollect only some of the numerous fallacies which have arisen from chemical theories.

AUTHOR.

If it be philosophical to argue against the utility of an instrument from a few instances of its abuse, I must certainly leave you in the undisturbed enjoyment of your assumed triumph; but remember the exclamation of Friar Lawrence—" *Virtue itself turns Vice, being misapplied.*"

PRACTITIONER.

You acknowledge then the extent of the evils which have flowed from your favourite science.

AUTHOR.

Most cheerfully. Nay, I go farther, I triumph in them. Pope has somewhere said that when a man owns himself to have been in error, he does but tell you that he is wiser than he was; and you must admit that it would be truly discouraging, if the Physiologist of the present day, with all the advantages of modern discovery, were unable to detect the errors of his predecessors.

PRACTITIONER.

You will scarcely venture to assert that the living power is not constantly opposed to chemical action.

AUTHOR.

And, for that reason, it is essential to learn the nature of chemical action, before we can attempt to appreciate the extent of that force which modifies or resists it. But there are changes perpetually going on in the animal body that are beyond the control of the living principle, and therefore the Physiologist, who is not a Chemist, will be utterly at a loss to comprehend them.

PRACTITIONER.

He will thus, at least, avoid error; and escape the danger of those false lights, which, like Will o' the Whisps, have ever led astray those who have pursued them.

AUTHOR.

Chemistry alone can enable you to avoid such false lights; the benighted traveller will best defeat the treachery of the Will o' the Whisp by carrying a lamp in his own hand.

PRACTITIONER.

And you are now endeavouring to seduce me from the safe path by the false glare of an allegory.

AUTHOR.

I will at once refute that charge by a plain practical illustration. I need not remind you that, for many years, the cause of all calculous disorders was referred to the impurities of water. This idea, so far from being the suggestion of chemical theory, arose, I believe, entirely from the occurrence of certain sediments observed in water, and supposed to resemble, in appearance, the calculi found in the human subject; and had not the Chemist refuted the error, the prejudice might still have continued, and the crust of the tea kettle have remained a source of perpetual terror. Is this no obligation conferred by Chemistry upon medicine?

PRACTITIONER.

I admit it. The fact is striking, and never occurred to me before.

AUTHOR.

Let us examine the subject a little farther. You must well know that old persons are extremely liable to be affected with calculous accumulations in the prostate gland. The Chemist examined their composition, and found them to consist of phosphate of lime combined with ammonia; by reasoning upon this fact, he was led to conclude that the matter was derived from the spontaneous decomposition of urine, which by yielding ammonia precipitated the phosphoric salt, and thus formed the concretion in question. The practice suggested by this discovery is obvious; to empty the bladder at intervals, and prevent the decomposition of urine in its cavity. The experiment succeeds, and much human misery has been prevented. Is this no obligation conferred upon Medicine by Chemistry?

PRACTITIONER.

I confess myself a convert upon these points; but I have another charge to bring against Chemistry, which, I fancy, you will not so easily refute. It has enabled the manufacturer to adulterate our drugs with greater art.

AUTHOR.

Then become a Chemist, and you will be enabled, by very simple means, to detect the fraud; in this case, Chemistry, like the spear of Telephus, carries upon its surface a balm for every wound it may inflict. Once for all, my good friend, I advise you to cease your hostilities; for like the Philistines of old your forges serve for no other purpose than to sharpen the defensive weapons of your adversaries.

PRACTITIONER.

I am content. I feel the great importance of the science from the few striking arguments you have used, and I am most anxious to impress the same feeling upon my son.

AUTHOR.

Before I quit the subject, I would say a few words regarding the impolicy of neglecting the study of Chemistry in the present enlightened age. This science, which during the last ten years has scarcely been known beyond the precincts of a small circle of philosophers, is rapidly extending itself through all ranks of the community; the higher classes of society are crouding to the lecture rooms of our public institutions, and the mechanic and artisan are becoming enlightened through the medium of chemical societies, and the extensive circulation of cheap publications. What would be the surprise of our fathers, could their spirits revisit us, to see a sectarian meeting house crouded to excess with mechanics, to hear a chemical lecture from the reading desk! In such a state of intellectual advancement, what will become of the medical practitioner, unless he keeps pace with the general progress? It is obvious that he must lose his place in the scale of society, and the credit and respectability of the profession must sink. The mere mechanic will be the better man, and he will look down with contempt upon the physician or apothecary, who being ignorant in Chemistry, must, as his reading will have iuformed him, be ignorant in the most essential parts of his avocation.

Practitioner.

Patients trouble themselves no more about the science of their doctors, than we do about the mathematical acquirements of our carpenters.

Author.

There is no analogy in the cases. The mechanic merely follows, by routine, the same work which has been performed by hundreds before him ; but the Physician, in every particular case, has a new problem to solve, for which he has to exert his reason, and to invent expedients for each exigency. He is not therefore to be compared to the working mechanic, but to the inventor of new machinery, and you will admit, I trust, that science is of some use to the Engineer.

Practitioner.

It is unnecessary to carry your arguments any farther, I have already confessed myself your convert, and am now anxiously waiting to learn the plan on which my son ought to commence the study.

Author.

Before he attends any lectures upon Chemistry, it will be advisable for him to read some elementary work, in order that he may gain a general notion of the nature and objects of the science. He may then attend a regular course of lectures, and he will derive the greatest advantage from repeating the experiments in his own chambers.

Practitioner.

Then he must procure apparatus, which will of course be attended with considerable expense; surely by attentively observing the experiments at the lectures this may be rendered unnecessary.

Author.

Without actual experiment, believe me, it is quite impossible for a student to acquire any solid knowledge of the science. " *Nihil est in intellectu quod non fuerit in sensu,*" was the motto which the celebrated Rouelle caused to be affixed in large characters in a conspicuous part of his laboratory, and I heartily concur in the justness of its application. With respect to the nature and extent of the apparatus which is required for the elucidation of philosophical principles there is a very general misconception. By means of a common wooden tub, a quantity of tobacco-pipes, and florence flasks, and a few dozen differently sized corks, with glasses and phials, I will undertake to illustrate all the leading facts in Chemistry.

Practitioner.

I am no less amused than gratified by the success with which you have combated my prejudices, and removed the difficulties which my ignorance of the sub-

ject had created. But you just now observed that
some elementary work should be attentively read before a student enters upon the lectures. May I request
you to direct our choice upon this occasion.

AUTHOR.

There are three works which every student, desirous
of becoming an accomplished Chemist, should possess.
Henry's Elements of Chemistry, Brande's Manual, and
Dr. Ure's Chemical Dictionary. As, however, no elementary work on Chemistry has hitherto been published for the exclusive use of the medical student, I
have undertaken to supply the deficiency; and in the
execution of this task, I have endeavoured to exclude
whatever appeared to me to have no direct application
to the profession; indeed the work is founded upon
the notes from which I formerly lectured; and as my
pupils were entirely medical, it was my care to collect
all the chemical facts of professional interest, to conduct the student to a knowledge of their principles by
the shortest path, and to remove from his road every
adventitious object that might obstruct his progress, or
unprofitably divert his attention.

PRACTITIONER.

You are, nevertheless, a bold man to add another
work on a science upon which so much has been already written.

AUTHOR.

It may be so. But you will say that I am still
bolder, when I tell you that it is my intention to in-

troduce the heads of the conversation we have just held together, in the place of a didactic Preface.

PRACTITIONER.

It will at least have novelty as its recommendation.

AUTHOR.

That I fear would not go far in disarming the wrath of the critic, unless it ensured advantages which the more usual style of a preface could not command.

PRACTITIONER.

But you will not surely venture to print your observations upon the present style of lecturing.

AUTHOR.

Why not ? " *Licet omnibus, licet etiam mihi, dignitatem ARTIS MEDICÆ tueri ; potestas modo veniendi in publicum sit, dicendi periculum non recuso.*"

PRACTITIONER.

But may they not give personal offence ?

AUTHOR.

I have no such intention. " *Quis rapiet ad se quod erit commune omnium ?* "

CONTENTS.

—✐✐✐—

PART I.

CRYSTALLIZATION

PART II.

ON ELEMENTARY BODIES, AND THE COM-
POUNDS WHICH RESULT FROM THEIR
COMBINATION WITH EACH OTHER

16. *Salts of platinum.*
MURIATE OF PLATINUM, 468.
SULPHATE OF PLATINUM.

PART III.

ELEMENTS OF CHEMICAL SCIENCE

AS IT RELATES TO THE DIFFERENT BRANCHES OF

MEDICINE.

———

1. Chemistry is that branch of Science which enables us to examine the constituent particles of bodies, with reference to their nature, proportions, and different modes of combination; and to investigate the laws of Attraction, Heat, and Electricity, by the operation of which these particles are perpetually undergoing change.

2. As the minute particles of matter are alone actuated by such forces, *Chemical* changes are not accompanied by *sensible* motions; a fact which enables us, very conveniently, to consider *Chemistry* as distinct from *Natural Philosophy*, for the phœnomena, which are conventionally referred to this latter province, are characterised by *apparent* motion. Thus, the science of Mechanics treats exclusively of the nature, production, and alteration of motion, and the doctrine of equilibrium; that of Hydrostatics, of the motions of fluids, and of the phœnomena which result from them; while Astronomy traces the motions of the heavenly bodies, determines their orbits, and measures. their velocity.

A

3. Nor are the results of these two classes of action less distinct from each other, than are the causes by which they are produced; that which attends the former being a change of properties, whereas that which distinguishes the latter is evidently, at most, a mere change of place; thus, if two highly caustic substances, as *Sulphuric acid*, and *Potass*, are made to act chemically upon each other, a *new* body is produced, which bears no analogy to either of the ingredients; and yet this change, great and striking as it may appear, is not accompanied by any motions which can be submitted to calculation or admeasurement.

4. But while we thus acknowledge the truth of such a distinction, for the sake of adopting a classification that may serve to subdivide the chain of human knowledge, which from the number and extent of its links can no longer retain its unity, it must nevertheless be acknowledged that, in treating either of the one, or the other, of these classes of phœnomena, we shall not unfrequently be compelled to exceed the exact boundaries of our definition, and even on some occasions to call in the aid of the one, for the full investigation or illustration of the other. Thus, for instance, on the very outset, as well as at the conclusion of many chemical inquiries, we shall require the use of the balance for estimating the absolute weight of the bodies submitted to examination; the theory and application of which, in their relations to Gravitation, must be considered as the exclusive object of Natural Philosophy; while, for the purpose of ascertaining the comparative densities, or *Specific Gravities*, of such substances, we shall require the assistance of Hydrostatics. In the same manner the various instruments employed in chemical experiments will be found to be either mechanical, hydrostatical, or hy-

draulical, in their principles of construction. * In the examination of the laws of chemical combination, we shall find that the same materials will, in general, unite in certain definite proportions, which cannot be changed, without producing corresponding changes in the characters of the compounds arising from their union—an investigation which involves the necessity of mechanical, or of mathematical knowledge; while the theory of Crystallization, and the examination of the various regular and determinate figures to which that process gives origin, require some acquaintance with the principles of geometry.

5. The same motives that suggested the advantage of separating Chemistry from the other branches of Philosophy, may now sanction the propriety of dividing Chemistry itself into as many ramifications as the distinct and prominent purposes to which it is subservient; for, when this science is discussed in accordance with its generally received definition, (1) it necessarily embraces a range of subjects so wide and diffusive, that the light which, if properly concentrated in a focus, might be thrown with so much effect, upon any particular branch of knowledge, is thus, as it were, enfeebled by the extent of its radiation. Such was the conviction that induced me to collate the chemical notes of my Lectures, and to present them in an embodied form to the medical student; and, as far as the intimate relation which subsists between the different parts of Chemistry would allow the separation, I have endeavoured to exclude whatever has not a direct

---

* As one of the objects of this work is to œconomise the labour of the medical student, and to supersede, as far as possible, the necessity of reference to a variety of books, I shall introduce as much Natural Philosophy as may be necessary for the elucidation of the principles upon which such instruments are constructed.

application to the study and practice of the profession.
Few circumstances have contributed more largely to
the general advancement of science than such a division
of labour, and I question whether to such a cause we
may not principally attribute that rapid progress of
philosophy which has so eminently distinguished the last
half century. Physiology is certainly much indebted
to an arrangement of this kind for its present extended
scale of improvement; for, although the highest importance had been attached to the study of the human
body from the earliest period, yet its functions were
never made a distinct and separate object of inquiry
until the beginning of the last century. It is true that
the writings of the ancient physicians, and of the earlier
among the moderns, abound in physiological speculations, but they are rarely brought forward in a connected or systematic form; so that we are obliged to
collect our knowledge of their tenets more from a number of scattered fragments, dispersed through works on
medicine and pathology, than from treatises expressly
devoted to the subject.*

6. The science of Chemistry may be said to be subservient to Medicine, in demonstrating the various
changes which occur in the component parts of the
animal body, under the different conditions of health
and disease; and in appreciating and explaining the
phœnomena accompanying such changes;—in inves-
tigating the composition and occasional deterioration
of the air we breathe, and of the various solid and fluid
substances which we employ as aliment or medicine;—
in suggesting processes of art by which natural bodies

---

* See " An Elementary System of Physiology, by John
Bostock, M.D. F.R.S., a work which I strenuously recommend
to the attention of the medical Student.

may be adapted, or new and artificial compounds produced, for administration as remedies;—in detecting the presence, and counteracting the effects, of various noxious substances which may, either from accident or design, become instrumental in impairing health, or in destroying life;—and lastly, in instructing the practitioner how he may best direct the admixture and combination of various remedies, without the risk of occasioning such changes in their properties, as may alter or invalidate their efficacy. It would appear therefore that the Physiologist, Pathologist, Physician, Toxicologist, and Pharmacologist, may each, in his turn, derive important information from the cultivation of Chemistry; and numerous examples, in illustration of this truth will occur to the reader during the progress of the present work.

7. The living body has been frequently compared to the Laboratory of the Chemist, in which those various compositions and decompositions are continually proceeding, upon which the phœnomena of its œconomy have been supposed to depend. It will be readily perceived that so general a proposition must involve many fatal fallacies, and their detection has accordingly furnished some specious arguments against the utility of chemistry, while the absurdities, into which its sanguine votaries have been betrayed, have too apparently sanctioned the propriety of the objection, "A reproach to a certain degree just," says Sir Humphry Davy, " has been thrown upon those doctrines known by the name of Chemical Physiology; for in the application of them, speculative philosophers have been guided rather by the analogies of words than of facts. Instead of endeavouring slowly to lift up the veil which conceals the wonderful phenomena of living nature, full of ardent imaginations, they have vainly and presump-

tuously attempted to tear it asunder." Before however, the philosopher can be fairly convinced of the
inaptitude of chemical reasoning to the objects of physiology, he must receive some arguments against the
use of this science, more potent than those which are
deduced only from its abuse. An objection, the most
specious perhaps that has ever been urged upon this
occasion, is founded upon the axiom, that animated
bodies are not only enabled to resist all the laws of
inanimate matter, but even to act on all around them
in a manner entirely contrary to those laws. To become satisfied of this universal truth, it is said that we
have only to consider these bodies in their active and
passive relations with the rest of nature; a subject
which has been beautifully illustrated by Cuvier in his
Lectures on Comparative Anatomy. " Let us," says
he, " contemplate a female in the prime of youth and
health. That elegant, voluptuous form,—that graceful
flexibility of motion,—that gentle warmth,—those
cheeks crimsoned with the roses of delight,—those
brilliant eyes, darting rays of love, or sparkling with
the fire of genius,—that countenance, enlivened by
sallies of wit, or animated by the glow of passion, seem
all united to form a most fascinating being. A moment
is sufficient to destroy the illusion. Motion and sense
often cease without any apparent cause. The body
loses its heat; the muscles become flat, and the angu-
lar prominences of the bones appear; the lustre of the
eye is gone; the cheeks and lips are livid. These,
however, are but preludes of changes still more horri-
ble. The flesh becomes successively blue, green, and
black. It attracts humidity, and while one portion
evaporates in infectious emanations, another dissolves
into a putrid sanies, which is also speedily dissipated. In short, after a few days there remains only a

small number of earthy and saline principles. The other elements are dispersed in air, and in water, to enter again into new combinations. It is evident that this separation is the natural effect of the action of the air, humidity, and heat;—in a word, of external matter upon the dead body; and that it has its cause in the elective attraction of those different agents for the elements of which the body is composed. That body, however, was equally surrounded by those agents while living, their affinities with its molecules were the same, and the latter would have yielded in the same manner during life, had not their cohesion been preserved by a power superior to that of those affinities, and which never ceased to act until the moment of death." It will not require much address to convert any objection against the utility of chemical knowledge, grounded upon such reasoning, into an argument in proof of its necessity and importance. If the energies of life are thus rendered manifest by the dominion which they exert over chemical and mechanical forces, it certainly behoves the physiologist to ascertain the nature and extent of such forces; for, without determining exactly the exceptions which an animated body enjoys over that which is inanimate, how can he expect to understand the nature of the power which governs it? as well might the mechanical philosopher attempt to measure the force of a machine without any estimate of the weight and nature of the materials which it is designed to actuate.

8. Nor must it be forgotten that in some of the functions of the living body, the vital energy would seem rather to correspond in its action with chemical affinity, than to oppose or supersede its influence; and several of the senses may be said to owe their energies to the perfection of organs which are entirely con-

structed upon philosophical principles. Thus are the
laws of optics and acoustics in active operation during
the exercise of the visual and auditory apparatus ; and
it is a question whether some chemical action is not
established by the agency of sapid bodies upon the
epidermis of the mucous membrane of the mouth; it
is, at least, seen evidently in some cases, as in the
effects of vinegar, the mineral acids, a great number of
salts, &c. By the same agents similar effects are produced upon dead bodies; and Dr. Majendie thinks
that to this species of combination the different kinds
of impression made by sapid bodies may be fairly attributed, as well as the variable duration of such impressions. Nor is it reasonable to deny that many of our
remedies may act by a chemical action on the alimentary canal; alkalies are thus frequently serviceable, by
clearing the *primæ viæ* of superfluous animal matter,
which they effect by forming with it a soluble compound. If the origin of animal heat cannot be satisfactorily traced to a strictly chemical source, its maintenance, distribution, and regulation may, at least, be
shewn to depend upon the agency of those laws which
alike govern the temperature of inert matter. Do we
not perceive that every animal, suffering from diminished temperature, instinctively diminishes the surface
of its body, which is in contact with the cooling medium? Man, under such circumstances, is seen to bend
the different parts of his limbs upon each other, and
to apply them forcibly to the trunk.* It will be also
seen, when we come to consider the nature of capillary
action, that many of the phœnomena of living bodies,
which have been erroneously attributed to the action

---

* Children and weak persons often take this position when in
bed. It would therefore be improper to confine young children
in swathing clothes, so as to prevent the necessary flexion.

of the living principle, may be satisfactorily explained
by the simple operation of this attractive force. The
absurdities of the chemical and mechanical sects have
undoubtedly driven the modern physiologist into a
mischievous scepticism with regard to the influence of
physical causes upon a living animal ; John Hunter
even, to associate whose name with error will be regarded by many as an act little short of impiety, has
repeatedly attributed to the specific effect of *life*,
actions that ought to be solely referred to the powers
belonging to inanimate matter. In the same manner,
an objection to impute to the physical property of
elasticity, certain phenomena exhibited by membranous
structure, has led Bordieu, * Bichat, † Blumenbach, ‡
and others, to assign to it a peculiar vital power whose
existence has neither been proved by experiment, nor
rendered probable by analogy.

9. But there is another point of view in which the
same question may be advantageously regarded ; the
pathologist will have to contemplate the living powers
in various states of languor and decay, when they will
be found incapable of wholly resisting the laws which
govern inanimate matter; and we shall learn, during
the progress of the present work, that in certain conditions of the human body, several of the fluids will
undergo the same chemical decompositions, as would
take place in the laboratory. The same observation
will apply to the agency of mechanical causes. In a
state of perfect health, the fluids of the body will not
descend to the inferior parts, agreeably to the law of
gravitation, because the vital power opposes itself to
this hydraulic phœnomenon, and with an energy, in

---

* Recherches sur le Tissu Muqueux, § 70.
† Traité des Membranes, p. 62, 101, 133.
‡ Institut. Physiolog. § 40, 59.

direct proportion, as it would seem, to the robust and
vigorous state of the individual; for, if the person be
reduced by disease, this tendency will be only imperfectly resisted; the feet in consequence will swell. The
following experiment of Richerand may be here related, to shew how greatly the power manifested in the
living body of resisting, with more or less success, the
influence of physical force, is enfeebled by disease. He
applied bags filled with very hot sand all along the leg
and foot of a man who had just undergone the operation for popliteal aneurism ; the artery was tied in two
places under the ham. Not only was the usual cold-
ness which follows an interruption of the circulation
thus prevented, but the extremity so managed acquired
a degree of heat much greater than the ordinary temperature of the body. The same apparatus, when
applied to a healthy limb, was unable to produce that
excess of caloric, obviously in consequence of the
energy of life opposing such an effect.

10. Mr. Earle has published an interesting paper *
to prove that, when a limb is deprived of its due share
of vitality, it is incapable of supporting any fixed temperature, and is peculiarly liable to partake of the heat
of surrounding media. The cases which are adduced
prove also that a member so circumstanced cannot,
without material injury, sustain a degree of heat which
would be perfectly harmless, or even agreeable, to a
healthy part; thus, the arm of a person became para-
lytic, in consequence of an injury of the axillary plexus
of nerves, from a fracture of the collar bone ; upon
keeping the limb, for nearly half an hour, in a tub of

---

* Cases and Observations, illustrating the influence of the
Nervous System, in regulating animal heat, by H. Earle, Esq.
Published in the 7th volume of the Transactions of the Medico-
Chirurgical Society.

warm grains ' which were previously ascertained by the other hand *not to be too hot*,' the whole hand became blistered in a most alarming manner, and sloughs formed at the extremities of the fingers. In a second case, the ulnar nerve had been divided by the surgeon, for the cure of a painful affection of the arm ; the consequence of which operation was, that the patient was incapable of washing in water at a temperature, that was quite harmless to every duly vitalized part, without suffering from vesication and sloughs.

11. It follows then that the PATHOLOGIST may frequently derive important conclusions from the doctrines of Chemistry, although we must deeply regret that this department of medical knowledge, like that of Physiology, should have suffered from too hasty generalization ; but let no one attempt to tolerate his apathy, or to encourage his despair, by a reference to failures which have arisen, on the one hand, from a deficiency of knowledge, and, on the other, from errors which are to be solely attributed to a perversion of it ; let him rather seek encouragement in the contemplation of those useful improvements which have been derived from the judicious application of chemical science, and which are daily augmenting the resources of the intelligent physician, and tending to diminish the aggregate of human suffering.

12. To the PHARMACOLOGIST the chemical history of the Materia Medica forms an indispensable subject of inquiry. It was a natural, and consequently an early conjecture, that substances which possessed an analogy to each other in their action on the living system, must bear a corresponding resemblance in their composition; whence chemical analysis was eagerly embraced as the means of obtaining a knowledge of their medicinal properties. There can be little doubt

of the justness of such a conclusion, as a general proposition, and although its application has frequently
failed, the failure must be rather attributed to our very
imperfect acquaintance with all the conditions of the
problem, than to any fallacy in the proposition itself.
Thus, in the earlier part of the seventeeth century we
find the Chemists universally engaged in the analysis
of the different vegetables used as remedies; many
hundred plants were accordingly, for this purpose,
submitted to examination, but not a single result was
obtained that could in any degree sanction the pretensions of this science to that practical utility which
theory had assigned to it; the most inert, and the most
virulent, vegetables were found to afford the same products.   That such a failure, however, was not attributable to the non-existence of those relations which
they had endeavoured to trace, is at once rendered
evident on comparing the successful results which have
been obtained, within the last few years, by pursuing
the same general principle of investigation by more
perfect, and less objectionable processes.   Had even
the experimentalists of the seventeenth century conducted their operations with all those essential precautions which it was impossible that the state of Chemistry at that period could have suggested, the manner in
which their analyses were performed was such as to
have precluded the chance of any useful result, for the
plants subjected to examination were indiscriminately
exposed to heat, and the products, so obtained by
their destruction, collected and rudely examined; now
it is quite clear that these products did not preexist in
the vegetable, but were formed by new combinations
of its elements; and, since these elements are in all
vegetables nearly the same, we cannot be surprised
that the experimentalists should have been incapable

of tracing the least connection between them and the qualities of the substance from which they were so obtained. With equal reason and success might they have attempted to appreciate the style of a literary production by ascertaining the letters of which its words were composed.

13. When we enter upon the subject of Vegetable Chemistry, the Student will learn that, by adopting a more refined method of experimenting, we have at length succeeded in obtaining the active principles of many of the most valuable vegetable remedies, and that the Chemist has been thus enabled to reconcile many anomalies, which were, previous to such discoveries, in apparent discordance with the conclusions of medical experience ; while, on the other hand, the practitioner has received from the laboratory an interesting confirmation of those views which he derived from observations made at the bed side; thus, it was stated in the first edition of my Pharmacologia,* that the result of the practice at the Westminster Hospital, in cases of Gout and Rheumatism, afforded considerable support to the conjecture of Mr. James Moore, respecting the composition of the *Eau Medicinale*, who considered the *Hellebore* as its active ingredient. As soon, however, as it was known that the *Colchicum autumnale* constituted its basis, the supposed error into which so many practitioners had been betrayed, furnished an ample theme for the pen of the cynic ; but the heavy charge under which we might have fallen for having carelessly observed, and ignorantly mistaken, the effects of such a medicine, has been completely cancelled by the discovery of the singular

* Published in the year 1812.

fact, that *Colchicum* and *Veratrum* alike owe their properties to one and the same alkaline principle.

14. To what extent our list of important remedies has been increased by the aid of Chemistry must become apparent on the slightest inspection of the Pharmacopœia; while the fatal blunders into which the practitioner who prescribes them may be betrayed from a deficient knowledge of the same branch of science, has been very fully, and I trust satisfactorily, pointed out in the various editions of my work on Pharmacology. The adulteration of Medicines is another subject for the comprehension of which a certain portion of chemical knowledge is essentially necessary; and, lastly, the investigation of Poisons derives most of its value from the accuracy and success with which this *master key* of science may be applied for disclosing the many obscure facts which are involved in the history of these agents. *

---

* Upon this subject the reader may consult our work on " Medical Jurisprudence," Vol. ii. *Art. Poisons.*

OF MATTER, AND ITS PROPERTIES.

15. Whatever is capable of acting upon our senses has been denominated MATTER.

16. This definition, however, is too general to be unexceptionable. Light affects our organs of vision, and Caloric and Electricity our sensations, and yet it is doubtful whether such effects are produced by the agency of distinct matter; they may be the result of certain forces, or of modifications of other bodies, or each may be an essence *sui generis*.

17. Matter may be more correctly defined to be that which is capable of occupying space, or which has the qualities of length, breadth, and thickness.

18. A BODY is any portion of matter.

19. Matter is said to possess certain essential or general properties, such as *Extension*, *Divisibility*, *Impenetrability*, *Porosity*, *Mobility*, and the power of *attracting*, or *being attracted*, to which may be added *Polarity*.

20. For this enumeration of properties we are indebted to the Natural Philosopher, who, unlike the Chemist, only regards *masses* of matter without a reference to its ultimate structure; it will accordingly

be found, that some confusion has arisen from an in-
discriminate application of the same terms to matter, in
a state of aggregation, as well as to the primitive materials of which such masses are composed. These difficulties will become apparent in the progress of the
enquiry, and will be successively combated as they
arise.

21. Besides the properties above enumerated, matter is said to possess certain *secondary*, or *contingent*
properties, such as *Elasticity, Fluidity*, &c. and which,
by their combination with the general properties, constitute the condition or state of bodies. It is by gaining or losing some of these secondary properties that
bodies change their state; thus, water may appear
under the form of ice, under that of a fluid, or of a
vapour, although it is always the same body.

22. When we speak of the *mechanical* properties of
matter, we mean those which are obvious upon any of
the mechanical operations of breaking, weighing, measuring, or the like, without any regard to the composition of the body under examination; thus, the general *mechanical* effects of water will be the same whether
it be taken from a river or a spring; and so of the
various gases or airs, although, if their composition be
examined, they will be found very different; such an
examination can only be effected by *chemical* means,
and the properties so discovered have therefore been
properly distinguished by the term *chemical* properties
of matter.

23. DIVISIBILITY is a property which belongs to
every substance which can be brought under the cognizance of our senses; but it by no means follows that
matter in its elementary state possesses it; indeed it is
more probable that at some term, however distant, the
resulting particles lapse into simple atoms incapable of

any further resolution.* If marble, or any brittle substance, be reduced to the most impalpable powder which art can produce, its original particles will not be bruised or affected; since, if this powder be examined by a microscope, each grain will be found a solid stone, similar in appearance to the block from whence it was broken; and, of course, if we possessed suitable implements, would admit of being again subdivided, or reduced to a still finer powder. To what extent this reduction might be carried before we arrived at the simple elementary atom, we shall probably never be able to conjecture, for the divisibility of matter, if not infinite, at least exceeds the utmost limits of our imagination. The marble steps of the great churches in Italy are worn by the incessant crawling of abject devotees; nay, the hands and feet of bronze statues are, in the lapse of ages, wasted away by the ardent kisses of innumerable pilgrims that resort to those shrines. What an evanescent pellicle of the metal, says Mr. Leslie,† must be abraded at each successive

---

* Amongst the different arguments which might be adduced in support of such an idea, none is more forcible than that which is afforded by Dr. Wollaston in his paper on " The finite extent of the Atmosphere." (Phil. Trans. 1822.) This distinguished philosopher considers the non-existence of a perceptible atmosphere around the Sun as a fact conclusive against this indefinite divisibility of matter; for, were it infinitely divisible, so also must be the extent of our atmosphere, which would thus pervade all space, and be gathered and condensed around the Sun, Moon, and Planets, in a proportion corresponding with the force of their respective attractions. The value of the argument cannot, however, be understood by the student, until he is made thoroughly acquainted with the nature and laws of gravitation, as explained in a future section of the work.

† Leslie's Elemen   Natural Philosophy.

contact! Thus again, a single grain of the *Sulphate of Copper* will communicate a fine azure tint to five gallons of water; in which case, the Copper must be, at least, attenuated ten million times, and yet each drop of the liquid may contain as many coloured particles distinguishable by our unassisted vision; and, if the experiment be extended by still farther dilution, so that the metal shall cease to be an object of sense, it may nevertheless be recognised by chemical tests. In the same manner, to what a most extraordinary degree of division are odorous bodies reducible? a single grain of Musk has been known to perfume a large room for the space of twenty years; at the very lowest computation, the Musk, in such a case, must have been subdivided into 320 quadrillions of particles, each of which was capable of affecting the olfactory organs. In like manner a lump of Assafœtida exposed to the open air, and filling the surrounding atmosphere with its effluvia, was found to have lost only a single grain in seven years.

23. Nor are we compelled to derive our illustrations of the almost infinite divisibility of matter from inanimate bodies only; the naturalist and physiologist will supply us with a multitude of striking examples from the vegetable and animal kingdoms. How extremely minute, for instance, must be the parts of the seed by which the peculiar plant constituting mouldiness, is propagated? for Reaumur found this production in the interior of an addled egg, whence the seeds must have passed through the pores of the shell! Mr. Leewenhoeck has informed us, that there are more animals in the putrefying milt of a cod-fish than there are men on the whole earth, and that a single grain of sand is larger than four millions of these creatures, so that thousands of them could be lifted on the point of

a needle, and yet each individual must be provided with a series of organs; of what inconceivable minuteness then must be the ultimate fibres of such organs ! But the infusory animalcules, as they are termed, display in their structure and functions, the most transcendent attenuation of matter. The *Vibrio Undula,* found in duck-weed, is computed to be ten thousand million times smaller than a hemp-seed. The *Vibrio Lincola* occurs in vegetable infusions, every drop of which contains myriads of those points. The *Monas Gelatinosa,* discovered in ditch-water, appears in the field of a microscope a mere atom endued with life, millions of which are seen playing, like the sun-beams, in a single drop of liquid.

24. The human structure likewise affords many wonderful instances of similar attenuation. The red globules of the blood have an irregular roundish shape, from the 2500th to the 3300th of an inch in diameter, with a dark spot in the centre of each. The globules of perspirable matter have been computed as, at least, ten times smaller than those of the blood, each being about the 5000th part of an inch in diameter, and when the quantity of perspirable matter which is daily discharged, and the number of pores through which it passes, are estimated,* it will follow, that no fewer

---

* It has been computed, that the skin is perforated by a thousand holes in the length of an inch. If we estimate the whole surface of the body of a middle sized man to be sixteen square feet it must contain not less than two millions, three hundred and four thousand pores. These pores are the mouths of so many excretory vessels, which perform the important function of Insensible Perspiration. The lungs discharge, every minute, six grains, and the surface of the skin, from three to twenty grains, the average over the whole body being about fifteen grains of lymph, consisting of water, with a very minute admixture of Salt, Acetic acid, and a trace of Iron. Leslie's Natural Philosophy.

than 400 of such globules must issue from each orifice every second.

25. POROSITY is a property belonging to all bo dies with which we are acquainted, and is so diffuse throughout them, that it is by no means an improbable supposition that their real bulk may bear no sensible pro-portion to the space which they appear to occupy. Upon the extent of this quality the relative densities of dif-ferent bodies depend; thus, for example, if it be sup-posed that a million particles of *Gold* are contained in a cubic inch of that metal, 500,000 particles of *Iron* might also be capable of occupying that same space, or 100,000 particles of *Wood*. In the iron and wood there must, therefore, be many more interstices, or pores, than in the gold, and, of course, the gold will be the heaviest, or most dense. This superior density and weight does not therefore arise from the individual atoms of gold being heavier than those of wood, but from a greater number of them being forced into the same space, for it is assumed, that the original particles of matter, although they may possess different forms and magnitudes, nevertheless possess the same relative weight or density.

This hypothesis is not unsupported by experiment. Gold, for instance, which is one of the heaviest of solids, may, by being dissolved in nitro-muriatic acid, and having its solution transferred to æther, be made to remain equally suspended in every part of the lightest of all visible fluids.

26. The term *Porosity*, then, can evidently be alone applied with propriety to aggregates; it neces-sarily implies division, and cannot therefore have any relation to atoms which are *indivisible*. By pursuing this subject farther than may be necessary to correct popular fallacy, we should only merge into metaphy-

sical speculations, which are generally little more than unsuccessful efforts to extend the boundaries of human knowledge beyond the reach of the human faculties.

27. The existence of porosity in every species of matter which can be subjected to our senses, is sufficiently proved by the universal compressibility of bodies. There is no substance, however dense, that may not be made either by pressure, or reduction of temperature, to occupy less space; and, were it possible to bring the ultimate atoms into absolute contact, the globe itself might probably be compressed into an extremely narrow compass.

28. Nor is the arrangement of the atoms of matter which is thus indicated by its porosity, less important than its universality is obvious. It is clear that, if the constituent particles had not been so disposed in relation to each other, as to have allowed free latitude of motion, natural bodies could never have undergone those changes in form and composition, upon which their utility in the scheme of creation entirely depends. It becomes a question even, whether they could have been susceptible of change of temperature, for, if we regard Caloric as material in its nature, there would, in such a case, have been no space to have allowed its ingress; and, if we consider it as a species of vibration, it is equally evident that the atoms without free motion could never have vibrated, for the act of vibration necessarily implies change of place.

29. IMPENETRABILITY. The existence of this property necessarily follows from the incontrovertible fact, that no two bodies can occupy the same place in the same precise instant of time.

30. The atoms of matter, whether they be so arranged as to constitute gases, liquids, or solids, are impenetrably hard; a property which Nature appears

to have wisely adopted, in order to insure eternity to
her works, and to render them incapable of waste or
decay; for, although to the superficial observer, matter
may in many instances seem to disappear, as in the
cases of burning and evaporation, yet the Chemist's art
distinctly proves that it is incapable of annihilation,
and that the original atoms, in all cases, still exist,
although by change of arrangement they are made to
assume different states.

31. Although the quality of impenetrability is thus
shewn to be inseparable from the idea of an elementary
atom, it nevertheless cannot be said to relate to aggregates or masses; for a solid or fluid of certain dimensions may be incorporated with other bodies, without
any change in its magnitude, but in this case, the
interstices admit the new matter, and a corresponding
increase of density arises. This phenomenon perpetually occurs in the processes of chemistry, and its
nature should be distinctly understood by the student;
it has been termed a *penetration of dimensions.* Thus,
for instance, if a pint of water and a pint of oil of
vitriol be mixed together, the mixture will not measure
a quart. The density of the compound is manifestly
increased by this circumstance; and it will be increased
in the same proportion as the bulk is diminished. In
like manner two equal quantities of water and strong
spirit, when mixed, will not produce double the first
quantity, though their joint or separate weights are not
disturbed. The phenomenon is one of considerable
importance, for it presents, as we shall hereafter find,
a considerable practical difficulty in obtaining correct
results in the application of the hydrometer. The
rationale of the phenomenon will be more readily understood from the annexed diagram.

A   B

O O O O O   + + + + +

O O O O O   + + + + +

O O O O O   + + + + +

O O O O O   + + + + +

C

+ O + O + O + O + O

O + O + O + O + O +

+ O + O + O + O + O

O + O + O + O + O +

If we suppose A and B to represent the atoms of water and alcohol, respectively, and C, their mixture, it will be seen that, in consequence of their greater approximation, the dimensions of the mixture will be less than the sum of those of the two ingredients when separate.

32. POLARITY, or a power of arrangement, must evidently be admitted as a property with which the atoms of matter are endowed. Each body would thus appear to possess a structure peculiar to itself; for example, a piece of iron, tin, or any other metal, or substance, will, when broken, always exhibit the same arrangement and disposition of parts, or *Grain*, as it is sometimes called. Upon this principle also are the phœnomena of Crystallization to be wholly explained. To the same property, in conjunction with the relative force of cohesion, we may refer that variety distinguishable in bodies with respect to their hardness, softness, ductility, elasticity, fragility, &c. We may imagine

for instance, that when the particles of bodies are disposed without any order, they cannot afford a strong resistance to a motion in any direction; but that, when they are regularly arranged in certain positions with respect to each other, any change of form must displace them in such a manner as to increase the distance of a whole rank at once, and hence, that they may be enabled to co-operate in resisting such a change; just as a disciplined regiment is capable of affording more resistance than an unorganised mob.

Before we proceed to the consideration of the elementary constitution of different bodies, it will be necessary for the student to become thoroughly acquainted with the effects of that universal force, to which all matter is subservient, and which is expressed by the term ATTRACTION.

ATTRACTION.

33. *Attraction* may be considered under several modifications, not as being in itself a variable and inconstant power, but as manifesting different phœnomena, in relation to the circumstances and conditions under which it operates. The nature and origin of this force are entirely unknown to us, but some of its more important laws have been discovered, and successfully applied in the investigation of the phœnomena of the Universe.

34. The following Tabular arrangement exhibits the different species of Attraction, which it will become our duty to investigate, *viz.*

ATTRACTION.

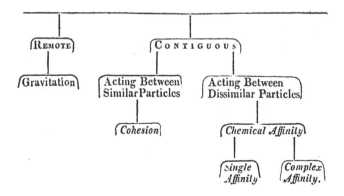

35. ATTRACTION may be defined, that force which causes distant bodies to approach each other, while it preserves those which are contiguous, in apparent contact.

36. It may therefore be considered as acting between large masses of matter at *sensible* distances, when it is termed the ATTRACTION OF GRAVITATION; or, as operating between the minute atoms of bodies, at *insensible* distances, when it is donominated CONTIGUOUS ATTRACTION. This latter force, moreover, admits of a further distinction, as it may act between *similar* or *dissimilar* particles of matter. In the former case, as for instance, where it operates between the particles of marble, it is termed the ATTRACTION OF AGGREGATION, or COHESION, since the effect of it is to produce an aggregate or mass; but, in the latter case, where it' brings together particles perfectly dissimilar in their nature, as those of an acid and an alkali, or an earth, it is termed CHEMICAL AFFINITY, and the result is a new body. It will be neecssary to consider the laws and phenomena of these forces with attention, for they constitute the basis of all our knowledge respecting the composition and affections of matter.

GRAVITATION.

37. This force acts on bodies remotely situated with respect to each other, *directly* in proportion to the quantity of matter, and *inversely* as the square of the distance. It is the operation of this power which maintains the moon in her orbit, and upholds the circulation of the whole system of planets around the sun, and which causes every body upon our globe to fall towards its centre, or in other words, in a line perpendicular to its surface.

38. Since the force of Gravitation is in proportion to the quantity of matter, it necessarily follows from the globular form of the earth, that it will act in the direction of a line passing through its centre; for the longest line that can be drawn is a diameter, and which must therefore pass through the greatest quantity of matter contained in any one direction, as the annexed diagram may more clearly illustrate.

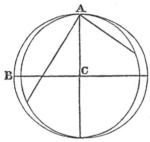

39. As the gravity of any mass is only the sum of that of its particles, it may be employed as the expression of quantity of matter, whence the theory of the balance becomes obvious.* Since, however, it has been just stated, that matter gravitates with a force which diminishes as the squares of the distance, it is evident that the weight of a body cannot be the same in all places and situations. It is stated by Professor Leslie that a lump of lead, which weighs a thousand pounds at the surface of our globe, would lose two pounds, as indicated by a spiral spring, if carried to the top of a mountain four miles high ; and, if it could be conveyed as deep into the bowels of the earth, it would lose one pound. The same mass transported from Edinburgh to the Pole would gain the addition of three pounds; but, if taken to the Equator, it would suffer a loss of four pounds and a quarter, since from the ellipticity of the earth, it would in such a situation

---

* *Weight* and *Gravity* are generally considered as one and the same thing; although some philosophers distinguish the latter as the quality inherent in the body, and the former as the same quality exerting itself according to its natural tendency.

be farther removed from the general mass that influences it.*

40.   As it appears, then, that the quantity of the bodies we employ in our various operations is most accurately determined by their force of gravitation, or *weight,* it becomes necessary to adopt some standard of comparison, by which such relations may be expressed; but as this has ever been derived from some initial quantity of arbitrary value, we find the weights and measures of different countries and ages variable and precarious.   At present we have two kinds of weight in common use; viz. *Troy,* and *Avoirdupois.* The former, which is of Norman origin, is used in the valuation of gold and silver, and in the composition of medicines.   The latter, which appears to have been introduced by the Romans, is the one universally employed in this country in every species of merchandize. The following TABLE exhibits the manner in which the Troy or Apothecaries pound, and the Avoirdupois pound are divided.

TROY WEIGHT.

Pound.	Ounces.	Drachms.	Scruples.	Grains.
1 =	12 =	96 =	228 =	5760
	1 =	8 =	24 =	480
		1 =	3 =	60
			1 =	20

AVOIRDUPOIS WEIGHT.

Pound.	Ounces.	Drachms.	Grains.
1 =	16 =	256 =	7000
	1 —	16 =	437·5
		1 =	27·975·

---

* The centrifugal force may also contribute something to this effect.

The Pound is usually expressed by the sign lb., and its subdivisions by the following symbols, viz.

The annexed figure represents an ounce; by omitting the upper part and commencing at A, a drachm; and the lower portion B. C. a scruple. A grain is expressed by the simple contraction gr.

The use of these symbols in medical prescriptions has been frequently condemned as being liable to occasion error; it is said that the characters for ounce and drachm are so similar that they may be easily written for each other; " a stroke of the pen too much may kill the patient, and a stroke too little may produce a medicine of no efficacy." To obviate this objection, it has been recommended to employ the contracted words *Unc.*, *dr.*, *scr.* : but let me ask whether such contractions are not equally liable to perversion in the hands of the careless prescriber? In support of this belief it is only necessary to state, that in the very work in which this fastidious objection is urged, *drs.* are more than once *printed* by mistake instead of *grs.*

41. In Chemical experiments, it will be found very convenient to admit no more than one description of weight. The grain is of such magnitude as to deserve the preference. It has been a mathematical problem to ascertain the least number of such weights which may be necessary for chemical purposes; but Dr. Ure very justly observes that the operator ought rather to seek the most convenient number for ascertaining his inquiries with accuracy and expedition. The error of adjustment is the least possible, when only one weight is in the scale; that is, a single weight of five grains is

twice as likely to be true, as two weights, one of three, and the other of two grains, put into the dish to supply the place of the single five; because each of these last has its own probability of error in adjustment. The most convenient set of weights for the laboratory is one that corresponds with our numerical system, thus 1000 grains, 900 grains, 800 grains, 700 grains, &c. down to $\frac{5}{10}$ of a grain. With these the Chemist will always have the same number of weights in his scales as there are figures in the number expressing the weights in grains; thus 742·5 grains will be weighed by the weights 700, 40, 2, and 5-10ths.

42. It must appear evident that all such standards of weights are merely arbitrary, and that were they, from any accident, to be lost,* it would be impossible to renew them; for since the foundation of the whole series was originally the weight of a certain number of grains of wheat,† it would be impossible to recover the initial quantity with such accuracy that the variation, however small, should not become fatal by repeated multiplication. In order to form correct and uniform standards of weight, it will be readily per-

---

* To obviate, as far as possible, the chance of such an occurrence, an ordinance was made in the reign of Henry VIII, that a set of the most esteemed weights and measures should be collected and lodged in the Exchequer, as standards for the whole country to abide by, and with which all future weights and measures should be compared and examined. This regulation has been preserved and attended to ever since, and duplicates of the standards have been very accurately formed, and are deposited at the Tower, with the Royal Society, and in other places of security.

† By an act of HenryIII. cap. 51, it is ordained that, in order to regulate the weights of the realm, that quantity of metal which will balance 32 grains of dry wheat picked from the midst of the ear, shall be called a penny-weight, that 20 such penny-weights shall make an ounce, and 12 ounces a pound.

ceived that we must recur to measure, for it is quite impossible to form any body into a weight without reference to its dimensions. The terms *Pound* and *Ounce* carry with them no specific ideas of their extent of weight to persons previously unacquainted with them; but, if it is stated, that four cubic inches of cast iron of a particular density would be equal in weight to a pound, or that a cubic foot of distilled, or pure rain water, weighs 1000 ounces, then such weights may be immediately formed by any one in possession of measures, and of course a standard lineal measure becomes an object of the greatest importance. In the earlier æra of England primitive measures were determined either from vegetable productions, or from parts or actions of the human body, * all of which were necessarily vague and inaccurate. In order to obtain an invariable and natural standard of lineal measure, two proposals have been offered; the one founded upon the mensuration of a degree of a great circle of the earth; the other, upon that of the length of the pendulum vibrating seconds. The former, although attended with much trouble and expense, has been accomplished and carried into practical effect by the French, and as they have founded upon it a scale which is frequently referred to in chemical works, a

---

* Thus was our inch derived from three barley corns laid end to end. The Saxons introduced a measure called the *Gyrd*, which corresponds with our yard, as a proper measure of unity, of which, fathoms, furlongs, and miles were made the multiples, while the foot, the span, the palm, and the inch might be considered as fractional parts of it. But in all such measures no certainty existed, since the yard was determined by Henry I. to be the length of his own arm, while the foot, the cubit, the ulna, or ell, the palm, the span, the hand, and many others, are evidently derived from the dimensions of the human body.

short account of the method will not be unacceptable to the student. After a series of most laborious and accurate observations carried on for many years by the first mathematicians of France, it was ascertained, that a quadrant of a meridian, extending from the pole to the equator, measures 5130740 toises, the ten millionth part of which was afterwards definitely decreed by the legislative body to be the *Metre*, or standard of unity, upon which they were to form all their other measures, whether greater or less; this Metre accords very nearly with the length of a pendulum vibrating seconds, as hereafter described, and may be considered as the present yard of the French, and as it ascends decimally, the next step, or degree, becomes the perch or *decametre*; the next, the mile, or *kilometre*; and then the *myriametre* or league; but the value of these several measures, as compared with those used in England, will be more clearly understood by referring to the Table inserted in the Appendix, and which was originally drawn up by Mr. Millington, and will be found to contain particulars not generally met with in Arithmetical Tables.

43. The method of using a Pendulum to obtain a standard of measurement is very simple; for it has been ascertained, that the stars perform an apparent journey round the earth once in 23 hours 56 minutes, with the greatest regularity. If then a small telescope be firmly and immoveably fixed against a wall in such a direction that any bright star may be seen through it, that star will pass the telescope once in that period, and if a clock be placed near it, having a pendulum beating seconds, that clock will indicate the above portion of time between every transit of the star, provided its pendulum be of the right length, and if not, it must be lengthened or shortened until such an ad-

justment is accomplished, which, in the same latitude can only occur when it is of one particular length. Thus then we may at once obtain a well defined standard of length, to which we could at any time resort, and which might be made the *Metre*, or base upon which other measures would be constructed; and having obtained an accurate standard of lineal admeasurement, solid measures, or measures of capacity, as well as weights, would arise out of it; and, as pure rain water, under equal temperatures, is less liable to a change of density than any other known substance, so it appears to be the best calculated for obtaining standard weights; thus, a cubic foot of pure water weighs about 1000 ounces avoirdupois, and either this measure, or the cube of the length of the pendulum, or an aliquot part of it, might be taken as the standard or base upon which to form larger and smaller weights, the same being whole, and not fractional parts of the first quantity, and taken at a certain point of the barometer and thermometer. In this, or a similar manner, might a series of measures and weights be established, which would be in the power of any one to adjust or examine with an apparatus of small expense, and without any serious loss of time, while the weights and measures we at present possess and use, are so very uncertain as to afford no greater proof of their accuracy, than the reliance which is placed on the correctness of their makers, unless, indeed, they have undergone the ordeal of a comparison with the national standards, which can never be expected in the great number that are made and sold.

44. The troy weight has been also adopted by the Edinburgh College for apportioning liquids, as well as solids, with a view to obviate the errors arising from the promiscuous use of weights and measures: but the

London and Dublin Colleges, with a great regard to practical convenience, order liquids to be measured; and for this purpose the London College employs measures derived from the wine gallon, which is subdivided for medical purposes, in the manner exhibited in the following TABLE, which, at the same time, represents the symbols employed for denoting the several measures.

A Gallon (*Congius*) cong : .... ⎫   ⎧ Eight Pints
A Pint (*Octarius*) O.............. ⎪   ⎪ Sixteen Fluid-ounces
A Fluid-ounce (*Fluid-uncia*) f ℥ .... ⎬ contains ⎨ Eight Fluid-drachms
A Fluid-drachm (*Fluid-drachma*) f ʒ ⎪   ⎪ Sixty Minims
A Minim (*Minima*) m.............. ⎭   ⎩ ——————

Table of the proportions of the Wine Gallon.

Gallon.	Pints.	Fluid-ounces.	Fluid-drachms.	Of Water.	
				Minims.	*Grains.†*
1 =	8 =	128 =	1024 =	61440 =	58016
	1 =	16 =	128 =	7680 =	7272
		1 =	8 =	480 =	454·5
			1 =	60 =	56·8

The London College have introduced the *Minim* measure as a substitute for the drop, the inaccuracy of which had been long experienced, and will be hereafter adverted to.

45. For measuring fluids, the graduated glass measures should be always preferred : they should be of different sizes, according to the quantities they are intended to measure. For chemical purposes fluids are measured either by cubic inches, or by ounce measures equal to the bulk of an ounce of water. The cubic inch is found to weigh 252·72 grains of water at 62°.

———————————

† According to Lane's Measures, sold at Apothecaries' Hall.

46. Although the method of ascertaining quantities of matter by the balance was known, and practised in the very earliest periods of the world, still such an instrument was for many ages merely employed as the measure of gross weight, without any reference to the density of the bodies so examined; and yet it is difficult to conceive that the obvious fact could have long escaped observation, that equal bulks of different substances, such, for instance, as wood and stone, differed essentially from each other in weight.* No application, however, was made of it, towards elucidating the properties of bodies, nor were any means devised to estimate or express such differences, until a comparatively late period.

---

* It may be here repeated, that it is not meant that the *elementary* particles of these bodies differ in weight.

OF SPECIFIC GRAVITY.

47. The comparative weight of a body has been termed its *specific gravity*; and, since the density of a body is as the quantity of matter contained in a given space, its *specific gravity* may be regarded as only another term for its density.

48. By this term, therefore, is meant the weight of a body compared with that of another whose magnitude is the same; and, for the accurate expression of such a relative quantity it became necessary to fix upon some substance as a standard. The philosophers of different nations have accordingly agreed to consider Distilled Water, at the temperature of 60° of Fahrenheit's thermometer, as the unit of comparison, or the datum from which all calculations of specific gravity should proceed, and is always called 1·000; thus, if a cubic inch of any solid body were found to be double the weight of a cubic inch of water, such a body would be specifically heavier than water, in the proportion of two to one, and its specific gravity would accordingly be set down thus, 2. If, again, its weight were equal to that of two and a half cubic inches of water, it would be specifically heavier than water in the proportion of two and a half, to one; and its specific gravity would in that case be set down as 2·5;

the fractional parts being always expressed by deci-
mals. The strongest Sulphuric acid of commerce is
very nearly nine-tenths specifically heavier than water,
and is stated to be 1·85.

49. Independent of the general advantages which
attend the adoption of water as the unit of comparison
upon these occasions, it fortunately happens that a
cubic foot of this fluid weighs exactly 1000 ounces,
avoirdupois, so that the weight of any given bulk of
material, of which the specific gravity is known, may
be readily computed, for the same figures that denote
the specific gravity must also express the number of
ounces in a cubic foot of the same substance; thus
will that measure of Sulphuric acid be found to weigh
1650 ounces.

50. Since it appears then, that specific gravity is
but another term to express the comparative density of
bodies; and, as the nature and properties of a great
variety of solids, as well as fluids, have an intimate
relation with that condition of matter, it is evident
that, in the examination of natural and artificial substances, a knowledge of their specific gravities is of
very considerable importance. Suppose the medical
practitioner to have received a parcel of *Glass of Antimony*, and that from the clumsy appearance of some
of the pieces, he suspects it to have been mixed with
Glass of Lead, he has only to ascertain their specific
gravities in order to confirm or falsify his conjecture,
for that of the Antimonial oxide never exceeds 4·95,
whereas that of the Lead is 6·95, or in round numbers
their comparative gravities are as 5 to 7. Thus again,
the officinal Solution of Potass (*Liquor Potassæ*),
may be obtained in too dilute a state, a fraud which
may be at once discovered by its specific gravity being
less than 1·056. On some occasions we may thus

arrive at a conclusion which might otherwise require
an elaborate chemical inquiry; thus we may fairly
infer that the *Tinctura Ferri Muriatis* contains its
proper proportion of Peroxide of Iron, by ascertaining
its specific gravity to be about ·994. In estimating the
strength of our spirituous preparations, the purity of
ammonia, and the concentration of the mineral acids,
the chemist must rely upon the respective indications
which are afforded by this mode of investigation. The
manufacturers of Epsom, Rochelle, Glauber, and other
salts, have no proper criterion for ascertaining the
strength of the several liquors which they have prepared for crystallization, but that of their specific
gravities.

51. Nor is this subject of less importance to the
physician than to the manufacturer, for in prescribing
various forms of medicine, an attention to the specific
gravities of the ingredients is necessary to direct him;
thus, for instance, in regulating the size of pills, unless
he takes this circumstance into consideration, he will
be unable to adjust their magnitude; for although he
may conveniently direct six or eight grains of mercury
in a single pill, the same weight of soap would exceed
the standard dimension. In like manner, when powders are added to liquid vehicles, their specific gravities
must be duly attended to, as I have already explained
in my PHARMACOLOGIA.

52. During the various changes which bodies undergo by combination with each other, a diminution
or increase in their specific gravities is not the least
remarkable or important; in some cases they are diminished, and in others increased, under such circumstances; it has, for instance, been already stated (31)
that alcohol and water, on admixture, will have a
volume less than the sum of their respective volumes;

and it will hereafter appear that the extent to which these changes take place through the agency of heat, constitutes a very important branch of inquiry.

53. Various organs of the human body also undergo alterations in structure, either by disease, or natural developement, by which their absolute weight is increased, at the same time that their specific gravities are diminished; and vice versa. To the Physiologist such changes are fraught with interest; thus, the pulmonary organs before birth are of such density as to sink in water, but no sooner has the function of respiration been established, than the lungs become so inflated that they are enabled to float in that medium, while their *absolute* weight, so far from suffering a corresponding diminution, is nearly doubled by the influx of new blood into their numerous vessels. * In certain diseases, on the contrary, the same organs lose a considerable portion of their weight, while they acquire an increase in density. It seems probable, also, that certain parts of the body increase in density, in proportion to the age of the subject, and were accurate tables formed so as to express such increments, some results of great physiological interest might possibly be elicited. The knowledge of the specific gravities of our fluids, under different circumstances of health and disease, would likewise be a desideratum in this branch of science.

54. In order to ascertain the specific gravity of any substance, it is evident from the definition (48) that we have only to estimate the gross weight of a given bulk, and to compare it with that of a similar bulk of any body which we may adopt as the unit of comparison;

---

* See our work on MEDICAL JURISPRUDENCE, Vol. iii. p. 109.

thus, if a cube of water weighs an ounce, and a similar cube of sulphur is found to weigh two ounces, the specific gravity of the latter, when compared with that of the former, is as 2 to 1. This was the method actually pursued by Lord Bacon, in the construction of his Table.* Having first formed a perfect cube of pure gold of an ounce weight, he caused cubes, of the same size, to be made of various other materials, when, by weighing these against each other, he at once ascertained their specific gravities. It is sufficiently evident, however, that this method, although perfectly correct in theory, must be attended with such inconvenience and difficulty as to be almost impracticable, owing to the trouble of forming solids of the same exact dimensions; while, in order to obtain the comparative weight of an irregular piece of metal and its equal bulk of water, it would be necessary to have an exact measure of its capacity to be filled a certain number of times, which would be difficult, or even impossible to effect with precision and certainty. The same difficulty will not oppose itself in the examination of liquids and certain powders, and it will accordingly appear hereafter that a method, founded upon this principle, is the one more generally adopted on such occasions.

1. METHODS OF ASCERTAINING THE SPECIFIC GRAVITIES OF SOLID BOBIES.

1. *The Sp. Grav. of the body being greater than that of water.*

55. It is a proposition in Hydrostatics that *the weight which a body loses when wholly immersed in a fluid,*

---

* The oldest Table of Specific Gravities now extant, and which may be seen in his Tract entitled " Historia Densi et Rari," printed in the second volume of his works. Folio, 1741.

is equal to the weight of an equal bulk of that fluid. This at once affords a simple and easy method of taking the specific gravity of any solid body, by weighing it, as it is termed, *Hydrostatically,* that is, by comparing the difference which there is in its weight when in air and in water, and dividing the absolute weight by the loss, when the quotient will be the specific gravity.

56. When we say that a body loses part of its weight in a fluid, we do not mean that its absolute weight is less than it was before, but that it is partly sustained by the reaction of the fluid under it, so that it requires a less power to support it.

57. The instrument by which the operation may be most conveniently performed is termed the HYDRO-STATIC BALANCE.

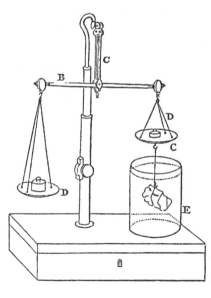

B, C, D, is a balance ;—E, a glass jar about six inches in height, which contains distilled or rain water. The mode of using this instrument is as follows,—Let the solid, say for example a piece of common brimstone, whose specific gravity is to be ascertained, be suspended by a fine hair,* silk, or thread, from the scale C, and weighed in air; suppose it to weigh 12 grains; let it next, still suspended to the balance, be immersed in the distilled water of the temperature of 60° *Fah.* as represented in the annexed figure; the scale containing the weight will now preponderate; add therefore to the scale C, as many grain weights as may be necessary to restore the equilibrium; suppose that six grains are necessary for this purpose, then this will indicate the amount of the weight lost in the water. We must now divide the real weight of the body in air, viz. 12 grains, by this loss, 6 grains, which gives us 2 as the specific gravity of the body under examination.

58. In conducting an experiment of this kind, it is desirable to obviate every chance of confusion, I am therefore induced to recommend to the student a method which I have usually adopted in registering the different steps of the process. It consists in drawing a cross, and putting down the results at the alternate angles, as represented in the annexed diagram; in which it will be seen that the figures are thus at once placed in the most convenient position for the necessary operations of subtraction and division.

---

* *Hair* is recommended as affording the most convenient mode of suspension, because it possesses the greatest strength, with the least bulk to affect the accuracy of the experiment; and it, moreover, is not liable to absorb any sensible quantity of water. Where greater strength is required, a single or double horse hair may be employed.

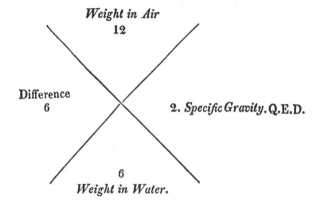

59. The rationale of this process will be easily understood by referring to the Hydrostatical proposition announced at the head of the present section, (55). In the first place, we ascertain the absolute weight of the body under examination; we next find the weight of a portion of distilled water equal in bulk to that of the body so weighed; and, lastly, by the operation of division, we compare the former with the latter, and thus obtain the comparative or specific gravity.

60. In conducting the above process there are several sources of error, against which it is necessary to guard the inexperienced operator; and there are, also, some circumstances, which, unless they be appreciated, and provided for, must render the experiment inconclusive, or altogether impracticable. These difficulties will be now considered in succession.

A. *The Temperature and Purity of the Water employed.*

61. Unless we use water at the standard temperature of 60°, we cannot expect to obtain a just result,

for it is stated by Mr. Nicholson that he has found, by experiment, the fifth decimal figure to change in *water* at every three degrees of Fahrenheit's thermometer. In like manner the presence of saline impurities will, by altering its specific gravity, necessarily affect the accuracy of every experiment connected with it.

> B. *The occurrence of accidental vacuities, or fissures, in the substance submitted to examination.*

62. This circumstance occurs more frequently than is generally suspected, and will go far to explain the discrepancy which exists in different works with regard to the specific gravities of numerous bodies; thus will the occurrence of pores render substances apparently lighter. It will, for the same reason, be necessary for the operator to take care that no air bubbles are attached to the substance, for they would have a tendency to buoy it up; should they occur they may be easily detached by means of a feather.

> C. *The Solid is too small to admit of suspension.*

63. In this case the operation must be performed through the intervention of the *glass bucket*, thus,

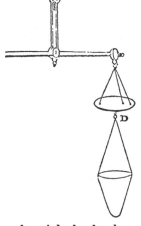

Suspend the glass bucket by a thread to the hook of the scale **D**, and find its weight in air, then place the substance which is to be tried in it, and weigh it again; the former weight subtracted from the latter leaves the weight of the substance in *air*; this being done, the same operation must be repeated in water; for which purpose, let the loaded bucket be weighed in water, then remove its contents and weigh the bucket alone in water;* subtract the latter weight from the former, and the remainder is the weight of the substance under examination in *water*; having thus obtained the weight of the body in *air* and in *water*, the operator will proceed in the solution of the problem as already directed (58).

D. *The Substance to be examined is soluble in water.*

64. Bodies of this description may be gently heated and covered with a thin coat of melted bees wax; thus defended, they may be plunged without any risk in distilled water. A slight allowance should be made for the buoyant influence of the coat of wax, which, however, must be very trifling, since this plastic matter

---

* The trouble of this part of the experiment is much abridged by having two weights, the one of which shall counterpoise the weight of the bucket *in air*, the other that of the bucket *in water*.

has very nearly the density of water itself. In some
cases the substance may be weighed in a fluid of known
specific gravity, in which it is not soluble, as in *Spirit
of Turpentine, Alcohol, Naptha,* &c.; or, if it be a
salt, in a saturated solution of the same; the specific
gravity of which media may be afterwards easily referred to that of distilled water, by the common rule of
proportion; thus say

> *As the Specific Gravity of Water*
> *Is to that of the Solution employed*
> *So is the Specific Gravity found*
> *To the true Specific Gravity.*

And, since the specific gravity of water is unity, it is
evident that we have only to multiply the sp. gr. of the
solution by that found, in order to arrive at the conclusion.

65. In like manner, we may easily find the specific
gravity which a body will assume when weighed in any
other medium than that of water, *viz.* by dividing its
true specific gravity by that of such a medium; thus,
a body whose specific gravity is 2·5 if weighed in
Sulphuric acid would give an apparent one of 1·333,
thus $\frac{2\cdot500}{1\cdot875} = 1\cdot333$. If in the sea water (1·026)
that of 2·436. The knowledge of this simple fact is
of great practical importance, as it will at once enable
the chemist, at sea, to use the water of the ocean, in
his experiments, with as much facility and accuracy,
as if he employed that which had been distilled.

66. This subject may perhaps receive a clearer explanation by the statement of the following problem.
Suppose a piece of Opium weighing 30 grains, is found
to weigh only 9·5 grains, when immersed in olive oil,
what is its specific gravity?

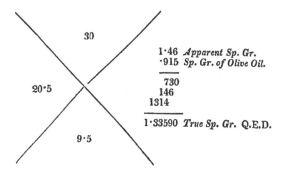

1·46 *Apparent Sp. Gr.*
·915 *Sp. Gr. of Olive Oil.*

730
146
1314

1·33590 *True Sp. Gr.* Q.E.D.

67. There is still another mode in which the specific gravity of solid bodies may be estimated, and which is founded upon the proposition that *a body immersed in a fluid displaces a quantity equal to its bulk.* If, then, into a bottle of water, the weight of which is accurately known, we carefully drop a hundred grains of the substance to be examined, we shall displace a quantity of water equal in bulk to these hundred grains, when by weighing the bottle we shall easily find, by the difference of weight, the specific gravity of the matter under examination. Suppose, for example, the bottle is found to weigh fifty grains more than it did when filled with water only, we may, in that case, infer that a hundred grains of the substance displaced only fifty grains of water, and that consequently its specific gravity is 2.

68. It was by some such method that Archimedes is said to have discovered the deceit which had been practised upon his relative Hiero the Second, king of Syracuse, by an unprincipled goldsmith, who having received a quantity of pure gold to fabricate a crown, as an offering to the gods, secreted a part of it, and substituted the same weight of silver.*

---

* Vitruvius. Lib. ix. c. 3.

2. *The Specific Gravity of the body is less than that of Water.*

69. It is obvious that, unless a substance will sink
in water, its specific gravity cannot be found by any
of the methods above described. In such a case we
may suspend the substance in company with some
heavier body ; and, having ascertained the exact weight
of the former in *air*, and that of the latter in *water*,
we are, by means of thread to fasten them together,
not so closely, however, as to exclude the water from
their contiguous surfaces, or to harbour bubbles of air
between them ; after this adjustment, the bodies are to
be weighed in water, when it will be found that their
weight will, together, be less than that of the heavier
body alone, in consequence of the latter being partly
buoyed up by the lighter substance to which .it is
attached. If we subtract the weight of the lighter
body from that of the heavier body, and add the remainder to that of the former in air, we shall obtain
the weight of a quantity of water equal in bulk to the
lighter body, and we have then only to divide the weight
of the lighter body in air by this last mentioned sum, in
order to obtain its specific gravity. This process will be
rendered more intelligible by the following example.
A piece of Elm wood, having been varnished in order to
prevent its absorbing any water, was found to weigh
920 grains in *air*. A piece of Lead, chosen as the
ballast, was ascertained to weigh 911·7 grains *in water*.
The Elm and the Lead were then tied together, and,
being suspended from the hook of the scale C in the
usual manner, were found to weigh *in water* only
331·7 grains, being 580 grains less than the weight of
the Lead alone ; therefore 580 were added to 920,

i. e. to the weight of the elm in air, which made up the sum of 1500; lastly, 920 were divided decimally by 1500, and the quotient ·6133 gave the specific gravity required.

70. This problem may be still more easily solved by means of a contrivance which has been termed the *Sinking Pulley*. A small pulley, moving with but little friction, is fastened to the bottom of the water jar, or to a weight sufficiently heavy to preserve it steadily in its position; and the hair attached to the substance must in this case pass downwards under the pulley, and rise again, so that its opposite end may be fixed to the hook of one of the scale pans, as represented in the annexed figure.

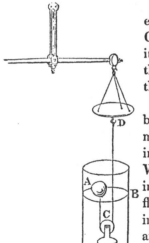

A, is the floating body to be examined; B, the jar of water; C, the pulley at the bottom of it; and D, the opposite end of the hair hooked on to one of the scale pans of the balance.

The substance A is first to be weighed in the ordinary manner, and afterwards placed in the jar as here represented. Water must then be poured in, until the substance by floating draws the scale beam into an horizontal position; after which weights must be placed in the opposite scale until the substance is sunk, or drawn under the water. Suppose, for instance, the body be a piece of Cork weighing 30 grains in air, then it will be found that about 150 grains must be placed in the opposite scale to sink it, and this weight must be added to the original weight of 30 grains, making 180 grains for the weight of a bulk of water equal to that of the cork. The original weight (30 grs.) must now be divided by that of the water, viz. 180, which of course can only be done fractionally, the result of which will shew that the specific gravity is negative, or less than the standard; it consequently cannot be expressed by an integer, but will come among the fractions, being only 0·24.

71. In conducting this investigation we may conveniently register the results in the manner already recommended (56) taking care to add, instead of subtracting, the second charge; thus,

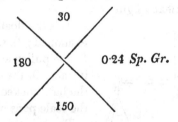

72. The following problem will afford a farther illustration of this subject—*If a piece of Potassium, weighing 30 grains, requires 220 grains to draw it under the surface of Naphtha, what is its specific gravity ?*

For the solution of this problem, we proceed as follows:

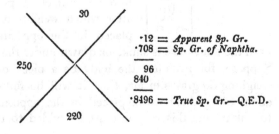

II. The Modes of ascertaining the Specific Gravity of Fluids.

73. There are several methods by which this problem may be solved; but that which is unquestionably the most convenient, and, at the same time, the most accurate, is performed by what is termed a *Gravity Bottle.** It is a glass bottle with a slender neck, and is furnished with a ground conical stopper, in the side of which there is a notch, or indentation, by which the operator is enabled to put in the stopper after the vessel has been completely filled, the redundant fluid escaping through this groove. Unless such a contrivance were adopted, it would be difficult to fill a bottle with liquid, without inclosing some bubbles of air.

The weight of this bottle, with its stopper, must be carefully noted in grains, or a weight may be at once procured, that shall always serve as a counterpoise. The bottle is then to be filled with distilled water, at 60° *Fah.* and its weight in grains, carefully noted. In taking the specific gravity of any other fluid, we have only to fill the gravity bottle with it, and then to ascertain the exact weight, in grains, which, if it be divided by the weight of the water contained in the bottle, will give a quotient which will be the specific gravity of the fluid in question. Suppose, for example, we wish to find the specific gravity of a solution of the *Carbonate of Potass,* and that the bottle, in which the experiment is to be

* Where such an apparatus is not at hand, any phial may be substituted, provided the orifice in the neck be small, and it be first ascertained what quantity of water it is capable of containing.

made, contains exactly 1750 grains of the solution, and
that the same quantity of water weighs no more than
1425 grains. In this case, we have only to divide the
greater by the lesser sum, and the quotient 1·222 will
be the specific gravity required. For conducting this
experiment with greater facility, a gravity bottle is now
usually sold, under the name of a " *Thousand Grain
bottle*," together with a weight which is an exact
counterpoise for it when filled with distilled water at 60°
Fah. This instrument consequently does not require
the aid of any computation, but is simply filled with
the fluid to be examined, and placed in one scale of the
balance, while its counterpoise is placed in the other.
If the contained fluid be lighter than water, it will
appear deficient in weight, and as many grains must be
added to the scale that contains it, as may be sufficient
to restore the balance. This shews, at once, that the
specific gravity of the fluid in question is *negative*, or
less than the standard, and, consequently, that it must
be expressed by a fractional number; but, should the
fluid be heavier than water, the bottle will preponderate, and weights must be put in the opposite scale,
when their amount being *positive*, must be added to
that of the standard. For example, if the bottle were
filled with Sulphuric Æther, it would require 739
grains to be placed in the same scale to restore the
balance, and, consequently, its specific gravity would
be expressed thus 0·739. Had it been filled with sea
water, which is rather more dense than that which is
distilled, 26 hundredths, or rather better than a quarter
of a grain must have been added in the opposite scale,
and which, as already explained, must be added to the
standard 1·000 to express the specific gravity of such
water, which would be stated thus, 1·026. Sulphuric
acid, again, being still heavier, would, in like manner,

require 875 grains, and must accordingly be expressed as 1·875.

74. By a similar process we may proceed to investigate the specific gravities of certain powders, always, however, taking care to ram them down so as to leave no interstices that might vitiate the accuracy of the results.

75. Another mode of estimating the specific gravity of fluids, is founded upon the same proposition as that to which the Hydrostatical balance above described (53) owes its utility, viz. *that a body, when weighed in any liquid, loses just so much of its weight as is equivalent to that of the same bulk of the liquid.* It is clear therefore, that, if we are required to compare the specific gravities of two liquids, we have only to take some solid body heavier than water, and to observe how much weight it loses in each, and we shall thus acquire the comparative weight of the same bulk, which is as the specific gravities of the fluids respectively; but, as the specific gravity of water is not, in such a case, expressed by unity, we must say

> *As the loss of weight in water*
> *Is to the loss of weight in the other fluid,*
> *So is Unity*
> *To a Fourth Proportional.*

In a word, divide the loss of weight in the other fluid by the loss of weight in the water, and the quotient will express the specific gravity of the former.

A ball, or pear-shaped lump of glass, suspended by a fine platinum wire, as represented in the annexed figure, is usually sold for the purpose of the above experiment; and, were it ground to such a size as to lose exactly a thousand, or ten thousand grains in distilled water, no computation would be required; its loss of weight by immersion, indicating at once the specific gravity of the fluid.

76. Although the processes above described are capable of ascertaining the specific gravity of any liquid with the utmost nicety, and are those employed in every philosophical investigation, yet, for the purposes of trade and commerce, some more expeditious and less operose method is desirable; to answer this object, instruments termed *Areometers** and *Hydrometers* have been invented.

77. The *Areometer*, or *Water-poise*, is an instrument of considerable antiquity. It was long supposed to have been invented, at the close of the fourth century, by Hypacia, the celebrated mathematician of Alexandria; but Usebe Salverti, in a memoir on this subject in the 22nd volume of the *Annales de Chimie*, has proved that it was invented by Archimedes. It consists of a graduated tube, with a bulb at one end, and so contrived by being balanced with Mercury, or Lead, that it may swim in the fluid in a perpendicular position. The specific gravity is indicated by the degree marked on the stem of the instrument, to which it sinks in the fluid to be examined, and this

---

* *Areometer*, from αραιος *thin*, and μετρον *measure*.

will consequently always be lower, in proportion as the liquid is lighter.*

78. The Areometer of Baumé, which has been long in common use in France, is of this construction, and the student will find a useful table in the Appendix, for reducing the degrees of this instrument to the common standard of specific gravity.

79. We have hitherto considered the numbers employed to denote the specific gravity of any body, as having a direct reference to distilled water as unity. To this rule, however, there are some important exceptions. In Great Britain and Ireland, the specific gravity of ardent spirits is expressed by other terms, and various instruments, called *Hydrometers*, have been used for this purpose; the degrees of which are calculated from the strength of an arbitrary spirit, which is termed by the Excise board *Proof* Spirit; its specific gravity being about 0·933. The relation which all other spirituous liquors have to this standard is expressed by saying that they are so much *above*, or *under, Hydrometer proof;* thus, when it is stated that a spirit is 20 *above proof*, it means that its strength is such, that one hundred gallons of it will admit an addition of twenty gallons of water, to reduce it to the strength of *proof*; and it is considered to be 20 per cent. *under proof*, when the same quantity contains 20 gallons of water more than is contained in Proof Spirit.

---

* *The immersed portion of a floating body*
Is to the whole body,
As the Specific Gravity of the Body
Is to the Specific Gravity of the Fluid.

Hence it follows, as a corollary, that, if the same body float upon two different fluids, the parts immersed will be inversely as the specific gravities of such fluids.

80. The ordinary form of the Hydrometer is shewn in the annexed figure.

p is a hollow ball of copper, of the shape and about the size of a hen's egg; i is a graduated scale of thin brass placed upon the top of the ball, and k a brass weight attached by a wire to the lower side of the ball, for the purpose of keeping the instrument in an upright position, while it is floating, and is so adjusted as to cause the top of the ball to stand but a very little distance above the surface of distilled water, when placed in the same; m is a weight, (of which there are several distinguished by different numbers) that drops on to a pin projecting above the top of the scale to complete the adjustment of the instrument; by which it is so much further depressed that the *Zero* point, or bottom of the scale, just coincides with the surface of the water; thus adjusted, if it be floated in a vessel containing spirit, or any other fluid of less specific gravity, and consequently of less buoyancy than water, it will sink in it to a certain extent, and indicate the difference by the surface of the fluid intersecting a higher point upon the graduated scale. Should the instrument be placed in a fluid of greater specific gravity, and buoyancy than water, so that the weight m will not be sufficient to depress the scale down to the surface, then that weight must be changed for one of higher power.

81. The principal source of fallacy, in the use of this instrument, arises from the susceptibility of Spirit, for the examination of which the Hydrometer is chiefly used, to expand and contract with slight change of temperature, by which its density is subject to constant

variation.* It is evident that such an inconvenience can be alone remedied by the constant use of the thermometer with the instrument, and as the divisions are made to correspond with the density of fluids at a temperature of 60° *Fahr.* so at this point the result, as indicated by the instrument, will not require correction. But, since the hydrometer is to be used in all seasons, and it would be impossible in the extensive concerns of commerce, to vary the temperature of the fluid under examination to some common standard, so corrections must be made either by tables, calculations, or change of weights, to suit the particular exigency of the case, and several modifications of the instrument have, accordingly, been contrived to facilitate the process.

82. CLARK'S HYDROMETER, which varied very little from the one already described, (80) except that the shifting weights were very numerous, and were applied at the situation k, within the fluid, instead of being placed upon the top of the scale, as at m, was adopted and used during a long period in the Customs, under the directions of an Act of the Legislature ; but, having been found intricate and troublesome, it was superseded in the year 1816, by SIKES'S HYDROMETER, which is the one now generally used in England, as ordered by an Act of Parliament (56th Geo. III. c. 140.) This instrument has but nine shifting weights, applicable upon the upper part of the stem, as in the figure above represented (80), and is used with a set of tables, or a sliding rule which is sold with it, for computing compensation for different temperatures. The scale is

---

* It has been found that a cubic inch of good Brandy is ten grains heavier in winter than in summer, or that 32 gallons of spirit, in the former season, will measure 33 in the latter : so that it is most profitable to buy such articles in the winter, and to sell them in the summer.

divided into ten principal divisions, each of which is
subdivided into five parts, and, by the separate application of the weights in succession, completes the
range of strengths from pure alcohol to water; each
weight being equivalent to ten principal divisions.
This Hydrometer, with the weight marked 60 screwed,
as at *k*, on to the lower stem, is so adjusted as to sink
to the line mark *p* on the scale of the instrument, when
placed in *proof* spirit, of the temperature of 51° *Fahr.*;
and by the addition of the square weight on the top,
it sinks to the same point in distilled water of the same
temperature. This weight being just one-twelfth part
of the entire weight of the whole hydrometer, together
with its bottom weight, No. 60, causes the scale to
shew the difference between water and proof spirit,
which the Act, above referred to, states shall weigh
exactly twelve-thirteenths of an equal bulk of distilled
water.

83. There is another practical difficulty opposed to
our obtaining correct results from the Hydrometer,
depending upon the fact, that the specific gravity of
mixtures is, in very few instances, the mean of the
separate specific gravities of their ingredients, as already explained (31); thus, if to 100 gallons of spirit
of wine found to be 66 above proof, we add 66 gallons
of water to reduce it to the proof state, the mixture
instead of producing 166, will only measure 162 gallons.

84. The Hydrometer of JONES is still more simple
than those already described, having but three shifting
weights, and the thermometer being attached to the
instrument with such a scale that the compensation for
temperature, and the allowance for the mutual *penetration*, or incorporation of the fluids, are given at
sight, without any tables, or requisite calculation.

85. The Thermometer-makers of London prepare small glass bubbles, like beads, but hermetically sealed, and formed of different weights, so that, while some will float upon the ordinary brandy of commerce, others will sink in it; and the specific gravities of these bubbles having been ascertained, are marked upon them respectively, by means of a diamond, and they thus become very useful instruments for trying the comparative goodness of spirits; for if a bubble be found which will just sink beneath the surface of good brandy, it will sink to the bottom of that which is still stronger, and float upon the surface of that which is not so strong. The figures 5, 10, 15, &c. marked upon these bubbles indicate per centage strengths, as in the hydrometer, but are not so much to be depended upon, unless used with a thermometer at the temperature at which they were originally formed. They answer, however, very well for the cursory examination of spirits, as above hinted at.

86. Where it is necessary to take the specific gravity of a fluid at a boiling temperature, such bubbles will be found very convenient. Mr. London, who has obtained a patent for purifying rock salt, has adopted the ingenious expedient of employing *two* bulbs of different gravities at once, by which he readily determines when his saline liquor is evaporated to the point which is necessary for his purpose; for if the *lixivium* be too dilute, both these bulbs will fall to the bottom; and, if too dense, both will swim; but when he perceives that the heaviest falls, and the other floats, he then knows that he may rely upon the liquor being of the medium specific gravity which he requires.

87. The methods of ascertaining the specific gravity of Gases will be considered hereafter, when the student

may be supposed to be better acquainted with those manipulations which the management of such bodies will require.

An ample Table, exhibiting the Specific Gravities of those different substances, which are most interesting to the medical student, will be found in the Appendix.

CONTIGUOUS ATTRACTION.

88. We have hitherto only considered Attraction, as exerted over *masses* of matter, at *sensible* distances; we have now to examine its influence over the minute atoms of bodies placed with respect to each other at *insensible* distances. Where these atoms are *similar* in their nature, the result of this power is simple aggregation; but where *dissimilar*, it gives origin to new and infinitely varied productions.

COHESION.

Synon. *The Attraction of Aggregation;—Cohesive Affinity;— Corpuscular Attraction; — Homogeneous Affinity.*

89. It may be defined, that force, or power, by which particles, or atoms of matter, of the *same* kind, attract each other, and produce an aggregate or mass.

90. This force is exceedingly various in different bodies, and even in different states of the same body. In solids its force is exerted with the greatest intensity; in liquids it acts with much less energy, and in aeriform bodies it is doubtful whether it exists at all; thus water, in a solid state, has considerable cohesion, which is much diminished when it becomes liquid, and is entirely destroyed, as soon as it is changed into vapour.

91. The force of cohesion in solid bodies is measured by the weight necessary to break them, or rather
to pull them asunder; thus iron is composed of particles cohering so strongly, that if a rod of this
metal be suspended in a perpendicular direction, it
will require an enormous weight to be attached to
the lower extremity in order to break it; a smaller
force is sufficient to overcome the cohesion of lead,
and a still smaller to separate the particles of chalk
from each other. We are indebted to Muschenbroeck
for the most complete set of experiments upon this
subject, and by which he has been enabled to construct
a Table expressive of the relative degrees of cohesion
possessed by different bodies.

92. The vital power has been said to modify cohesion, as it does every other physical force; this
assertion, however, can only be supported with reference to muscular structure; numerous experiments
have shewn that there exists a greater degree of cohesion between the particles of the muscular fibre
during life than immediately after death. Sir Gilbert
Blane made the following experiment upon the flexor
muscle of the thumb of a man, five hours after death,
while the parts were yet warm and flexible. All the
parts of the joints having been separated, except the
tendon, a weight was hung to it, so as to act in
the natural direction, and was increased gradually till
the muscle broke, which happened when *twenty-six*
pounds had been appended; whereas, he found that a
man of the same age, and the same apparent size and
strength, with the subject of the preceding experiment,
could with ease lift *thirty-eight* pounds by the volun-
tary exertion of the same muscle. * Similar remarks

---

* Select Dissertations on several subjects of Medical Science,
by Sir Gilbert Blane, Bart. p. 237.

have been made by Carlisle, * and by Bichat; † and, with a view to the same conclusion it has been observed, that when a vessel is ruptured during life, it is the tendinous part which is disposed to give way, while, on the contrary, after death, the fleshy part is always weaker than the tendon. ‡

93. In liquids, the force of cohesion is demonstrated by the spherical figure which they assume, when suffered to fall through the air, or to form drops. The drop is spherical, because each particle of the fluid exerts an equal force in every direction, drawing other particles towards it on every side, as far as its power extends; and it follows, upon the principles of Mechanics, that the equilibrium of the attractive forces can only take place when the mass has received a globular form. To the operation of the same force is owing the property possessed by all liquids of remaining heaped up to a sensible height above the brims of the vessels which contain them, whether formed of glass, or of metal.

94. The force of cohesion varies in different liquids, as it does in different solids, and hence the size of their respective drops must also vary. We perceive, therefore, how extremely incorrect it would be to assume a drop as, in all cases, equivalent to a grain (45); thus a drop of alcohol, the most ordinary solvent in tinctures, is not only much lighter, but even much smaller, than one of water. This consideration sanctions the propriety of the introduction of a measure that may supersede the necessity of apportioning the dose of a liquid by the ordinary mode of dropping.

---

* Phil. Trans. for 1805, p. 3.
† Anat. Gen. t. ii. p. 398.
‡ Upon this subject, see Bostock's Physiology, p. 217.

95. It is difficult to measure directly, the cohesion of liquid bodies; but it is evident from what has been above stated, that an approximation may be derived from the magnitude of drops, and the thickness of liquid sheets, heaped upon an horizontal surface.

96. As the Attraction of Cohesion does not extend to any sensible distance, (88) it must follow that whenever the parts of any substance are separated, or broken, it will be difficult to reunite them. If, however, they can be brought into sufficient approximation, then this attraction will operate, and their union take place; thus, two leaden bullets, having each a flat surface of a quarter of an inch in diameter, if scraped smooth, will on being forcibly put together, cohere so strongly as sometimes to require a force of 100 lb. to separate them. As the constitution of liquids will allow a more perfect contact, separate portions may be made to cohere still more perfectly; in this way, if we place two or more globules of Mercury on a dry glass, or earthen plate, and push them gently towards each other, the globules will attract each other, and form one mass or sphere, greater in bulk, but precisely the same in nature; but if these globules should have been previously moistened, the necessary approximation will be prevented, by the intervention of a film of water, and no adhesion will take place.

97. An important modification of this force occasions liquids to rise in small tubes, and as this phœnomenon is most conspicuous, when the width of the bore is so small as to resemble that of a hair, it has been denominated CAPILLARY ATTRACTION, for the height to which the fluid rises in such tubes is always inversely as their diameters. The popular mode of explaining the fact is to refer the suspension of the slender column of water to the attraction of the in-

terior ring of glass immediately above it, but Mr.
Leslie* very shrewdly inquires why the ring just below
the summit of the column should not attract it equally
downwards? and such opposite forces producing a
perfect equilibrium, the water would merely preserve
its level, and shew no disposition to rise in the tube.
The chief obstacle in explaining the mode of capillary
action, continues this acute philosopher, arises from
the prejudice, that a *vertical* attraction is necessary to
account for the elevation of the liquid; yet such undoubtedly is not the primary direction of the force
evolved; for the action of the glass being evidently
confined within very narrow limits, this virtue must be
diffused over the internal surface of the tube, and must
hence exert itself *laterally*, or at right angles to the
sides. Nor is it difficult to conceive how a lateral
action may yet cause the perpendicular ascent; for it
is a fundamental property of fluids, that any force im-
pressed in *one* direction may be propagated equally in
every direction. The tendency of the fluid, then, to
approach the glass will occasion it to spread over the
internal cavity of the tube, and, consequently, to
mount upwards.

98. Capillary action is not confined to glass tubes;
but is exerted among all substances which are perforated by pores, or subdivided by fissures or interstices. It is this attraction, for instance, which causes
water to rise in sponge, cloth, sugar, sand, &c.

99. To the Physiologist Capillary attraction is a
phœnomenon of very great interest, for on its power
depend chiefly the functions of the excretory vascular
system in plants and animals; thus, says Professor
Leslie, if the pores of the human skin were no finer

---

* Elements of Natural Philosophy.

than the three-thousandth part of an inch in diameter,
they would yet be sufficient to support lymph to an
altitude of 120 inches, or 10 feet, or much higher indeed than is required for any individual. The rejection
of the perspirable matter from these external mouths
must occasion a continued flow of the liquid from the
lower and wider trunks of the capillary vessels, aided
no doubt by a connected chain of alternate contractions and dilatations, extending through their muscular
structure. The pores in the leaves of trees and tall
plants must be still finer, seldom perhaps exceeding
the ten-thousandth part of an inch. As fast as the
humidity is exhaled into the atmosphere, it is constantly supplied by the ascent of sap from the roots.
Dr. Hales attempted, by an ingeniously devised experiment, to demonstrate the power of the vegetating
principle, by measuring the *force* with which the sap
ascended in the ramifying vessels of a growing plant,
but had the same experiment been repeated, with a
dead branch, the same result would have followed,
provided the evaporation from its extreme surface had
been sufficiently copious.

100. To the Chemist the force of cohesion is more
immediately interesting, as it is continually opposed to
the action of chemical affinity; for the more strongly
the particles of any body are united by this power, the
less are they disposed to enter into combination with
other bodies; hence certain *mechanical* operations,
such as *rasping, grinding, pulverising,* and other
modes of division, are generally employed as preliminary steps to chemical processes; the application of
heat* will also diminish the cohesion of bodies, and is

---

* For this reason the size of a drop of liquid will vary at different temperatures, a fact which offers an additional difficulty in

therefore frequently useful on such occasions. In these respects the digestive action of the stomach may be said to resemble a chemical process, for where the *ingesta* are not finely divided by mastication, the operation is tardily or imperfectly performed, and the teeth, accordingly, perform the same duty in the œconomy of the animal, as the mortar, or mill, in the laboratory of the Chemist.

101. It will be here necessary to describe more fully the nature and importance of these mechanical operations of *division*, and *separation*, to which we have just alluded. In the former class are included the processes of *Pulverization, Trituration, Levigation, Granulation, Rasping,* and *Grating.* In the latter, those of *Sifting, Elutriation, Decantation, Filtration, Expression, Coagulation,* and *Despumation.*

102. *Pulverization* is that process by which friable and brittle substances are reduced to powder; and is generally performed by means of *pestles* and *mortars.*

103. These instruments are constructed of various materials, according to the nature of the substances they are intended to pulverize; it being necessary that they should at once resist both mechanical force, and the chemical action of the bodies they may contain. Wood, iron, marble, siliceous stones, porcelain, Wedgwood's ware, and glass, are all employed for such an object. Thus, if our purpose be to submit astringent matter to their action, we must carefully avoid an iron vessel, or the mass will combine with the metal; if for instance, the *Confection of Roses* were to be beaten in an iron mortar, we should soon obtain a compound as black as jet; so again, substances containing *Calomel*

---

any attempt to measure quantity by the number of drops. Upon the same principle we may be enabled to appreciate some of the effects of temperature in affecting the functions of animals.

should be rubbed in mortars of glass or earthenware, since this mercurial salt is decomposed by iron, lead, or copper; nor ought marble, or metallic mortars to be used for acid substances.

104. Mortars are also required to be of various sizes. The largest are usually made of cast iron, as represented in figure A, fitted with wooden covers, so

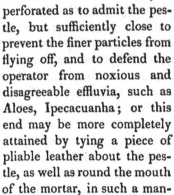

A

perforated as to admit the pestle, but sufficiently close to prevent the finer particles from flying off, and to defend the operator from noxious and disagreeable effluvia, such as Aloes, Ipecacuanha; or this end may be more completely attained by tying a piece of pliable leather about the pestle, as well as round the mouth of the mortar, in such a manner that the former shall have free motion. So penetrating and acrid, however, are some of the subtances to be thus treated, as Euphorbium, Cantharides, &c. that it will be necessary for the operator to cover his mouth and nostrils with a wet cloth, and to stand with his back to a current of air, that the particles which arise may be carried off from him. To lessen the labour, the pestle is often attached by a cord to the end of a flexible wooden beam, placed horizontally over the mortar, the elasticity of which elevates the pestle to the proper height after each stroke is made.

105. The most useful mortars for smaller articles, and for the purpose of dispensing medicines, are those of Wedgewood ware, as they are smooth, hard, and resist the action of any chemical re-agent.

106. Of whatever materials mortars are constructed,

their internal bottom ought to be made in the form of a hollow sphere, and their sides should have such a degree of inclination as to make the substances they contain fall back to the bottom, when the pestle is lifted, but not so perpendicular as to collect them too much together, otherwise too large a quantity would rest below the pestle, and prevent its operation. For the same reason, too great a proportion of the substance to be pulverized ought not to be put into the mortar at once; and the particles already reduced to powder should, from to time, be separated by means of *sieves*, to be hereafter described.*

107. In many cases, the subject to be pulverised requires some previous treatment, to render it obedient to the pestle; thus, vegetable matters require to be dried before they can be pulverized; and wood, roots, and barks, should be previously cut, chipped, or rasped. When roots are very fibrous, as for example, those of ginger, it is necessary to cut them *diagonally*, in order to prevent the powders from being full of hair-like filaments. Resinous substances, which soften at a moderate temperature, should be pulverized in cold weather, and only gently beaten to prevent them from running into a paste, instead of forming a powder.

108. In some cases the disintegration of a solid body is much accelerated, and extended by the addition of other materials, hence the pharmaceutical aphorism of Gaubius, ' *Celerior atque facilior succedat composita quam simplex pulverisatio*,'' thus the pulverization of camphor is assisted by a few drops of spirit; that of aromatic oily substances, as nutmeg, mace, &c. by sugar; upon the same principle, the Pharmacopœia directs the trituration of Aloes with

* Lavoisier's Elements of Chemistry, translated by Kerr, vol. ii. p. 68.

clean white sand, in the process for preparing *Vinum Aloes,* to prevent the particles from running together into masses. This subject, however, will be found more fully treated of in my Pharmacologia (*Vol.* 1. *art. Pulveres*). Metals which are scarcely brittle enough to be powdered, and are yet too soft to be filed, as for instance Zinc, may be powdered while hot in a heated iron mortar; or metals may be rendered brittle by alloying them with mercury, but as such bodies are not required to be reduced to the state of very fine powder for pharmaceutical purposes, these processes are rarely, if ever performed.

109. The degree of fineness to which substances should be reduced by pulverization, in order to obtain their utmost efficacy, is an important question, and I have already endeavoured to establish some precepts upon this subject, for the guidance of the practitioner.*

110. *Trituration* is intended to produce the same effect as pulverization, but in a greater degree. It is effected by a rotatory motion of the pestle, and should be perfomed in mortars of Wedgewood's ware, or glass, as here represented.

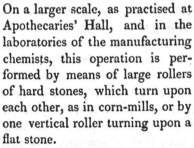

On a larger scale, as practised at Apothecaries' Hall, and in the laboratories of the manufacturing chemists, this operation is performed by means of large rollers of hard stones, which turn upon each other, as in corn-mills, or by one vertical roller turning upon a flat stone.

111. *Levigation* is a process similar to trituration, except that the rubbing is assisted by the addition of a liquid in which the

---

* PHARMACOLOGIA. Edition 6th. Art. *Pulveres.*

solid under operation is not soluble. Water, or spirit is usually employed on such occasions, and in particular cases, viscid and fatty matters, such as honey, lard, &c. The substance to be levigated is spread on a flat table of porphyry, or some other hard stone, as represented in the annexed figure, and is then rubbed with

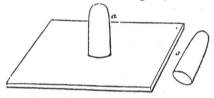

a *Muller* of the same materials, either of a pyramidal shape, as shewn in the wood cut (*a*), or a portion of a large sphere. A thin spatula of ivory, horn, wood, or iron, is employed to bring back the materials from the edges of the table, to which the operation of the *muller* continually drives them. In some cases, however, levigation may be performed in a common stone mortar, although with less facility than by the former method, where from the flatness of the two surfaces, the paste cannot elude the pressure.

112. *Granulation* is employed for the mechanical division of metals, and of phosphorus. It is performed either by pouring the melted metal into water, or by agitating it in a box, until the moment of congelation, at which instant it becomes converted into a coarse powder. For the granulation of phosphorus the former process can be alone employed.

113. Substances are, moreover, reduced to the state of coarse powder by rasping, and filing; and the softer vegetable bodies are made into a pulp by means of the grater.

114. Such being the different modes, by which solid bodies may be reduced to powder, in order to facilitate their combination with other substances, or to adapt them for administration as remedies, we have

next to describe the processes by which the parts of
substances may be separated from each other by different mechanical means.

115. *Sifting.* None of the mechanical operations
employed for reducing bodies to powder, are capable
of producing it of an equal degree of fineness throughout; the powder obtained by the longest and most
accurate trituration being still an assemblage of particles of various sizes. The coarser of these are removed,
so as only to leave the finer and more homogenous particles, by means of sieves of different degree of fineness, adapted to the particular purposes for which they
are intended. This coarser part is then again submitted to the action of the pestle; and thus, by degrees,
will the whole assume an uniform fineness. Sieves are
composed of iron-wire, or of hair cloth, or gauze, or
sometimes of parchment pierced with round holes of an
appropriate size. When very subtle materials are to
be sifted, which are easily dispersed, or when the finer
parts of the powder may be injurious to the operator, a
Compound Sieve is made use of, as here represented,

which consists of the simple sieve, *c,*
with a deeper rim; a lid, *b,* covered with
leather; and a receiver, *d,* with leather
stretched across one end, and made sufficiently wide to admit the lower portion
of the sieve to enter, and fit tightly within it. When these are put together,
the finest powder may be separated by
them without any loss or inconvenience
to the operator.

116. *Elutriation,* or *Washing.* By this operation
we are enabled to procure powders of a greater, and
more uniform fineness, than by the sieve; but it can

only be employed with such substances as are not acted upon by the liquid vehicle. The powdered sub⁅stance is mixed and agitated with water, or any other convenient fluid; the liquor is allowed to settle for a few moments, and is then *decanted* off; * the coarser powder remains at the bottom of the vessel, and the finer passes over with the liquid. By repeated de⁅cantations in this manner, various sediments are ob⁅tained of different degrees of fineness; the last sedi⁅ment, or that which remains longest suspended in the liquor, being the finest. By a method of this kind the *Creta Præparata* of our Pharmacopœia is obtained in a very impalpable form. This process may also be used with great advantage for separating substances of different degrees of specific gravity, though of the same fineness.

117. Where it becomes an object to decant the liquor without disturbing the sediment, a *Syphon* may be very conveniently employed; and as this instrument is of great value to the chemist, I shall take this oppor⁅tunity to describe its nature and operation.

118. The Syphon is any bent tube having its two legs either of equal, or unequal length, as here repre⁅

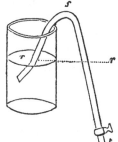

sented. If the two legs of the tube be of equal length, and it be filled with water, and then inverted with the two open ends downwards, and held level in that position, the water will re⁅main suspended in it; for the atmosphere will press equally on the surface of the water in each

* In performing this operation, especially if the fluid be de⁅canted from a wide-mouthed vessel, a glass rod should be applied to the rim, which will have the effect of directing the liquor in a regular stream.

end, and support them. But if, now, the Syphon be
a little inclined to one side, or, if the legs be of unequal
length, as above represented, which is the same thing,
so that the orifice of one end be lower than that of the
other; then the equilibrium is destroyed, and the water
will all descend out by the lower end, *t*, and rise up
in the higher. For, the air pressing equally, but the
two ends weighing unequally, a motion must commence where the power is greatest, and so continue
till all the water has run out by the lower end. And
if the shorter leg be immersed into a vessel of liquid,
and the Syphon be set running as above, it will continue to run until all the water be exhausted, or, at
least, as low as that end of the Syphon. Or it may be
made to run without first filling the Syphon, as above
described, by only inverting it, with its shorter leg in
the liquid; then with the mouth applied to the lower
orifice, *t*, sucking the air out, when the fluid will
presently follow, being forced up into the Syphon by
the pressure of the air on the water in the vessel. It
must, however, be here stated, that the operation of
this instrument does not depend upon the actual length
of its legs being unequal, but upon the virtual or *effective* legs being so, and the *effective* legs are from the
surface of the fluid to the bend or point *s*, and from
thence to the extreme point of discharge at *t*; consequently, if the leg *s r* were to be made indefinitely
longer than *s t*, the Syphon would nevertheless run,
provided the fluid at *r* was higher than the end *t*; but
it would only continue to run until the surface of that
fluid should be depressed down to the level of the
point *t*, when it would stop. The action of the Syphon,
then, is simply as follows; drawing the air out of the
end *t* produces a vacuum in the tube; the pressure of
the atmosphere on the surface of the water *r* therefore

causes it to rise and fill the tube as far as *s*, and after gaining that summit the liquid continues to flow over it down to *t*; the Syphon being thus filled with water resolves itself into two opposing or counteracting columns which occupy the two legs of the instrument, and as fluids press according to their perpendicular height, if *s t* be the longest its column will preponderate and overcome that in *r s*, and by running out at *t* will tend to reproduce or maintain a vacuum in the upper bend *s*, which is constantly prevented by a new supply of water being forced ·up at *r*, by atmospheric pressure; consequently the flow will be constant so long as there is water at the end *r* above the level of the discharging point· *t*, but as soon as that water becomes depressed to the level of that point which it may be supposed is raised to the dotted line *r*, then *r s*, and *s r*, being equal in length, will balance each other, and no more water can flow. As the Syphon is dependent on the pressure of the atmosphere for its action, the bend *s* at its top must in no case exceed 33 feet from the surface of the water to be raised, or it will be incapable of acting.*

119. The ordinary form, and mode of using the Syphon, or *Crane* as it is called in commercial lan-

---

* A column of mercury 31 inches high is a balance for, or is the greatest height of, that metal which the pressure of the atmosphere can sustain when at its greatest natural density, and as this density diminishes it can sometimes support no more than 28 inches of mercury in the barometer tube. Now as the specific gravity of mercury is 13·56, or rather more than 13 times and a half as heavy as water, it follows that, if the pressure of the atmosphere can support 31 inches of the former, it will also support above 13½ times as high a column of the latter, or one of about 420 inches, or 35 feet in perpendicular height. (*Millington's Natural Philosophy.*)

guage, are shewn in the annexed figure.

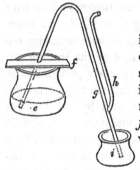

The inconvenience of applying the mouth to the extremity of the leg is prevented by the supplementary tube *h*. The instrument is supported by means of the perforated board *f*, at the proper depth in the vessel *e*.

120. A very simple and useful application of the principle upon which the Syphon operates, may be made by means of strips of linen rag, or skeins of cotton, wetted and hung over the side of the vessel containing a fluid, in such a manner as that one end of the rag or cotton may be immersed in the liquid, and the other end may remain without, below the surface.

121. *Filtration.* An operation, by means of which a fluid is mechanically separated from consistent particles merely mixed with it. It does not differ from *Straining*, except perhaps in degree.

122. A filtre is a species of very fine sieve, which is permeable to the particles of fluids, but through which those of the finest solids are incapable of passing; hence its use in separating impalpable powders from suspension in fluids.

123. The instruments for this purpose are constructed of various materials, and in their selection care must be taken that they be not acted upon by the substances for which they are employed.

For many purposes in pharmacy, as for the straining of saccharine and mucilaginous liquors, flannel, or fine woollen cloth, or linen, is chiefly used; the filtre is usually formed in a conical shape, as possessing the advantage of uniting all the liquor, which drains through, into a point, where it may be readily collected in a receiving vessel (A), thus—

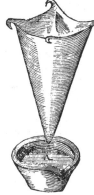

Flannel filtres are particularly eligible also where our object is to preserve the solid residue, but when the filtered liquor is the valuable product, linen is generally preferable, as it absorbs less of the fluid, which is thus obtained also in a more limpid state.

124. For smaller processes, and where it is essential to have the filtres perfectly clean, unsized paper is substituted. A square piece of this paper, of a size proportionate to the quantity of the substance to be filtered is taken, and first doubled from corner to corner into a triangle, which by second doubling forms again a smaller triangle, and this when opened constitutes a paper cone, as here exhibited (Fig. 1) which is to be supported in a glass funnel (Fig. 2) before the liquor is poured into it. It is of advantage to introduce glass rods, pieces of straw, or quills, between the paper and funnel, to prevent them from adhering too closely ; for this reason *ribbed* (3) are preferable to plain funnels.

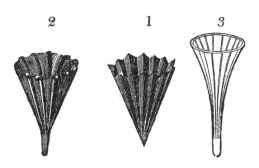

2 1 3

125. Very acrid liquors, such as acid, and alkaline solutions, act too powerfully on the ordinary materials of filtres, and are accordingly filtered through strata of differently sized particles of siliceous matter; for this purpose, a glass funnel is filled with powdered quartz, or flint glass; a few of the larger pieces being placed in the neck, which are covered with the smaller fragments, the finest, or white sand, being placed over all. The liquid to be filtered is poured gently on the surface, and soon passes through it, leaving the impurities behind. The porosity of this filtre, however, retains much of the liquor; but it may be obtained by gently pouring on it an equal quantity of distilled water; the liquor will then pass through, and the water will be retained in its place.

126. In many instances the first portions of fluid that pass through the filtre are turbid, and require to be returned, until the pores of the filtre are sufficiently obstructed to permit the most limpid part only of the liquor to pass. In cases where the solid residue is small, and it is requisite to collect the whole of it, it is useful to have a small glass tube, drawn to a capillary point at one extremity, and having a bulb in its centre, as represented in the margin. By filling which with distilled water, and putting the larger end into the mouth, the force of the breath can direct a small strong stream of water round the sides of the paper in the funnel, which will wash down to its bottom all the minute particles of solid matter lodged on its sides.

127. The filtration of some liquids is assisted by heat; in such cases a funnel with a double case may be constructed so as to hold boiling water. Where several filtrations are in progress at the same time, a

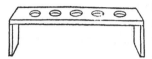

stand, similar to the one here represented, will be found very convenient for the reception of the funnels.

128. *Expression* is a species of filtration, assisted by mechanical force. It is principally employed to obtain the juices of fresh vegetables, and the unctuous vegetable oils. The subject of the operation is first bruised, or coarsely ground, then inclosed in a hair-cloth bag, and subjected to violent pressure between the plates of a screw press.* The bags should be nearly filled; and the pressure should be gentle at first, and gradually increased.

129. Vegetables intended for this operation ought to be perfectly fresh, and freed from all impurities. In general, they should be expressed as soon as they are bruised, as the bruising disposes them more readily to ferment; but sub-acid fruits yield more juice, and of a finer quality, when the bruised fruit is allowed to stand for some days in an earthen or wooden vessel. To some vegetables, that are not very juicy, the addition of a little water is necessary.

130. For unctuous seeds iron plates are used; and it is customary not only to heat the plates, but to warm the bruised seeds in a kettle over the fire; it would, however, be better to expose them in a bag to the steam of boiling water.

131. *Despumation.* Some fluids are so thick and clammy as to be unable to pass through any filtre, in

---

* For this, and similar purposes, where great pressure is required, Mr. Joseph Bramah's admirable *Hydraulic Press* may be employed with great advantage. This instrument is constructed upon the same principles as those which explain the well known *Hydrostatical Paradox.*

which case they are clarified by the process of *despu-*
mation, which consists in heating the liquor, and carefully removing the scum as it is thrown up on the
surface.

132. *Coagulation.* When the process of despumation is effected by means of glutinous, or albuminous
matter, as the white of egg, it is called *coagulation*;
when the substance is not spirituous, as syrups, for
example, the albumen which is mixed with the fluid
coagulates when it is boiled, and entangling the impurities of the fluid, rises with them to its surface in
the form of scum; but spirituous liquors may be clarified with Isinglass, without the assistance of heat, the
alcohol coagulating it so as to form a scum, which, by
descending slowly to the bottom of the vessel, carries
along with it all the impurities. In this manner, it
will be readily perceived that the network of isinglass
merely acts the part of a filtre, but with this difference,
that in this case the filtre passes through the liquor,
instead of the liquor through the filtre.

133. Fluids, of different specific gravities, when
mixed together may be separated by means of the
Separatory funnel which is here represented.
It is principally employed for separating the
Essential oils from the water with which they
are entangled during their distillation. The
funnel is first stopped at the bottom, and then
filled with the mixed fluids, the heaviest of
which gradually subsides into the narrow part
below; and, when the cork at the bottom is
removed, and the stopper above a little loos-
ened, it flows out; by which contrivance the lighter
is easily obtained in a separate state.

In some cases a *Separatory* is employed; a vessel of which the annexed is a sketch. The mixture is introduced through the central opening, which is then to be stopped. By inclining the bottle to one side, and at the same time closing the orifice on the opposite side with the finger, the fluids will separate; and when the finger is removed, the heavier will run through the lower orifice, or neck, before any of the lighter escape. If this apparatus is not to be procured, the separation of the fluids may be very easily performed by means of the common glass funnel, regulating the egress of the liquid by means of the finger held at the lower extremity of the tube.

CHEMICAL AFFINITY.

----

134. This species of Attraction, like that of Cohesion, is effective only at insensible distances; but it differs from the latter force, in being exerted between the particles, or atoms of bodies of *different* kinds, and in producing by its agency, not a mere aggregate possessed of the same properties as the separate parts, and differing only in its greater quantity or mass, but a *new* compound, in which the properties of the components have either wholly, or in part, disappeared, and been superseded by new qualities.

135. Where two heterogeneous or dissimilar bodies, are placed within the sphere of each others attractions, and form a direct union, the phœnomenon is termed COMBINATION, the substance so produced, a COMPOUND, and the bodies from which it has been formed, its CONSTITUENT PRINCIPLES, or COMPONENT PARTS.

136. The process by which a compound body is resolved into its constituent parts is in chemical language termed ANALYSIS; whereas that, by which a compound is formed by combination, is denominated SYNTHESIS. And under these are comprehended the greater part of the operations of Chemistry.

137. That chemical affinity is effective only at *insensible* distances may be demonstrated by the following experiments.

Experiment 1. Into a deep ale-glass, or glass jar, pour two fluid-drachms of a solution of the Subcarbonate of Potass *(Liquor Potassæ Sub-carbonatis.* P. L.) diluted with about fourteen fluiddrachms of distilled water. Under this introduce, very cautiously, half a fluid-ounce of water holding a drachm of common salt in solution; and again, under both these, two fluid-ounces of Sulphuric acid which has been previously diluted with an equal quantity of water. The delivery of these different fluids may be effected by the *dropping tube* above described (126), or by means of a common glass tube open at each extremity, which, having been plunged in the liquid, must be withdrawn with the thumb closely applied to the upper orifice; when, by immersing it to the bottom of the jar, and then removing the thumb, the fluid will be deposited in the manner required. If this arrangement be carefully accomplished, we shall perceive, notwithstanding the powerful attraction which subsists between the alkali and acid, that no action will take place; because their particles are separated from each other by a thin stratum of brine; as soon, however, as this arrangement is disturbed by agitation, a brisk effervescence will commence, and a chemical combination take place.*

---

* The same phœnomenon, to a certain extent, will take place whenever a lighter fluid is made to repose upon the surface of a denser one, and for which it has an attraction. In such a case those particles, which are in immediate contact, will combine,

Exp. 2.—Take two grains of *Oxymuriate,* or *Chloride of Potass,* and one grain of the *flowers of Sulphur,* mix them cautiously, and no action will ensue; rub them, however, in a mortar, so as to bring the particles into nearer contact, and a smart detonating noise will be produced. Continue to rub the mixture hard, and the reports will be frequently repeated, accompanied with vivid flashes of light.

Exp. 3.—Mix a little *Acetate of Lead* with an equal quantity of *Sulphate of Zinc,* both in fine powder; stir them together with a piece of wood or glass, and no chemical change will be perceptible; but if they be now rubbed together in a mortar, the two solids will operate upon each other, and *a fluid will be produced.*

138.   In many cases the necessary contiguity of the particles cannot be effected without the processes of solution and fusion; thus, for instance, the most intimate incorporation of Sulphur and Potass, by trituration, will only produce a *mixture* of these bodies; for, if a portion of the powder be thrown into water, the Sulphur will separate and rise to the surface, and the Potass be dissolved; but if these same substances be previously melted together, their union will be rendered so perfect, that the compound will be completely soluble; again, if we mix together, however intimately,

---

and the compound, thus produced, having no farther affinity for the upper or lower liquid, become a stratum of separation. This quiescence, however, will be only of short duration; for, since the specific gravity of the new compound will rarely suffer it to retain its situation, fresh portions of the two fluids will successively come in contact, until a general combination throughout the whole is effected.

the powdered crystals of *dry* Tartaric acid, and Carbonate of Soda, no action will be perceived, but the moment water is added, and a solution thus effected, they will act chemically upon each other, and a brisk effervescence will arise.

139. In order to shew that the result of chemical combination is the formation of a *new* body, in which the properties of the components have either wholly, or in part, disappeared, and been superseded by others, numerous experiments might be related, but the following have been selected as being at once striking, and easy of performance.

Experiment 4. — Two Odorous bodies produce a compound destitute of Odour. Water impregnated with *Ammonia (Liquor Ammoniæ.* P. L.) and concentrated *Muriatic Acid,* are fluids of a very pungent odour; mix them together in proper proportions, and a fluid will result entirely devoid of smell.

Exp. 5.—Two Fluids produce a solid. Into a saturated solution of *Muriate of lime,* let fall, gradually, concentrated *Sulphuric acid*; a quantity of pungent vapour *(Muriatic acid gas* and *water)* will immediately arise, and an almost solid compound *(Sulphate of lime)* be produced.

Exp. 6.—Two Solids produce a Fluid. This has been already exemplified by experiment 3; or we may triturate together equal parts of crystallized *Nitrate of Ammonia* and *Sulphate of Soda,* when the saline mixture will shortly assume the liquid form.—Or,

Exp. 7.—Take an Amalgam of Lead, and another of Bismuth; let these two *solid* bodies be mixed by triture, and they will become liquid.

Exp. 8.—*Two Liquids produce a Gas.* Mix about five fluid-drachms of Rectified Spirit, with three fluid-drachms of Nitric acid, when, in a few seconds, a gas will be evolved, which may be inflamed by holding a lighted taper to the mouth of the flask from which it escapes.

Exp. 9.—*Two Gases produce a Solid;* thus, when muriatic gas and ammoniacal gas are mingled together, they produce the solid salt called *Sal Ammoniac,* or Muriate of Ammonia. This experiment may be easily performed by putting an equal portion of Sal Ammoniac into two wine-glasses, and then adding quick-lime to the one, and sulphuric acid to the other; from the former ammoniacal, from the latter muriatic gas, will be evolved. As soon as the two vessels are made to approach each other, the gases will intermingle and combine, and display the formation of a solid product, by the appearance of white fumes of great density, thus,

Exp. 10.—*Two colourless fluids produce a compound of exquisite colour.* Into a wine-glass, containing water, add a few drops of Prussiate of Potass; and into another, a little of the dilute solution of Sulphate of Iron; on mixing these two *colourless* liquids together, a bright deep blue colour will be instantly produced, which is the true "*Prussian Blue.*"—Or,

Exp. 11.—Introduce a solution of Prussiate of Potass into one glass, and into a second, that of Nitrate of Bismuth; on mixing together these liquids, a *yellow* compound will result.

Exp. 12.—*The same Vegetable infusion will yield three different colours, on the addition of three colourless fluids.* Pour boiling water upon a few slices of red cabbage, and, when cold, decant the clear infusion; distribute the liquor in three wine glasses. To the first add a solution of Alum; to the second, that of Potass; and to the third, a few drops of Muriatic acid. The liquid in the first glass will be thus made to assume a *purple*, that in the second, a *bright green*, and that in the third, a beautiful *crimson* colour.

Exp. 13.—*Two blue liquids become red by combination.* Pour a little tincture of Litmus into a wine-glass; and into another, some diluted Sulphate of Indigo; and the compound will have a perfectly *red* colour.

Exp. 14.—*Two Corrosive substances produce a mild compound.* If we add together Sulphuric acid and Potass, two highly caustic bodies, we shall produce Sulphate of Potass, a mild and almost tasteless compound.

Exp. 15.—*Infusible bodies form fusible Compounds.* Thus, neither Clay, Silex, nor Lime, will melt

singly, but when these three earths are mixed in
due proportion, a highly fusible compound is the
result. In the same manner, Bismuth, Lead, and
Tin, when separately heated, require a comparatively high temperature to effect their fusion;
whereas the alloy formed by their combination
will melt in boiling water.

140. The experiments above related illustrate, in a
very striking manner, the important changes which
bodies undergo by chemical combination with each
other; but it does not necessarily follow that every
compound is marked by an equally great change; in
some cases of combination, the resulting body will
differ but little in appearance or qualities, from the
ingredients of which it is composed. This fact is exemplified by the phœnomena of Solution, a process
which offers, perhaps, the simplest instance of Chemical
Affinity, and may with great propriety and advantage
be investigated in this part of the work, as the process
is, in many cases, essentially necessary to prepare
bodies for those complicated changes which we shall
hereafter endeavour to explain; indeed so essential
was this preliminary condition considered, that the
older chemists regarded it as indispensable to chemical
action; hence was deduced the axiom " *Corpora non
agunt nisi sint soluta,*" a principle, however, to which
the progress of chemical science has discovered many
exceptions; it has been found, for instance, that many
solid bodies will act upon each other, chemically, both
at a low and high temperature ; thus two simple earths,
if heated together, will combine, although the temperature be much lower than that which would be necessary for the fusion of either separately. This consideration induced Morveau to modify the axiom, and

to express it more justly in the following terms, " *There is no chemical action between bodies, if one of them is not sufficiently fluid*" (or rather, if its aggregation is not sufficiently weakened,) " *that its particles may yield to that affinity, which tends to bring them into contact with each other.*"

Solution.

141. By this term Chemists understand a certain operation, by which a solid body combines with a fluid, in such a manner that the compound retains the form of a transparent, and *permanent* fluid; that is to say, incapable of separation by any *mechanical* process. The condition is essential to the phænomenon, and serves, at once, to distinguish it from simple mixture, or mere mechanical diffusion, as will more clearly appear by the following experiment.

Exp. 16.—Take a portion of the earth termed *Magnesia*, and diffuse it by agitation in a quantity of water; a white turbid mixture will be formed, which, upon standing sufficiently long at rest, will deposit a powder, the fluid above remaining transparent. In this case no solution has taken place; the magnesia and the water have been merely *mechanically* mixed, since on standing they mutually separate; each obeying its specific gravity. If, however, to this turbid mixture a few drops of nitric acid be added, it will immediately become transparent, and the magnesia can no longer be separated by any mechanical process, because a new compound has been thus formed, and that compound is *soluble*.

142. In ordinary cases the distinction between *Solution* and *Suspension* is so well marked that the student cannot fail to recognize it, but in others it is not quite so unequivocal, and to those endowed with a tolerable share of knowledge it has even appeared doubtful. It is sometimes difficult to pronounce decidedly upon the existence of perfect transparency, but the fact of *permanency* can never deceive us, for in the case of mere mechanical diffusion, although it may not always be impaired by rest, it will not resist the ordeal of the filtre. In explanation of this truth we may adduce the beautiful golden-coloured liquor which is produced by pouring a solution of sulphuretted hydrogen into one of arsenious acid, and which, at first sight, we should not hesitate to pronounce a perfect solution; but by rest, or filtration, a yellow powder will be separated, and the liquid become colourless. So again, certain vegetable matters, when diffused in water, will, from the extreme fineness of their particles, offer so small an obstacle to the passage of light, as scarcely to affect the transparency of the medium in which they float, but, by rest or filtration, such a source of fallacy is easily detected. The medical student is earnestly advised to preserve these distinctions in his remembrance, as they will form the basis of some important precepts respecting the efficacy of certain remedies, and the expediency of some particular processes instituted for their preparation.

143. The term *Solution* is not only applied to the result of the process just described, but to the process itself; thus, if common salt be thrown into water it will disappear, or undergo the process of solution, while the result, so obtained, is called a solution of that body.

144. The process is, doubtless, nothing more than an effect of Chemical Affinity, exerted between the fluid and the body to be dissolved.* In chemical language the fluid body is termed the *Solvent*, or *Menstruum*,† the solid, the *Solvend*. This distinction, however, must not be understood as conveying the idea, that the one body is more active in the operation than the other; for the attraction by which it is produced is exerted mutually.

145. In order to understand perfectly the phænomenon in question, it is necessary to consider the solid body as under the influence of two antagonist forces, *viz.* the *Cohesive* attraction of its own particles, on the one hand, tending to preserve it in a solid state, and its *chemical Affinity* for the fluid, on the other, exerting its influence to bring it into a state of solution. In proportion to the relative energies of these opposing forces, the body will be easy or difficult of solution; and whatever external circumstances are capable of modifying such powers, will exert a corresponding influence upon its solubility; thus, for instance, as cohesion is diminished or overcome by the processes of *rasping, grinding, pulverizing,* &c. as already explained (100), the Chemist frequently avails himself of such expedients to facilitate and expedite solution.

---

* We are indebted to Sir Isaac Newton for the first just view of the nature of this operation. He rejected the vague explanation, which supposed it to depend on the figures of the particles, by which the one body was fitted to break down the other; and ascribed it solely to an attraction subsisting between the particles of the two bodies, by which they were united.

† *Menstruum.* The term derives its origin from a conceit of the older Chemists, that a mixed body could not be completely dissolved in a less period than forty days, which they called a *Philosophical Month.*

Exp. 17.—Introduce a lump of fluate of lime (*Fluor-spar*) into concentrated sulphuric acid; no sensible action will take place; but reduce the same substance to powder, and bring it into contact with the acid, and considerable action will ensue. The knowledge of this simple fact may be extensively applied for the explanation of numerous phænqmena in the animate, as well as inanimate kingdoms of nature; thus will it explain the influence of aggregation in modifying the sapidity of some, and the medicinal activity * of other substances.

146. Agitation favours solution by removing from the solid the portion of the fluid already saturated with it, and bringing a new portion to act upon it.

Exp. 18.—Into a wine-glassful of water, tinged blue with the infusion of cabbage, let fall a small lump of solid tartaric acid. The acid, if left at rest even during some hours, will only change to red that portion of the infusion which is in immediate contact with it; but stir the liquor, and the whole will immediately become red.

147. Heat, by diminishing the cohesion of a solid, necessarily increases and extends the operation of solution; whence hot liquids, generally considered, are more powerful solvents than cold ones.

Exp. 19.—To four fluid-ounces of water, at the ordinary temperature of the atmosphere, add three ounces of sulphate of soda (*Glauber's Salt*) in powder; a portion only will be dissolved; but apply heat, and the whole of that quantity will disappear.

* This subject has been very fully investigated in my Pharma cologia, Edit. 6. p. 319.

148. The additional quantity, however, so dissolved will again separate as the solution cools, although there are some exceptions to this general law; thus cold water will dissolve but a very minute quantity of white arsenic, but if it be boiled on that substance, the solution when cooled, will be found to retain a more considerable proportion.

149. In some cases the solubility of a substance is not affected by temperature, as in the instance of common salt, in which water acquires no increase of its solvent power by the application of heat; nor is the solvent with respect to every body equally promoted by an increase of temperature; water, for example, will dissolve five, six, or seven times more of some substances, when boiling, than when cold; of others, three or four times; and so on. These differences exist principally with respect to those substances termed salts.

150. From the view which has been thus taken of the phænomenon of Solution, it will be readily perceived that, in most cases, there must exist some certain point, at which the force of affinity between the solid and fluid will be balanced by the cohesive attraction of the solid, beyond which it is impossible for solution to proceed; this point is called *Saturation*, and the resulting compound, a *Saturated* solution. This *point of saturation*, however, in any fluid is very different with respect to different solids; of some it will dissolve its own weight; of others, one half, or one fourth; and of others, not more perhaps than one or two hundredth parts. With respect to some substances, the solvent power appears to be almost without a limit.

151. After a menstruum has been fully saturated with one body, it may nevertheless be capable of dissolving another: and when saturated even with a

second, may still retain the power of dissolving a third ;
but rarely in such large quantities as it would do were
it pure.

> *Exp.* 20.—Saturate a given portion of water with
> common salt; taking care that a small quantity
> remains undissolved at the bottom of the vessel ;
> if some nitre be now added, we shall find that not
> only this second body will be dissolved, but that
> the residual salt of the first solution will also dis-
> appear on agitation.

152. The explanation of this apparent paradox is
to be found in the simple fact, that new compounds
acquire new powers as solvents.

153. The law is important to the pharmaceutic and
medical chemist, and admits of some highly useful applications in practice. We are, for instance, thus
enabled to purify various salts ; *Sulphate of Magnesia*, for example, as it usually occurs in commerce, is
generally contaminated with a muriate of that earth,
to remove which we have only to wash the crystals in
a saturated solution of the former salt, which, although
incapable of dissolving any portion of them, will nevertheless carry off the muriate.

154. *Lixiviation* is merely solution performed with
a particular view. If we have a mixture of two kinds
of matter, one of which is soluble in water, the other
insoluble ; by the affusion of a sufficient quantity of
that fluid, the former will be dissolved, and the latter
at length remain pure. The solution so obtained is
termed a *Ley*.

155. This operation is usually performed on a large
scale, in tubs or vats, having a hole near the bottom
containing a wooden spigot and faucet. A layer of

straw is placed at the bottom of the vessel A B C D, over
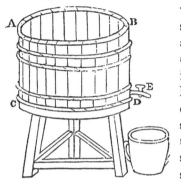
which the substance is
spread, and covered by
a cloth; after which hot
or cold water, accord-
ing as the salt is more or
less soluble, is poured
on. The water, which
soon takes up some of
the soluble parts of the
saline body, is after a
short time drawn off by
the spigot, E; and a fresh portion of water is successively added and drawn off, until the whole of the soluble matter be dissolved. The straw in this operation
acts the part of a filtre; while the cloth prevents the
water from making a hollow in the ingredients, when
it is poured on, by which it might escape without acting on the whole of the ingredients.

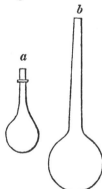

156. The vessels usually employed
by the chemist, in the operation of
solution, are small flasks or *matras-
ses* (a), the common Florence flask
is very convenient for such purposes,
or larger vessels (b), termed *Bolt-
heads*.

157. To such an extent are the medicinal properties of a body dependant upon its solubility, that the
physician may even adopt the maxim long since received by the chemist; viz: " *Corpora non agunt*

nisi sint soluta ; " and, although there may be many
exceptions to this general proposition, yet they will be
found neither more numerous, nor more inexplicable
than those which limit the chemical acceptation of the
same axiom.

158. The pharmaceutical operations of *Maceration* ;
Digestion ; *Infusion* ; *Decoction*, &c. are merely processes of solution, adapted to the particular nature of
the body to be dissolved, and to the objects for which
it is performed.

159. *Maceration* is an operation which is chiefly
performed on vegetable matter, by which the soluble
parts are dissolved by steeping them in cold water, or
in a spirituous menstruum. It is frequently useful in
preparing bodies for the operations of infusion and decoction, which are always rendered more effective by
the previous maceration of the materials.

160. *Digestion* is similar to maceration, except
that a gentle heat is employed to aid the solvent power
of the fluid. The process is usually conducted in a
glass matrass, and the evaporation of the liquid im-
peded by stopping the mouth of the vessel slightly with
a plug of tow, or by tying over it a slip of wet bladder
perforated with small holes. Where the menstruum is
of value, such as alcohol, a long open glass tube may be
affixed to the mouth of the matrass ; by which means
any part of the liquid, resolved into steam by the heat,
will be condensed, and returned into the vessel containing the materials. The matrass may be heated
either by a common fire, or through the medium of
water (*a water bath*) or sand (*a sand bath*).

161. *Infusion.* The object of this operation is to
extract from vegetable matter such principles as would
be lost or changed by digestion or decoction, or such

as, from their extreme solubility, are readily separated. See *Infusa*, Pharmacologia, Vol. ii.

162. *Decoction*, or boiling, is intended to answer the same purpose as infusion, but in a more extended degree. See *Decocta*, Pharmacologia, Vol. ii.

163. Although it has been shewn in the preceding pages, that the cohesive attraction is overcome by the opposing force of Chemical Affinity ; still it must not be considered that this power is wholly annihilated by it ; for it continues to exist, and is constantly tending to reunite the particles which are dissolved, and to restore their aggregation ; for, no sooner is a portion of the solvent removed by evaporation, than they become approximated, and are again brought within the limits of their mutual attraction, when they reunite, and the solid reappears.

164. The fluid menstruum may be either removed by *spontaneous* evaporation, occasioned by the simple exposure of a considerable surface to the atmosphere ; or by that which is produced by the application of heat. For this purpose, basins, or evaporating dishes of biscuit porcelain, made by Wedgewood, are usually employed ; and which are flat-bottomed, shallow vessels, of such a figure as to present a wide surface to the air.*

165. If the solid matter thus recovered from a solution, be of a vegetable nature it is termed an *Extract*, and the process is termed *Inspissation* ; whereas saline bodies, so obtained, assume regular forms denominated crystals, and the process in that case is called *Crystallization*.

---

* For chemical experiments, very good evaporating dishes may be made of the bottoms of broken flasks, retorts, &c. which may be cut round and smooth, by means of a hot iron or ring of wire, or by tying a piece of string moistened with spirit of turpentine, and then inflaming it. Watch glasses will also furnish a very convenient apparatus for such purposes.

166. As the fluid, which is dissipated during the process of evaporation, is entirely lost, and sacrificed for the sake of the fixed substance with which it is combined, evaporation is only employed in the cases where the liquid is but of little value, such as water; but where a solid is to be recovered from a more valuable liquid, as alcohol, for instance, a process termed *Distillation*, is instituted, whereby the chemist is enabled to carry on his operation in close vessels, so as to collect the fluid that is volatilized, and to preserve it without loss.

167. For the accomplishment of this object various forms of apparatus have been contrived. The common *Still*, as here represented, is the one more generally employed for the preparation of spirits, distilled waters, &c.

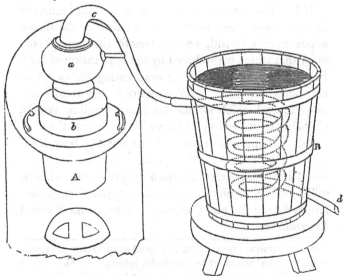

This apparatus consists of the following parts. The *Boiler, b,* the body of which, A, is partly sunk in the furnace; *a*, the head or *capital*, from the top of which rises the curved pipe, *c*, which enters the spiral tube, or *worm*, placed in a tub of cold water, B, termed the *Refrigeratory.* The Still is usually constructed of copper, but the *worm* is of pewter. The body, head, and worm require to be *luted** together, in order to

---

* LUTES, or CEMENTS, are used either to close the joinings of chemical vessels, to prevent the escape of vapours or gases; or, to protect vessels from the action of the fire which might otherwise crack, fuse, or calcine them. In this latter case the operation is termed *lorication,* or *coating.* To prevent the escape of the vapours of water, spirit, and liquors not corrosive, the simple application of slips of moistened bladder will answer well for glass; and paper with good paste for metal. Bladder to be very adhesive should be soaked some time in water moderately warm, till it feels clammy; it then sticks very well; but, if it be smeared with white of egg, instead of water, it will adhere still closer.

There is a great variety of receipts for the composition of Lutes, all of which are referable to one of the three following classes, *viz.*

I. *Unctuous and Resinous Lutes.*—1. Melt eight parts of beeswax with one of turpentine, and according as it may be required to be more or less consistent or pliable, add different proportions of common powdered rosin, and some brick-dust. This Lute adheres very closely to glass, and is not easily penetrated by acrid vapours; it has the advantage also of being very plastic and manageable; but it cannot sustain a heat above 140°.—2. Melt spermaceti, and while it is hot, throw in bits of caoutchouc. This is an excellent lute where much heat is not employed.—3. A solution of shell-lac in alcohol, added to a solution of isinglass in proof spirit, forms a cement that will resist moisture.—4. The lute best calculated for confining acid vapours for any length of time is termed the *Fat lute*, and is made by taking any quantity of tobacco-pipe clay, thoroughly dry, but not burnt, powdering it in an iron mortar, mixing it gradually with drying linseed oil, and then beating them together for a long time to the consistence of thick paste. Much manual labour is required, and it should be continued until the mass no longer adheres to the pestle. The edges of the glass or vessel, to which it is to be applied, should be per-

prevent the escape of any portion of the volatile product. The substances to be acted upon, having been
introduced into the boiler, and submitted to heat, are
soon volatilized, and raised into the *head,* whence they
pass into the *worm,* where they are condensed, and
issue in drops from the lower end of the pipe *d.* By degrees the water in the refrigeratory becomes warm, and
requires to be replaced by a fresh portion; and thence

---

fectly dry. Good glazier's putty, which is made of chalk, beat
up with drying linseed oil, much resembles the *fat* lute in quality.

II. *Mucilaginous and Gelatinous Lutes.*— 1. Linseed meal,
kneaded up with water to a sufficient consistence, and applied
tolerably thick over the joinings of the vessels; or Almond meal,
treated in a similar way, form very convenient lutes, which dry
and become firm in a short time.—2. Smear slips of linen on both
sides with white of egg, then apply these neatly to the joinings,
and when applied shake loosely over them some finely powdered
quick-lime. This lute dries speedily, is extremely hard and
cohesive, impervious to water, and impenetrable by most kinds
of vapour.

III. *Earthy Lutes.*—These are intended for operations which
require a high temperature.— 1. Mix burnt gypsum (*Plaister of
Paris*) in powder, with water to the consistence of a thick cream,
and apply it immediately. This forms a lute which sets as soon
as it is applied, and is firm; but a slight blow will easily crack it.
2. A very valuable fire lute may be made of about one part of
glass of borax, five parts of brick-dust, and five parts of clay,
finely powdered together, and mixed with a little water when used.

In every instance, where a lute or coating is applied, it is ad-
viseable to allow it to dry before the operation is commenced;
and even the *fat* lute, by exposure to the air during one or two
days after its application, is much improved in firmness. The
clay and sand lute is perfectly useless, except it be previously
quite dry. In applying a lute, the part immediately over the
juncture should swell outwards, and its diameter should be gradually diminished on each side, as may be seen in the figure
representing the distillation of Nitric acid, where the luting is
shewn applied to the joining of the retort and receiver.

the necessity of the tub being furnished with a stop-cock, by which the heated water may be drawn off, without disturbing the apparatus.

168. From the above description it will appear evident, that the common still can never be employed for the volatilization of substances that act on copper, or other metals; on such occasions a glass vessel termed an *Alembic,* * has been frequently used, and is represented by the accompanying sketch. It consists of

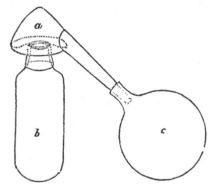

two parts ; the body, *b*, for containing the materials, and the head *a*, by which the vapour is condensed, and from which it is carried by a pipe into a receiver *c*. It will be perceived that the head is of a conical figure, and has its external circumference or base depressed lower than its neck, so that the vapours which rise, and are condensed against its sides,

---

* The Alembic appears to have been the most ancient form of a distilling apparatus. Both Dioscorides and Pliny mention the αμβίξ (*Ambix*) which is described by the latter of these writers, as a hemispherical iron cover, luted upon the earthen pots in which Mercury was procured by the distillation of Cinnabar; it is, however, probable that the Ambix was in the time of Pliny a mere plain Still, without any beak or gutter, since he mentions the mercury being *wiped off* in small drops from the inside of the vessel, the necessity of which manipulation would be superseded by the invention of a beak. The Alchemists having adopted this instrument, prefixed the Arabian article *al* to its name, and made considerable alterations in its form.

run down into the circular channel formed by its depressed part, from whence they are conveyed by the
nose or beak into the receiving vessel. The characteristic difference then between an Alembic and a Still
is in the construction of the head or capital, which in
the former is contrived not merely to collect, but to
condense the vapour; whereas, the corresponding part
of the latter serves only to collect the vapour, which
is transmitted in an elastic state through the beak,
and condensed in the worm.

169. In the English* laboratories the use of the
Alembic is now almost superseded by that of the Retort and Receiver, whose construction and applications
we have next to consider.

170. *The Retort* is a bottle with a long neck, so
bent as to make with the belly an angle of about sixty
degrees. From
this form it has
probably derived
its name. The
most capacious
part of the vessel
is termed its *belly*; its upper part is called an *arch*
or *roof*; and the bent part, the *neck*. A Retort may
be either plain, as represented above, or stoppered, as
shewn in the following figure, in this latter case it is
said to be *tubulated*.

171. To the Retort, a *Receiver* is a necessary ap-
pendage; and this may either be plain, as engraved
below (*c*) or tubulated. To some a pipe is added, as
may be seen in the figure representing the apparatus
for the distillation of Nitric acid. Such a receiver is
principally useful for enabling us to remove the dis-

* On the Continent, especially in France, the Alembic continues to be the favourite vessel for distillations in the large way.

tilled liquid, at different periods of the process, and is termed a *quilled receiver*.

172. In order to facilitate the condensation of the vapour, the neck of the retort is occasionally lengthened by an intermediate tube (*b*) termed an *Adopter*, the wider end of which slips over the retort neck, while its narrow extremity is admitted into the mouth of the receiver, as here exhibited.

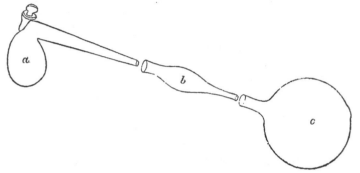

173. For the convenient introduction of the liquid into the retort several instruments have been contrived. When it is to be added at distant intervals during the process, the best contrivance is the one here exhibited, which consits of a bent tube, *a*, with a funnel at the upper end. When the whole is to be introduced at once, it is either done through the tubulure by the common funnel; or, if into a plain retort, through a funnel of a construction similar to the one here represented;

which will enable the operator to pour in the liquid without touching the inside of the retort

neck, which by being soiled would contaminate the
result of the process.

174. Retorts are made of earthenware, and of green,
or of white glass, according to the operations for which
they are destined. They are also heated in several
different modes. When the vessel is of earthenware,
and the substance to be distilled requires a strong heat
to raise it into vapour, the naked fire is applied. Glass
retorts are usually placed in heated sand; and, when
of a small size, the flame of an argand lamp, cautiously
regulated, may be very safely used. If the nature of
the operation be such as to require a glass vessel, and
a high temperature, the retort must be coated.*

175. In many processes, a large proportion of the
extricated matter is partly condensable into a liquid
and partly a gas, which is not condensed without the
presence of a considerable proportion of water. Unless
therefore, some means were provided either to effect
this condensation, or to allow the escape of the gas,
the apparatus would necessarily be burst in pieces.
To prevent such an accident, a small opening was
generally left, either in the joinings of the vessels, or
in the receiver, which could be kept shut, and occa-

---

* For this purpose a mixture of moist common clay, or loam,
with sand, and cut shreds of tow or flax, may be employed. The
following is the mode of applying it. After kneading the coating
material, so as to render it very plastic, let it be spread out on a
flat table, and lay the bottom of the retort in the middle of the
mass; then turn up the edges of the cake, 6o as to bring it round
the whole of the vessel, pressing it down in every part with the
fingers, until it applies uniformly and closely. If the distillation
be performed by a sand heat, the coating need not be applied
higher than that part of the retort which is bedded in sand; but
if the process be performed in a furnace the whole body of the
retort, and that part of the neck also which is exposed to the fire,
must be carefully coated.

sionally opened when the quantity of confined vapour
was supposed to be such as might endanger the rupture
of the vessels. By this contrivance, however, much
condensable vapour necesssarily escaped, and a large
proportion of the products of the distillation was lost.
This defect of the old apparatus was first attempted to
be remedied by Glauber, whose plan was improved by
Woulfe, the inventor of the apparatus here represented,
and which still bears his name.

176. It consists of a series of receiving vessels containing water.

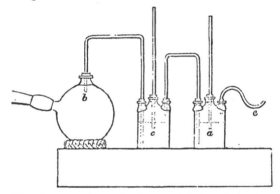

The first receiver (*b*) has a right angled glass tube,
open at both ends, fixed into its tubulure ; and the
other extremity of the tube is made to terminate beneath the surface of distilled water, contained, as high
as the horizontal dotted line in the three-necked bottle, (*c*). From another neck of this bottle, a second
pipe proceeds, which ends, like the first, under water,
contained in a second bottle (*a*). To the central neck
a straight tube, open at both ends is fixed, so that its
lower end may be a little beneath the surface of the
liquid. Of these bottles any number may be employed
that is thought necessary. The materials being intro-

duced into the retort, the arrangement completed, and. the joints secured, the distillation is commenced. The condensable vapour collects. in a liquid form in the. balloon *b*, while the involved gas passes through the. bent tube, beneath the surface of the water in *c*, which continues to absorb it till saturated. When the water of the first bottle can absorb no more, the gas passes, uncondensed, through the second right-angled tube, into the water of the second bottle, which in its turn becomes saturated. Any gas that may be produced, which is not absorbable by water, escapes through the bent tube *e*, and may be collected, if necessary. The use of the perpendicular tubes in the middle necks is to prevent the interruption of the process by an accident, which would perpetually occur without such a contrivance; for if, in consequence of a diminished temperature, an absorption or condensation of gas should take place in the retort, and of course in the balloon *b*, it must necessarily ensue that the water of the bottles *c* and *a*, would be forced by the pressure of the atmosphere, into the balloon, and possibly into the retort, which might cause a dangerous explosion; * but, with the addition of the central tubes, a sufficient quantity of air would, under such circumstances, rush through them to supply·any accidental vacuum. This inconvenience, however, is still more easily obviated by *Welther's tube of Safety*, which supersedes the necessity of three necked bottles.

---

* This transfer of a liquid from one vessel to another, in consequence of the formation of an imperfect vacuum, is termed *absorption*, and is highly annoying to the inexperienced operator.

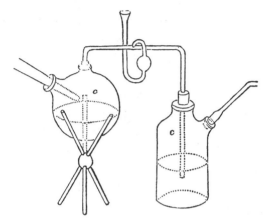

177. The apparatus being thus adjusted, as shewn by the above figure, a small quantity of water or mercury, is poured into the funnel, so as to about half fill the ball. When any absorption happens, the fluid rises in the ball, till none remains in the tube, when a quantity of air immediately rushes in and supplies the partial vacuum in *c*. On the other hand no gas can escape, under ordinary circumstances; because any pressure from within is instantly followed by the formation of a high column of liquid in the perpendicular part, which resists the egress of gas.

178. Very useful modifications of this apparatus have been contrived by Mr. Pepys, * Mr. Knight, † Dr. Murray, ‡ Dr. Hamilton, and Mr. Burkitt. ‖ There is also an American invention for the same pur-

* Philosophical Magazine, vol. xx.

† Nicholson's Journal, vol. iii.

‡ Murray's Chemistry, and Hamilton's Translation of Berthollet on Dyeing.

‖ Nicholson's Journal, 4to. vol. v. 349.

pose, which has considerable merit. † The arrangement by Mr. Pepys is represented in the annexed figure.

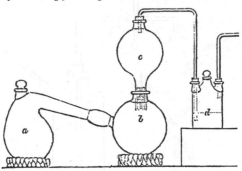

The receiver *b* is surmounted by a vessel *c* accurately
ground to it, and furnished with a glass valve, which
allows gas to pass freely into the vessel *c*, but prevents
the water which *c* contains from falling into the receiver *b*.

179. *Rectification.* This is the repeated distillation of any distilled product when it is not perfectly
pure. This second operation is conducted at a lower
temperature, so that the more volatile parts only are
raised, and carried over into the receiver, leaving the
impurities behind.

180. *Dephlegmation* or *Concentration.* The process of rectification is distinguished by these terms,
when the fluid is simply rendered stronger by the operation, as in the case of alcohol, by bringing over the
spirit, and leaving the superfluous water.

181. *Abstraction* is a term employed to denote that
the liquid is redistilled off from some fresh substance ;
as for instance in the rectification of acetic acid, where
Barytes is introduced into the still, in order to remove
any sulphuric acid that may be present.

182. *Cohobation* is a name given to rectification, when the product is redistilled from a fresh parcel of the same materials.

183. By the process of evaporation then, under which subject we have included the history of Distillation, the fluid part of a solution may be driven off, and separated from the fixed matter with which it was combined. We have next to consider under what circumstances, and in what forms, the *solid* constituent may be recovered.

184. It must be distinctly understood that, although in the process of solution, the cohesive attraction is overcome by the counteracting and superior power of chemical affinity, it is nevertheless not extinguished, but constantly manifesting its existence by tending to reunite the particles so dissolved; for no sooner is a portion of the menstruum removed by evaporation, or by other means to be hereafter described, than the particles reunite, and the solid reappears. In the accomplishment of this effect, the particles of certain bodies, as if actuated by a species of polarity, assemble in groups of a determinate figure, while others aggregate in confused masses.

185. The process, by which the particles of a body are enabled to arrange themselves into determinate forms, is termed CRYSTALLIZATION; while the regular bodies so produced are called CRYSTALS.

186. To impart freedom of motion to the particles of a solid body, we must confer upon it either the liquid, or aëriform condition; and it is equally evident that, in order to allow such particles to arrange themselves with regularity, or to *crystallize*, the body must be allowed to return to its solid form slowly, and without disturbance or interruption.

187. Solution and Fusion are the means by which the body is rendered fluid: evaporation and slow cool-

ing, or a gradual abstraction of the solvent by the action of chemical affinity, are those by which it is made to crystallize.

188. In some cases, heat at once annuls the cohesive force, and the substance is converted into vapour, which on being slowly condensed, will assume a regular crystalline form; it is thus that *Benzoic acid*; *Camphor; Arsenic; Iodine*, and various other solids to be hereafter examined, are made to assume the form of regular crystals; (*see Sublimation*).

189. It has been ascertained by observation and experiment, that every substance, in crystallizing, has a tendency to assume a particular geometric figure; thus, common salt crystallizes in cubes; Epsom salt, in four-sided prisms; Alum, in octohedrons; and so on. To the mineralogist and chemist, therefore, an intimate acquaintance with the crystalline forms and modifications of natural bodies is a subject of the highest importance, as enabling them to ascertain, directly, that which, independent of such aids, could only be arrived at by indirect and circuitous processes. Crystals to the mineralogist are what flowers are to the botanist, by which the former is enabled to deduce the chemical nature of a mineral from its mechanical figure, just as the latter pronounces from the structure of the corolla the class and order of the plant to which it belongs. The accomplishment of such an end, however, involves more extensive knowledge than a mere acquaintance with the external form under inspection, for the figures of crystals are liable to be altered by the influence of various circumstances upon the crystallizing process; and hence the geometric forms which the same substances present will often bear little or no *apparent* resemblance to each other; and yet, great as these diversities of figure may appear on a superficial view, they will be found capable of being resolved into

a small number of simple forms which, for each individual, will be *always the same.* The consideration of the geometric principles upon which such phænomena depend, and the explanation of the production of the variety of *secondary* forms, with which Nature and Art have alike furnished us, constitute a peculiar and elaborate branch of science termed CRYSTALLOGRAPHY,* the elucidation of which would be wholly incompatible with the plan and objects of an elementary work; it is only necessary to state, that there are not more than six *primitive* forms, and that the *secondary* crystals are supposed to arise from *decrements* of particles taking place on different edges and angles of the former. Thus a cube, having a series of decreasing layers of cubic particles upon each of its six faces will become a dodecaëdron, if the decrement be upon its edges; but an octoëdron, if upon the angles; and by irregular, intermediate, and mixed *decrements*, an infinite variety of secondary forms would ensue.

190. There are two methods by which the *primary* form of a crystal may be obtained;—by mechanical force, applied in the direction of its layers, which is termed by the mineralogist *Cleavage*,† and by the

---

* In Pharmacy a knowledge of Crystallography is turned to very valuable account, in estimating the purity of our saline medicines; and Mr. PHILLIPS has rendered the profession an important service, by the introduction of figures, illustrative of their crystalline forms, in his excellent translation of the London Pharmacopœia.

To those who are desirous of cultivating this branch of science, a work lately published by H. JAMES BROOKE, Esq. entitled " A FAMILIAR INTRODUCTION TO CRYSTALLOGRAPHY," is strongly recommended.

† It had been long known to jewellers and lapidaries, that every crystal may, by means of proper instruments, be split in certain directions, so as to present plane and smooth surfaces; while if split in other directions, the fracture is rugged, and the mere effect of violence, instead of being guided by the natural joining of a crystal. This simple fact has proved in the hands of Abbe Haüy, a key to the whole theory of crystallization.

action of fluid menstrua, by which its structure is gradually developed by the process of solution.　The former, however, is the method usually adopted by the mineralogist; the latter ought perhaps to be regarded as a subject rather of speculative interest, than of practical utility.　We are indebted to Mr. Daniel † for a knowledge of the phænomena which it presents, and which he has illustrated by many beautiful and satisfactory experiments.　A shapeless mass of alum, for instance, weighing about 1500 grains, being immersed in 15 ounce measures of water, and set by, in a quiet place, for a period of three or four weeks, will be found to have been more dissolved towards the upper than the lower part, and to have assumed a pyramidal form.　On further examination the surface of the mass

will be found to have been unequally acted upon by the fluid, and will present the form of octohedrons, and sections of octohedrons, carved, as it were, in high relief and of various dimensions, as represented in the annexed figure.　Borax, thus treated, will in the course of six weeks exhibit eight-sided prisms with various terminations; and other salts may be made to unfold their external structure by the same slow agency of water. The carbonates of lime, strontia, and baryta, will give also distinct results, when acted upon by weak acids; and even amorphous masses of those metals, which have a tendency to assume a crystalline form, such as bismuth, antimony, and nickel, when exposed to very dilute nitric acid, present, at the end of a few days, distinct crystalline forms. Mr. Faraday also found that a large crystal of sulphuret of antimony, when introduced into a portion of fused sulphuret, began to melt down, but not uniformly, crystals more than half an inch in length being left projecting from it, and thus affording an admirable illustration of Mr. Daniel's mode of displaying crystalline texture.

We may now proceed to consider the artificial process of crystallization in its practical connection with the objects of pharmacy.

191. It will follow from what has been already stated, that, in order to obtain well formed crystals from saline solutions three essential circumstances are required; viz. *Time, Space,* and *Repose.* By *time* the superabundant fluid is slowly dissipated, so as to allow the particles of the salt to approach each other by insensible degrees, and without any sudden shock; in which case they unite according to their constant laws, and form a regular crystal; indeed it is a general rule, that the slower the formation of a crystal the more perfect will be its form; the larger, its size; and the harder and more transparent, its texture; while, on the contrary, too speedy an abstraction of the separating fluid will force the particles to come together suddenly, and, as it were, by the first faces that offer; in which case the crystallization is irregular, and the figure of the crystal indeterminate; and, if the abstrac-

tion be altogether sudden, the body will ever form a concrete mass with scarcely a vestige of crystalline appearance. *Space,* or sufficient latitude for motion, is also a very necessary condition; for, if Nature be restrained in her operations, the products of her labour will exhibit marks of constraint. *A state of repose* in the fluid is absolutely necessary for obtaining regular forms; all symmetrical arrangement is opposed by agitation, and a crop of crystals obtained under such circumstances would necessarily be confused and irregular.

192. The whole art of crystallizing substances is founded upon these obvious truths, as will be fully illustrated in the sequel; although on many occasions, to ensure perfect success a certain address is required in the manipulation, which has enabled particular manufacturers to produce articles very superior to ordinary specimens.

193. In the act of separating from the water in which they were dissolved, the crystals of almost every salt carry with them a quantity of water, which is essential to the regularity of their form, as well as to their transparency and density, and which cannot be expelled without reducing them to shapeless masses. It is termed their *water of Crystallization,* and its proportion will be found to vary very essentially in the different salts; in some instances constituting more than half their weight, as in the case of sulphate of soda, carbonate of soda, nitrate of ammonia, &c. while in others it is extremely small; and yet, however it may differ in different salts, it always bears, in the same salt, the same definite ratio to the solid saline matter; thus in crystallized bicarbonate of potass, (*Potassæ Carbonas.* P. L.) to every three proportionals of salt there is one of water; while in the carbonate of soda (*Sodæ Sub-carbonas.* P. L.)

to every two proportionals of salt there are eleven of water. This water appears to be in a state of combination with the salt, and not simply interposed between its laminæ.

194. Where the water of crystallization exists in a large proportion, and the solubility of the salt is greatly increased by elevation of temperature, the application of heat is often sufficient to liquify the crystals, producing what is termed *Watery fusion.*

195. The water of crystallization is retained in different salts with very different degrees of force. Some crystals lose it by mere exposure to the atmosphere, in consequence of which they pass into the state of a dry powder, and are said to *effloresce* ;* while others, on the contrary, not only retain it very strongly, but even attract more; and, on exposure to the atmosphere, become liquid, or *deliquesce.*

196. The means by which a solution is made to surrender its saline charge in a crystalline form are twofold, *viz.* by REFRIGERATION and EVAPORATION ; and since the manner and degree in which such processes are to be directed must, in each case, depend upon the connection which subsists between solubility and temperature, it will be convenient to consider the subject under these relations.

CASE I. IN WHICH THE SOLUBILITY OF THE SALT IS INCREASED BY HEAT.

197. When the salt to be crystallized is considerably more soluble in hot than in cold water, as sulphate of

---

* This circumstance is very annoying to the collectors of crystals ; such as are liable to this change should be preserved in jars containing a portion of water ; or, where the salt is not acted upon by oil, the crystals may be effectually defended by merely smearing their surfaces with that substance.

soda, nitrate of potass, &c. it is only necessary to saturate hot water with it, and to set it aside to cool; by
which we shall obtain a considerable crop of crystals.
It might, indeed, have been readily anticipated, that a
salt of this kind would under such circumstances easily
crystallize, for since it ceases to be equally soluble in
water, of which the temperature is diminished, that
portion of it which owed its solubility to heat must
necessarily separate as the liquor cools; and when this
is effected, the solvent will only retain such a quantity
as cold water would dissolve. In conducting this process, however, the refrigeration of the fluid ought to be
gradual, for the more slowly the water cools, the more
regular will be the crystals obtained, as the saline particles will thus, with greater certainty, be made to approach each other by those faces which are most suitable (191); whereas, if the boiling solution be suddenly cooled, it will let fall, in a shapeless mass, all
the exces of salt that was dissolved by the agency of
heat. It will be expedient, therefore, to place the solution, while very hot, on a sand bath, or in a warm
place, and to lower its temperature by slow degrees;
the vessel should, at the same time, be covered with a
cloth to prevent the access of cold air. After the solution is perfectly cold, and all the excess of crystalline
matter is separated, the residual liquor may be treated
as directed in the following case.

CASE II. IN WHICH THE SOLUBILITY OF THE SALT
IS NOT INCREASED BY HEAT.

198. If we have a saturated solution of a salt which
is not more soluble in hot than in cold water, it is evident that refrigeration will not advance its crystalliza-

tion. In such a case we must proceed to abstract a portion of the menstruum; and this may be accomplished in two ways, *viz.*

A. *By evaporation through the agency of Heat.*

199. In common cases, all that will be required is to continue the process of evaporation, until a drop of the solution when placed upon a cold body shews a disposition to crystallize; or, at farthest, until we observe a thin film of saline matter, called a *Pellicle*, creeping over the surface of the liquid; a phænomenon which indicates that the attraction of the saline particles for each other, is becoming superior to their attraction for the water; and that the solution, therefore, when left undisturbed, will crystallize.* After the first crop of crystals has been thus made to separate, the evaporation may be repeated, and another crop obtained, and so on, until by a succession of evaporations and coolings, the greatest part of the saline charge is surrendered in a crystalline state.

200. If the crystallizable salt be perfectly pure, the whole of its solution may be thus crystallized; but it often happens that, if two or more salts exist in the same menstruum, after crystals have been obtained by several evaporations and coolings, the remaining portion of the fluid, although saturated with saline matter, will nevertheless refuse to crystallize; in which state it is technically denominated *mother water.*

201. This occurs where both of the salts have only a slight tendency to crystallize, and are of nearly equal solubility, such, for instance, as Nitrate of soda, and

---

* In some cases the operator derives his indication from the specific gravity of the solution under treatment, as already explained (86).

Sulphate of soda; whereas, if the two salts be endowed with a strong degree of crystalline power, and possess, moreover, different ratios of solubility, as in the case of Nitrate of potass, and Sulphate of potass, they may be readily removed by crystallization, without leaving any *mother water*.

202. The above process affords a convenient mode of separating salts, which coexist in the same solution; for on carefully reducing the quantity of the solvent by evaporation, the salt whose particles have the greatest cohesion will crystallize first. If both salts are more soluble in hot than in cold water, the crystals will not appear till the liquid cools. But if one of them, like common salt, is equally soluble in hot and in cold water, crystals will appear, even during the act of evaporation. In this way we may completely separate nitre from common salt, the crystals of the latter being formed during evaporation; while those of nitre do not appear till some time after the fluid has cooled.

203. In some cases, the affinity of a salt for its solvent is so great, that it will not separate from it in the form of crystals; but will yet crystallize from another fluid, which is capable of dissolving it, and for which it has a weaker affinity; thus, for instance, potass cannot be made to crystallize from its watery solution, but will yet separate, in a regular form, from its solution in alcohol.

B. *By spontaneous Evaporation.*

204. For this purpose, we must expose the saline solution to the temperature of the atmosphere, in vessels having a considerable extent of surface, and slightly covered with fine paper, or gauze, which may prevent any dust from falling into the liquor. without opposing

the progress of evaporation. For this operation, it will be advantageous to select a separate chamber well ventilated, and sufficiently light (215). The solution is thus left exposed to the air, till crystals are perceived in it, which sometimes will not take place in less than four, five, or six weeks, or, with some salts, for even a still longer period. This is by far the most successful and certain of all the processes for obtaining perfect crystals of considerable size, and is that which ought always to be preferred where time and circumstances will allow the election.

205. After the operation, however, has continued for some time in progress, the active vigilance of the operator becomes necessary; for when the quantity of salt held in solution becomes much diminished, the liquid will begin to act upon the separated crystals, and to re-dissolve them. The action is, at first, perceptible on their angles and edges; they become blunted, and gradually lose their shape altogether. Whenever this operation commences, the liquid must be poured off, and a fresh portion of the solution substituted, otherwise the crystals will be infallibly destroyed.

206. The student may exemplify these facts by the following simple and striking experiment.

Exp. 21.—Introduce a fully saturated solution of common salt* into a medicine phial of four ounces, so as to occupy about two-thirds of it, and cover its mouth loosely with a piece of filtering paper; then place it upon a shelf exposed to light (215). In the course of a fortnight, a cube of salt will be

---

* As common culinary salt usually contains a portion of muriate of magnesia, it will be necessary to remove this foreign ingredient, by precipitating it from the solution by carbonate of soda and then to filter the liquid.

distinguished at the bottom of the phial, which will gradually augment in size, until it attains a certain magnitude, when, if attentively watched, its angles will be observed to become less acute, and the crystal will gradually lose its figure.

207. There is yet to be mentioned another mode, by which evaporation may be accomplished; but, except for philosophical objects, I am not aware that it has ever been applied for the purpose of crystallization. It consists in exposing the solution, in a capsule of glass, to the absorbing action of Sulphuric acid, under the receiver of the air-pump. A process which will be better understood when we come to treat of Mr. Leslie's elegant method of producing artificial congelation. We may next proceed to examine the external and contingent circumstances which are capable of influencing the formation of crystals.

a. The form of the Vessel.

208. The shape of the evaporating dish produces a considerable variety in the figure and mode of concretion; for, since crystals extend themselves more in a horizontal, than in a vertical direction, it is evident that by attaching themselves to the oblique, irregular, and uneven sides of the vessels they must always become more or less irregular. The circumstance of crystals extending more rapidly at the bottom of a tall vessel than nearer the surface, will admit of a simple explanation from the fact of the integrant particles being denser than the solution from which they are separated, falling down, and so augmenting by their continual accretion the crystals below.

b. The presence of some foreign body which may act as a nucleus.

209. To prevent the crystals from adhering to the sides of the vessel containing the solution, some substance, such as pieces of thread, string, or wood, are usually introduced, for the purpose of collecting them. In this case the crystals are precipitated on the threads, and as the surface thus presented has very little extent, they have commonly the greatest regularity of figure ; it is thus that we meet with well formed crystals of sugar-candy, verdigris, &c. in the different shops, on strings and sticks. The student may satisfy himself of the value of such an expedient by the following experiment.

 Take a cold saturated solution of any crystallizable salt, pour it into a capacious jar, and suspend in it a line of silk, or horse hair, attached on its upper part to a piece of cork that will float on the surface of the liquid, and steadied by a shot affixed to the other extremity; as exhibited in the annexed figure. In a short time the line will become studded with well defined crystals.

210. This property of crystals, to deposit themselves around a nucleus, has, of late, been frequently illustrated by placing various bodies in saline solutions, by which they become so studded with crystals as to resemble natural productions of great rarity and beauty.

211. But a still more effectual way of inducing regular crystallization is to immerse in the solution a crystal of the same kind, which by becoming, as it

were, a rallying point for all the particles, considerably accelerates the process of aggregation. And it is moreover stated that, if there be two salts in solution, that one will most readily separate, of which the crystal has been introduced.

c. The influence of atmospheric pressure.

212. That the excess of air may exert an important influence on the process of crystallization is, at once, rendered evident by the following interesting experiment.

Exp. 22.—Prepare a concentrated solution of Glauber's salt, by adding portions of it gradually to water kept boiling, until this fluid will dissolve no more.* Pour the solution while boiling hot into common phials, previously warmed, and immediately cork them, or tie slips of wetted bladder over their orifices, so as to exclude the access of common air from the solution. This being done, set the phials by in a quiet place, without agitation. The solution will now cool to the temperature of the air, and remain perfectly fluid, but the moment the cork has been drawn, or the bladder punctured, and atmospheric air is admitted, it begins to crystallize on its upper surface, the crystallization shoots downwards in a few seconds, like a dense white cloud, and so much heat becomes evolved, as to make the phial very sensibly warm to the hand. When the crystallization is accomplished, the whole mass is so completely solidified, that, on inverting the phial, not a drop of

---

* An ounce and a half of water will thus dissolve two ounces of salt.

it falls out. The student however will remark, that its crystalline structure is very confused in consequence of the rapidity with which it took place (191). Should the salt not crystallize immediately on the admission of air, it will be only necessary to produce a slight agitation in the fluid (214) to insure the result.

213. The phænomenon is explained by Dr. Murray* in the following manner. When the saturated solution of the salt is enclosed in the vessel, and the pressure of the atmosphere is excluded, the particles in solution may be conceived as placed at distances too great to admit of the attraction of cohesion being asserted, so as to cause them to unite and crystallize. But when the pressure of the air, or any equivalent pressure, is brought to act on the surface of the fluid, its particles, as well as the particles of the solid contained in it, are forced nearer to each other; the distances between them are lessened; the attraction of cohesion is exerted, and the crystallization commences. The small crystals that are thus formed at the surface, afford solid points, from which other crystals are formed; and this proceeds rapidly through the whole fluid.

d. The effect of Agitation.

214. The absence of external motion has been stated (191) to be essential to the production of perfect crystals; there are, however, certain cases in which the process may be promoted by exciting a slight tremulous motion in the fluid, such as may be communicated by striking lightly, with the bottom of the vessel, the table upon which it rests. We may conceive that agitation

† System of Chemistry, vol. 1. p. 87.

not only can assist the saline particles in disengaging themselves from the aqueous ones, which may still oppose a small obstacle to their union, but we may imagine that they can occasion, at the same time, such motions as will lead to those positions which give the greatest advantage to crystalline attraction,*

e.   The effect of Light.

215. That the process of crystallization is influenced by light, is at once rendered evident by the appearance of crystals in the bottles of druggists, when placed in the window, and which are always most copious upon the surface exposed to light. If we place a solution of nitre in a room which has the light only admitted through a small hole in the window-shutter, crystals will form most abundantly upon the side of the basin exposed to the aperture through which the light enters; and Chaptal found † that by using a solution of a metallic salt, and shading the greater part of the vessel with black silk, capillary crystals shoot up the uncovered side, and that the extent of the exposed part is distinctly marked by the limit of the crystallization.

216. The phænomenon, termed *Saline Vegetation,* has been also shewn by Chaptal † to depend upon the influence of light. It consists in the creeping up of the salt around the edge of the vessel, and is very embarrassing to the operator, since it has been found that, when this operation is suffered to proceed, the spon-

---

* It is well known that in calm weather, water may be reduced several degrees below 32°, without being frozen ; but, by the least agitation, as by that occasioned by a slight breeze, a crust of ice will form immediately on its surface.

† Journal de Physique, T. iv. p. 300.

taneous evaporation of the fluid affords very few crystals; all the saline matter having extended itself over
the sides of the vessel. The easiest mode of preventing
this occurrence is to smear the edge of the vessel with
a little sweet oil.

f. The effect of Electricity.

217. To a peculiar electrical state of the atmosphere
the operator has often attributed his want of success in
a crystallizing process. It has been repeatedly observed
that saline solutions which have not yielded crystals,
after having been sufficiently concentrated, and left
to undisturbed repose for many days, have suddenly
deposited an abundant crop, during, or immediately
after, a thunder-storm.

218. Having thus enumerated the principal circumstances by which the process of crystallization is liable
to be disturbed, we may here introduce to the notice
of the practical chemist some account of the method
proposed by Le Blanc, * for obtaining perfect crystals
of almost any size. The process is as follows;—Let
the salt to be crystallized be dissolved in water; con-
centrate the solution slowly by evaporation, to such a
degree that it shall crystallize on cooling, which may
be known by suffering a drop of it to cool on a plate
of glass, or other substance. This being done, let the
solution be put aside, and when perfectly cold, pour off
the liquid portion from the mass of crystals at the
bottom, and put it into a flat-bottomed vessel. After
having stood for some days, solitary crystals will be
performed.. This having been accomplished, crystals
will begin to form at some distance from each other,
which gradually increase in size; select the most regu-

---

* Journal de Physique, T. iv. p. 300.

lar of these, and place them in a flat-bottomed vessel
at some distance from each other, and pour over them
a quantity of the concentrated liquid obtained by evaporating a solution of the salt, till it crystallizes on
cooling. Alter the position of every crystal once at
least every day, with a glass rod, that all the faces may
be alternately exposed to the action of the liquid; for
the side on which the crystal rests, or is in contact with
the vessel, never receives any increment. The crystals
will thus gradually increase in size. When they have
arrived at such a magnitude that their figure can be
easily distinguished, let the most perfect of them be
selected, or those having the exact form which we are
desirous of obtaining; and put them separately in a
vessel filled with a portion of the same liquid, and let
them be turned, in the manner directed, several times
a-day. By these means they may ultimately be obtained of almost any magnitude we may desire. In
the advanced stage of the process, for reasons already
explained (205), we must carefully watch their progress, or they will be entirely demolished.

219. From the view which has been just taken of
the subject of CRYSTALLIZATION, it will appear that
the structure of crystals bears no resemblance whatever
to *organization*. Nothing can be more different than
the increase or accretion of a crystal, and the growth
of an organic being;* the one takes place by the mere
juxta-position of new particles, mechanically or chemically applied to its exterior surface; whereas the
other increases in dimension by appropriating different

---

* Some physiologists of the French and German schools have
maintained a different opinion, but it may be traced to the erroneous definition of Linnæus, who has said " a Mineral *grows*;
a Vegetable *grows* and *lives*; and an Animal *grows*, *lives*, and
feels."

materials for its subsistence. At the same time it would be too much to assert, that there is no formation in the animal body which derives any assistance from the operation of those affinities which bestow impulse upon the particles of inanimate matter.*

ELECTIVE AFFINITY.

220. The student having learnt the nature of that species of attraction, by which two bodies of a different nature combine with each other, so as to lose their individuality, and to produce a compound with new characters, we are now prepared to conduct him one step farther in the inquiry, and to demonstrate the im-

---

* In the formation of bone, for instance, although it may be most agreeable to the laws of the animal œconomy to suppose that the minute arteries pour out the earthy matter from their extremities, yet, as Dr. Bostock justly observes, there are some circumstances respecting the mechanical mode in which the operation is effected, which it is not easy to reconcile to this supposition. We may conceive that from some unknown cause, the arteries of certain parts of the cartilage acquire a disposition to deposit the earthy matter which they contain; yet how can these depositions acquire the particular form which they exhibit? they do not seem to follow the direction of the arteries which accompany them in their course, but they sometimes assume the radiated structure, branching off from a common centre; or, at other times, compose masses of parallel lines, shooting out in a course contrary to that of the vessels which are supposed to carry the materials for their formation. If, adds Dr. Bostock, we might be permitted to indulge a conjecture upon the subject, it would seem as if certain portions of the cartilage acquired an affinity for the earthy matter, and that, when the deposit began to be formed, it received accessions of new matter in consequence of the affinity between the particles of the phosphate of lime producing a species of CRYSTALLIZATION.

portant truth that *the attractions exerted by any body towards others, are different in their force with respect to each.*

> *Exp. 23.*—Mix together equal weights of magnesia and of quicklime, in fine powder, and add diluted nitric acid. After some hours, it will be found that a considerable part of the lime has been dissolved, but that the whole of the magnesia has remained untouched. Hence it is clear that nitric acid has a stronger attraction for lime, than for magnesia.

221. In such a case the nitric acid has been metaphorically represented as making an election, whence this species of Affinity is denominated *Elective* Affinity.

222. Upon the discovery of this important law, it occurred to Geoffroy, a French chemist, that tables might be constructed, which should exhibit the relative forces of attraction of any body towards others. These have been since extended to such a degree as to comprehend the greater part of the combinations and decompositions in chemistry. The substance, whose affinities are to be thus expressed, is merely placed at the head of a column, separated from the rest by a horizontal line. Beneath this line are arranged the different substances for which it has any attraction, in an order corresponding with that of their respective forces of affinity; the substance which it attracts most powerfully being placed nearest to it, and that for which it has the least affinity, at the bottom of the column. The following series exhibiting the affinities of Muriatic acid for the alkalies, and alkaline earths, may serve as an example.

Muriatic Acid.

Baryta.
Potass.
Soda.
Lime.
Ammonia.
Magnesia.

223. In consequence of the same body thus uniting with others, with different degrees of force, we are enabled to decompose a compound by adding any substance which has an attraction to one of them superior to that by which they were held united. The two bodies, between which there is the strongest attraction, combine, and the third is separated. This may be easily demonstrated by a modification of the last experiment, viz.

Exp. 24.—Heat together, in a flask, nitric acid, and magnesia; these substances will combine, and a *nitrate of magnesia* result; at the same time, make a solution of lime in water, by agitating some powdered quick-lime in distilled water. Let the solution of lime be poured into that of magnesia, when a white powder will separate, and gradually fall to the bottom of the vessel. This powder is found to be magnesia, which is thus disunited from nitric acid, in consequence of the stronger attraction of lime for that acid.

224. The following series of experiments, which the student may easily perform, are admirably calculated to afford a farther illustration of this law of elective affinity.

Exp. 25.—Dissolve pure silver in nitric acid, or make a solution of nitrate of silver in distilled water. To this add mercury, which will be immediately dissolved, and the silver disengaged. The supernatant fluid will then be *a solution of mercury in nitric acid.*

Exp. 26.—To the above solution of mercury in nitric acid, present a piece of sheet lead; the latter metal will be dissolved, and the mercury become disengaged. The fluid will then be *a solution of lead in nitric acid.*

Exp. 27.—If in this solution of lead, a thin slice of copper be suspended, the copper will be dissolved, and the lead become disengaged. The fluid now is *a solution of copper in nitric acid.*

Exp. 28.—In this solution of copper, let a thin sheet of iron be immersed; in a short time the iron will disappear, and be replaced by metallic copper, and we have now a solution of *iron in nitric acid.*

Exp. 29.—Let a piece of zinc be next presented to the solution; the iron will be thus separated, and the *zinc remain in solution.*

Exp. 30.—To the zinc solution, we may now add ammonia, which will instantly combine with the nitric acid, and the zinc be thus separated; the solution will in this case alone contain *nitrate of ammonia.*

Exp. 31.—Into this solution of nitrate of ammonia pour some lime-water, the ammonia will be instantly disengaged (manifesting itself by a pungent odour) and the solution will be *nitrate of lime.*

Exp. 32.—If to this solution we add oxalic acid, the lime will be precipitated, and what now remains will be merely nitric acid.

We shall offer one more experiment in illustration of the present subject.

> *Exp.* 33.—Add to common writing ink a few drops of nitric acid; its blackness will instantly disappear. Into this colourless liquid drop a small quantity of potash in solution, and the blackness will be restored. Writing ink owes its hue to a compound of iron and gallic acid; upon the addition of the nitric acid, the iron forsakes the gallic acid, for which it has a weaker affinity, and a nitrate of that metal is produced; no sooner, however, is the potass presented to the solution than it carries off the nitric acid, and the iron thus abandoned returns to its former associate, the gallic acid.

225. When a body is liberated from a compound by the agency of elective affinity, it may either pass off in the state of gas, as in experiment 30, or remain in solution,* or be separated as an insoluble body; in which latter case it is said to be *precipitated,* the substance employed to produce the decomposition is termed the *Precipitant,* while the fluid which remains after the operation is called the *super-natant* liquor.

226. In those cases where the developed constituent assumes an elastic form, it must be obtained through the medium of a distilling apparatus connected with Woulfe's bottles, as already explained (176), or by means of a *hydro-pneumatic trough,* as will be hereafter described.

---

* It is essential to impress this fact upon the mind of the medical student; for a popular opinion has existed, that no change of qualities can occur, upon admixture, unless precipitation takes place: see Pharmacologia, Edit. 6. vol. 1. p. 342.

227. As *Precipitation* is an operation of the highest
importance, both in a chemical and pharmaceutical
point of view, it will be necessary in this place to enter
into some details, in explanation of its nature and uses.
It is employed to separate solids from the solutions in
which they are contained; to produce new combinations, which cannot be readily formed by the direct
union of their constituents; to purify solutions from
precipitable impurities; and to reduce a body to a
finer state of division than the most laboured mechanical process can effect.   To this latter circumstance,
precipitates frequently owe their medicinal activity, as
I have endeavoured to explain in another work ;† they
are, moreover, in this attenuated condition, more easily
disposed to enter into new combinations; thus, for
instance, Silica, although reduced to the finest powder
by levigation, may be boiled for some time in liquid
potass without being rendered sensibly soluble; but,
when first precipitated from a state of chemical solution, it is not only readily soluble in that menstruum,
but is even capable of being acted upon by certain
acids.

228.   In some cases the precipitate is separated by
the *precipitant*, in consequence of the latter having a
greater affinity for the liquid, and thence weakening
its attraction to the substance which is held in solution,
as occurs when water is added to spirit of camphor, or,
alcohol to the solutions of certain salts.   In other cases,
the precipitate is an insoluble compound formed by
the union of the added substance with that which was
previously held in solution, as takes place upon the
addition of sulphuric acid to a solution of baryta.

229.   The specific gravity and form of the precipitate will also vary considerably in different bodies, and

---

† PHARMACOLOGIA, Edit. 6. vol. 1. p. 318.

even in the same body under different circumstances. When the separated body is very light and rises to the surface of the liquid it is termed a *cream* ; thus, by the addition of any acid to a solution of soap, the alkali unites with the acid and the oil, thus separated, swims on the surface. In certain cases, where the process of precipitation is slow, the separating particles are enabled to obey the law of their polarity, and to arrange themselves in such a manner as to form regular solids ; a fact which serves still further to illustrate those views which have been already explained respecting the law of cohesion, and to connect the process of precipitation with that of crystallization.

    Exp. 34.—Into a saturated solution of sulphate of magnesia pour a portion of alcohol, and the salt will be precipitated in a *crystalline* form.

230. The vessels usually employed for precipitation are tall jars, which are sometimes narrower at bottom

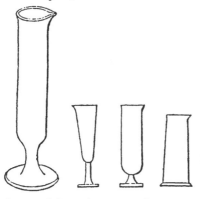

than at the mouth, so as to allow the precipitate to collect by subsidence, and the supernatant liquor to be afterwards decanted off. In other cases, such a form is troublesome, on account of the difficulty with which the precipitate is removed.

231. When the chief object of the process is to obtain the precipitate in a pure form, it is necessary to wash it, after it is separated by filtration, and to dry it

by a heat not exceeding 212°; for the accomplishment of which an extremely useful apparatus is sold by the makers of chemical instruments, and is represented in the annexed figure.

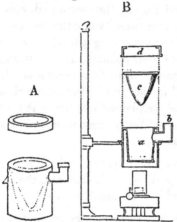

A exhibits the vessel, with its different parts *in situ.*

B, shews the same apparatus supported by the ring of a stand over an argand lamp; its parts being detached in order to render the description of them more perspicuous. The vessel *a* is of sheet iron, or copper, japanned and hard soldered; *c* is a conical vessel of very thin glass, having a rim, which prevents it, when in its place, from entirely slipping into *a*; and *d* is a moveable ring, which keeps the vessel *c* in its place. When the apparatus is in use, water is poured into *a*, and the vessel *c*, containing the substance to be dried, is immersed in the water, and secured by the ring *d*; the whole apparatus is then suspended over an argand lamp. The steam escapes by means of the chimney *b*, through which a little hot water may be occasionally poured, to supply the waste by evaporation. Where our object is to ascertain with accuracy the weight of any precipitate, as in the case of an analysis, if we previously estimate the weight of the filtre employed, we may at once arrive at the conclusion without incurring the trouble and fallacy of separating the matter deposited upon it.

232. The changes which occur, through the agency of that species of elective attraction which we have just considered, have been ingeniously represented by diagrams originally contrived by Bergman. The following scheme illustrative of the decomposition of Spirit of Camphor, by water, may serve as an example.

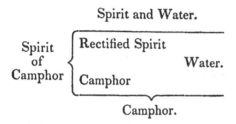

The original compound (Spirit of Camphor) is placed on the outside, and to the left of the vertical bracket; the included space contains the original principles of the compound (Rectified Spirit, and Camphor), and also the body (water) which is added to produce decomposition. Above and below the horizontal lines are placed the new results of their action. The point of the lower horizontal line being turned downwards denotes that the Camphor falls down, or is *precipitated*; while the upper line, being perfectly straight, shews that the new compound (Water and Spirit) remains in solution. Had both the bodies remained in solution, they would both have been placed *above* the upper line. Had both been precipitated, they would have been placed *beneath* the lower one. Had either one or both escaped in a volatile form, such a result would have been expressed by placing the volatilized body above the diagram, and turning up the middle of the upper line; thus if we add Sulphuric acid to Carbonate of lime, a Sulphate of lime is precipitated,

and the Carbonic acid escapes in a gaseous form, and
is represented as follows.

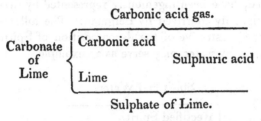

Carbonic acid gas.

Carbonate of Lime { Carbonic acid   Sulphuric acid
                    Lime

Sulphate of Lime.

In the construction of these diagrams Mr. Phillips *
has lately introduced an improvement which greatly
enhances their utility, and of which I shall avail myself in the following pages. It consists in printing the
new compounds formed during the process, or the constituents which assume a fresh state, in *italics*. In the
above diagram, for instance, *Sulphate of lime* would
have been thus denoted, as being a new compound;
whereas the only change which the Carbonic acid undergoes being from the state of solid combination to
that of an uncombined elastic fluid would have been
described as follows,—Carbonic acid *gas*.

233. In compliance with that principle which should
regulate the arrangement of every elementary work, the
student has been first made acquainted with the most
simple operation of elective affinity, in which a body
acts upon a compound of two ingredients, and unites
with one of its constituents, leaving the other at liberty;
and which is termed *Single* or *Simple* affinity, in contradistinction to the more complicated case which we
have next to consider.

234. *Double Elective Attraction*, or *Complex Affinity*, takes place when two bodies, each consisting of
two principles, are presented to each other, and mutu-

* Phillips's Translation of the London Pharmacopœia.

ally exchange a principle of each; by which means two new bodies, or compounds, are produced, of a different nature from the original compounds. In this case it frequently happens, that the compound of two principles cannot be destroyed, either by a third or fourth separately applied; whereas, if this third and fourth be combined, and placed in contact with the former compound, a decomposition, or a change of principles, will ensue. For instance, if to a solution of sulphate of soda we add lime-water, no decomposition takes place, because the attraction of the sulphuric acid is stronger for the soda than for the lime; so again if muriatic acid be added to the same salt no change is induced, since the sulphuric acid attracts soda more powerfully than the muriatic acid. But if the lime and muriatic acid previously combined (muriate of lime) be mixed with the sulphate of soda, a double decomposition is effected. The lime, quitting the muriatic acid, unites with the sulphuric; and the soda being separated from the sulphuric acid, combines with the muriatic. These decompositions, like those produced by single affinity, may be expressed by diagrams.

Muriate of Soda.

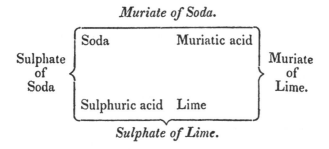

Sulphate of Soda

Soda   Muriatic acid

Sulphuric acid   Lime

Muriate of Lime.

Sulphate of Lime.

On the outside of the vertical brackets are placed the original compounds (Sulphate of soda, and Muriate of lime) above and below the horizontal lines, the new compounds produced (*Muriate of Soda*, and *Sulphate of Lime*), the upper line being straight indicates that the *muriate of lime* remains in solution; the dip of the lower line, that the *sulphate of lime* is precipitated.

235. In all such cases of decomposition, it is evident that two distinct series of attractions must be in conflicting operation; those tending to preserve the original compounds, and to which Mr. Kirwan has bestowed the expressive name of *Quiescent* affinities, and those which tend to disunite them, and to raise up new compounds by a fresh combination of their ingredients; to which latter forces he has applied the term of *Divellent* affinities. It is plain therefore that a double decomposition can only be effected where the sum of the *divellent* is superior to that of the *quiescent* attractions. Taking the instance above adduced, Dr. Henry * has placed this subject in a very clear point of view, thus—

$$\text{The attraction of Lime to Muriatic acid} .. = 104$$
$$\text{Soda to Sulphuric acid} .. = 78$$

$$\text{\textit{Quiescent affinities}} = 182$$

$$\text{The attraction of Soda to Muriatic acid} ... = 115$$
$$\text{Lime to Sulphuric acid} .. = 71$$

$$\text{\textit{Divellent affinities}} = 186$$

The original compounds, therefore, are preserved by a force equivalent to 182, and the tendencies to produce new compounds are represented by the number 186. The *divellent* affinities are consequently predominant. Dr. Cullen, with whom the happy idea of representing what passes in these complicated changes by diagrams, appears to have originated, proposed as

---

† Elements of Experimental Chemistry, Edit. 9.

an illustration, two cross sticks, moveable on a pivot (*a*) placed at their point of intersection, thus—

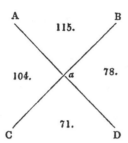

If, on mixing the compounds denoted by A C, B D, the attractions of A to B, and of C to D, overcome the *quiescent* attractions A C, B D; the resulting decomposition is at once represented by bringing together the extremities A B, C D; by which, also, the production of the two *new* compounds will be denoted by the conjunction of these letters.

236. It must be acknowledged, that the theory of chemical affinity, as above displayed, is highly captivating, as well from its extreme simplicity, as from the apparant facility with which it may be brought to bear upon questions of practical inquiry; but the student has yet to learn, that there exist numerous extraneous circumstances which are constantly operating in modifying and even reversing the laws of chemical affinity. These circumstances are, *Quantity of matter*;— *Cohesion*;—*Insolubility*;—*Specific gravity*;— *Elasticity*;— *Efflorescence*;— *Temperature*;— and *Mechanical pressure.* It will be necessary to examine each of these powers in succession, in order to establish some conclusions with respect to their force and value.

237. 1. *Quantity of matter.* M. Berthollet, to whom we are indebted for the first distinct views of

the relations of the force of affinity to quantity, conceives that these relations are universal, and that elective attraction cannot strictly be said to exist. This
assertion, however, has been very ably comb'atted by
Sir H. Davy,* who has pointed out several sources of
fallacy which had escaped the observation of Berthollet.
Although many of the cases which have been adduced
in support of the fact, that *excess in quantity of matter will compensate for deficiency of affinity*, may undoubtedly admit of a different explanation, still it
cannot be denied, that chemical decompositions are
frequently influenced by the ponderable quantities of
the substances placed within the sphere of action;
whence manufacturers operating on a large scale will
frequently obtain results which never arise in the progress of a scientific experiment.

238. It is probable that, in some cases, the influence
of quantity may be referred to a mechanical cause,
enabling the particles to have a readier access to those
for which they have an affinity.

239. The following experiment is well calculated
to illustrate the effect of quantity in favouring decomposition.

Exp. 35.—Mingle together, in a mortar, one part of
chloride of sodium (common salt) with half its
weight of red oxide of lead (*Litharge*, or *Red
Lead*) and add sufficient water to form a thin
paste. The oxide of lead, on examining the mixture after twenty-four hours, will be found not to
have detached the muriatic acid from the soda; for
the strong taste of that alkali will not be apparent.
Increase the weight of the oxide of lead to three
or four times that of the salt; and, after the same

---

* Elements of Chemical Philosophy.

interval, the mixture will exhibit, by its taste, proofs of uncombined soda. This proves, that the larger quantity of the oxide must have detached a considerable portion of muriatic acid from the soda, though the oxide has a weaker affinity for that acid than the soda possesses.

240. 2. *Cohesion.* The influence of this force has been already considered (100), and we shall hereafter have occasion to adduce farther evidence of its importance.

241. 3. *Insolubility.* This might, perhaps, be included under the consideration of the cohesive force, of which it is but an effect (145); the extent, however, to which it may modify chemical affinity requires farther explanation. If there be two bodies, one of which is soluble, and the other insoluble, but each possessing a nearly equal affinity to a third; upon bringing such substances into the sphere of attraction, the soluble body will possess great advantages over its antagonist, for its cohesion, trifling even at the outset, must be reduced to almost nothing by solution, while that of the insoluble body will remain the same. The whole of the soluble body will, moreover, exert its affinity at once, whereas a part only of the insoluble body can oppose its force. Hence it is evident that the former may attach to itself the greatest proportion of the third body, although it should possess even a weaker affinity than the latter, to the subject of combination. In some cases, however, *insolubility* may turn the balance in favour of the affinity of one body, when opposed to that of another; thus, if to the soluble compound, sulphate of soda, we add baryta, the new compound (sulphate of baryta) being precipitated the instant it is formed, and, thus removed from

the sphere of action, will escape from the dominion
which the soda might otherwise exert over it, in consequence of its greater quantity, or mass.

242. 4. *Great Specific Gravity.* This will necessarily concur with insolubility, in impeding combination in the one case, and in aiding the operation of
affinity, in the other, as explained in the preceding
section.

243. 5. *Elasticity*, or *Volatility*, will operate by
separating the particles of bodies so widely, as to remove them out of the sphere of their mutual attraction.

244. 6. *Efflorescence*, like Precipitation, tends to
remove one of the bodies from the operation of affinity.
It was first observed by Scheele, that if in a paste, composed of several saline substances, decomposition is
going on, one of the resulting compounds often rises
through the mass, and forms an efflorescence on its
surface.

245. 7. *Temperature.* This is, by far, the most
important of all the extraneous forces which are capable of modifying chemical affinity; in fact, the other
forces fall more or less under its dominion; that cohesion is thus diminished has been shewn under the history of Solution (147); and it is equally obvious that
the solubility, specific gravity, and elasticity of bodies
will be affected by the same power. The following
experiment will shew how far temperature, by favouring volatility, may modify, or even reverse the order of
chemical affinity.

    Exp. 36. — To a solution of nitrate of potass add
    alcohol, the spirit will immediately unite with the
    water, and the salt be precipitated. If the tem-
    perature be now raised, the alcohol will rise, on
    account of its volatility, and the nitre be re-dis-
    solved ;—or,

Exp. 37.—Into a solution of carbonate of ammonia, pour one of muriate of lime; a double decomposition instantly takes place, carbonate of
lime falls down, and muriate of ammonia floats
above. Let this liquid mixture be now boiled
for some time; exhalation of ammoniacal gas will
be perceived, and the carbonate of lime will be
re-dissolved, as may be shewn by the farther addition of carbonate of ammonia, which will cause
an earthy precipitate from the liquid which, prior
to ebullition, was merely muriate of ammonia.

246. Another beautiful illustration of this power is
afforded by the formation and decomposition of the
red oxide of mercury (*Hydrargyri Oxydum rubrum,*
p. l.) In this case, by exposing mercury to the air in
a degree of heat very nearly equal to that at which it
volatilizes, it will absorb oxygen, and become converted into the red oxide; but, if the heat be still
farther augmented, the oxygen will assume the elastic
state, and fly off, leaving the mercury in its original
state.

247. 8. *Electricity.* The influence of this power
over the affinities of bodies, will require a separate
chapter for its consideration.

248. 9. *Mechanical Pressure.* The effects of this
force are chiefly manifested in producing the combination of aëriform bodies, either with each other, or with
liquids or solids. The theory of its action has been
already explained (137).

249. Such are the principal circumstances by which
chemical affinity is modified; and after an impartial
review of their nature and influence, we may safely
conclude, that affinity is, nevertheless, *elective,* acting

in the different chemical bodies with gradations of attractive force; but which is, at the same time, liable to
be modified by quantity of matter, temperature, cohesion, solubility, and other conditions of the bodies
in action.

250. There remains to be noticed a case of attraction, which has been termed *Disposing Affinity*; in
which, two bodies incapable of combining, are made
to unite chemically by the addition of a third substance, although it may have no apparent affinity for
either. This may be illustrated by an example. Water
is a compound of oxygen and hydrogen; for the former of which Phosphorus has an attraction, but not so
powerful as to enable it to decompose water; if, however, the phosphorus be combined with lime, it instantly decomposes water with rapidity, by attracting
its oxygen, and thus liberating the hydrogen, with
which a portion of the phosphorus enters into union.
In like manner, iron has an attraction for oxygen; but
so little superior to that of hydrogen for the same body,
that it is unable to decompose water, at a low temperature, or, at least, with any energy; but, if a small
quantity of sulphuric acid be added, the decomposition instantly proceeds with very considerable rapidity.
In the one case, the Lime, in the other, the Sulphuric
acid, are said to exert a *Disposing* affinity.

251. Notwithstanding the light which late discoveries have thrown upon the general subject of chemical attraction, it must be confessed that the explanations
which have hitherto been offered of the nature of disposing affinity are, to say the least, extremely unsatisfactory. In some cases, the phænomenon may evidently be referred to Electro-chemical agencies; in
others, to the operation of a force, which Berthollet

has termed *resulting* * affinity, and which may be stated, in a few words, to be the combined action of different affinities existing in one compound; in many instances, however, it cannot be fairly explained upon any known principle of combination.

REAGENTS, or TESTS.

252. From what has been already stated upon the subject of chemical affinity, it will follow, that the application of certain bodies to different solutions will, by producing changes which are striking to the senses, detect the presence of very minute proportions of particular ingredients. Such bodies are termed *Tests*, or *Reagents*, and prove valuable instruments in the hands of the Chemist.† To illustrate their agency, let us

---

* Berthollet conceives that the affinities of a compound are not newly acquired, but are merely the modified affinities of its constituents, the action of which, in their separate state, was counteracted by the prevalence of opposing forces. By combination, he supposes that these forces are so far overcome, that the affinities of the constituents are enabled to exert themselves. To such affinities he has given the name of *elementary affinities*, while those possessed by the compound he calls *resulting affinities*.

† Every medical practitioner ought to be furnished with a chest containing a certain number of tests, in order that he may apply them, as occasions will require, for the examination of unknown substances;—the discovery of poisons, and the detection of those numerous adulterations to which his medicines are so unfortunately exposed. The period has not long passed, when it was supposed that, for the purposes of experimental inquiry, a laboratory regularly fitted up with an expensive suite of apparatus was indispensable. The great improvements, however, which have been lately made in the application and management of chemical Reagents have superseded this necessity ; and it is now universally

K

take a portion of spring water, whose purity it is an
object to ascertain; if we introduce a few drops of a
solution of the *Nitrate of Silver,* and a white precipitate follows, we may conclude from the phænomenon
that the water in question contains a muriatic salt or

---

admitted that every analysis may be performed, as far as practical
utility is concerned, by the assistance of a few phials of tests,
with the addition, perhaps, of some glass tubes, florence flasks,
and tobacco pipes. In the directions which I shall hereafter offer
to the student, for the construction of an *æconomical Laboratory,*
the necessary tests shall be enumerated. In this place it will be
only necessary to make some remarks on the modes of manipulation, and which are no less essential to abridge labour, than to
ensure accuracy. The most convenient bottle for containing a
chemical test is a common ounce phial, through the cork of which
should be introduced a piece of glass tube of small bore, two or
three inches long, and bent at one end to an obtuse angle, as here

represented. On inverting the phial, and
grasping the bottom part of it, the
warmth of the hand expells a few drops,
which may be directed upon any minute
object. When the flow ceases, it may
be easily renewed by setting down the
bottle, for a moment, with its mouth
upwards (which admits a fresh supply of cool air), and then proceeding as before. In some cases where extremely minute quantities are required, as in the application of concentrated alkalies
or acids, a bottle with an elongated stopper, as exhibited in the

annexed sketch, will be found very convenient for enabling the operator to take
up a single drop, and to allow it to fall
upon any body under examination.
The liquid to be examined is frequently introduced into glasses,
or tubes, but as this arrangement necessarily requires a certain
quantity of materials, a slip of glass may be very conveniently
substituted, or on some occasions, a piece of common writing
paper. Where our object, however, is to collect the precipitate
for the purpose of ascertaining its weight, or of submitting it to

some carbonated alkali or earth; such an effect, however, from these latter bodies may be at once prevented by the previous addition of some nitric acid. To shew the extreme delicacy of nitrate of silver, as a test for any muriatic salt, it may be stated, that if two glasses be filled with distilled water, and the finger is merely dipped into one of them, the silver test will actually indicate the impurity thus introduced. If the water should contain any sulphuric salt, the *Nitrate of Baryta* will indicate its presence by a white precipitate, which will not disappear on the addition of nitric acid, as it would, were the precipitant a carbonate. The presence or absence of lime may be inferred from the effect occasioned in the water by *Oxalate of Ammonia*, which will precipitate this earth, although it should exist in extremely minute quantities. These few examples are sufficient for present illustration, others will occur in our progress.

On the Proportions in which Bodies combine; and on the Atomic Theory.

253. Having considered the nature of chemical combination, and the circumstances by which it is directed and influenced, we may next examine the *proportions* in which different substances unite with each other, in order to produce various compounds.

---

other processes, we are necessarily compelled to work on a larger scale. For merely determining the composition of a liquid, without any reference to the proportions of its ingredients, the slip of glass will answer every intention; and it is quite surprising to what a degree of accuracy an experienced eye will arrive in deducing a conclusion from the colour, density, and other appearances of the precipitate thus produced.

254. In some cases bodies appear to unite in every proportion, without any assignable limit, as water with sulphuric acid, or alcohol with water, &c. In others, they unite in every proportion, as far as a certain point, beyond which no farther combination can take place, a fact which has been already exemplified in the history of saline solutions (150.) In all such cases the combination of the constituents is very feeble, and is easily subverted; the compound, moreover, thus produced, is not characterized by any remarkable change in properties, but possesses qualities which belong to the components in their separate state; thus dilute sulphuric acid, or spirit and water, exhibit no sensible characters which may not be discovered in the single ingredients; and the same observation will apply to the aqueous solutions of salts.

255. To express the point at which bodies cease to combine with each other, we use the term *Saturation* where there is no remarkable alteration in properties, (150), and that of *Neutralization* where the opposing properties of the ingredients disappear.

256. In every case of energetic combination, in which the qualities of the components are no longer to be distinguished in the compound, the constituents unite in proportions that are always definite; thus, when they combine so as to form one compound only, that compound under whatever circumstances it may be produced, always contains, with the most rigid accuracy, the same relative proportions of its component parts; and, where two bodies unite in more than one proportion, the second, third, &c. proportions are multiples, or sub-multiples of the first; thus, for instance, there are two well known combinations of iron

with oxygen,* constituting the *black* (Protoxide) and the *red* (Peroxide) oxides; now, if we consider the oxygen in the former as 2, that in the latter must be considered as 3; in like manner, there are two oxides of Mercury, the *black* (Protoxide—*Hydrargyri oxydum cinereum.* P. L.) and the *red* (Peroxide—*Oxydum Rubrum.* P. L.), and the latter appears to contain just twice as much oxygen as the former. We may derive another illustration from the compounds of Lead; this metal combines with oxygen in three different proportions; the first compound consists of 100 metal + 8 oxygen; in the second the same proportion of lead is combined with 12 of oxygen, and in the third with 16. When an alkaline substance combines with more than one proportion of acid, the same circumstances seem to occur; this is beautifully shewn by a simple experiment first made by Dr. Wollaston, *viz.*

> *Exp.* 38.—Let a given weight of the salt called carbonate of potass (*Potassæ Carbonas.* P. L.) be introduced into an inverted tube filled with mercury, and a quantity of sulphuric acid sufficient to cover it, be admitted; a certain volume of carbonic acid gas will be disengaged; let an equal weight of the same salt be heated to redness, by which it will become a *sub*-carbonate, and let this sub-carbonate be treated in the same manner, when it will be found to give off exactly *half* as much carbonic acid gas; the fact is, that the former salt consists of 70 parts of potass + 60 of carbonic acid; and the latter of 70 of potass + 30 of carbonic acid.

---

* In these illustrations it is impossible to avoid anticipating a certain degree of chemical knowledge, with which the student may be supposed to be not yet acquainted. In every case where it was possible to make a different arrangement it has been studiously adhered to.

257. But in no instance is the law of definite proportions more satisfactorily demonstrated than in the combination of gaseous bodies with each other, which will hereafter be shewn to unite, invariably, in simple ratios of volumes.

258. The idea of definite proportions appears to have first struck the mind of Richter; and Mr. Higgins, in his work on Phlogiston, subsequently maintained the opinion that bodies unite chemically, atom to atom. But the generalization of the doctrine was reserved for the genius of Mr. John Dalton, who deservedly enjoys the glory of having permanently established a theory, the dèvelopement of which, to use the expression of an eminent philosopher,* must be considered as the greatest step which Chemistry has yet made as a science; enabling us to establish principles of rigid accuracy as the foundation of our reasoning, and to call in the assistance of mathematics to promote the progress of a science which has hitherto eluded the aid of that unrivalled instrument of improvement. Already has the application of this theory shed a flood of lustre upon our science, not only by correcting former analyses, but by leading to the discovery of many unknown combinations, whose existence might, otherwise, have never even been suspected. It will, for instance, be easily perceived, that if we have a series of compounds, the proportions of whose constituents are respectively $1 + 1$; $1 + 3$; $1 + 5$; and so on, we are naturally led to enquire whether intermediate compounds may not be formed of the same bodies in the proportions of $1 + 2$, and $1 + 4$, to complete the series; and in

---

* On the Daltonian Theory of Definite Proportions in chemical combinations: by Thomas Thomson, M.D. F.R.S. &c. in a paper published in the second volume of the Annals of Philosophy.

many cases the existence of such combinations has been actually discovered; thus, there were four known compounds of Nitrogen and Oxygen, viz. *Nitrous Oxide,* consisting of 1 atom of nitrogen + 1 of oxygen ; *Nitric Oxide,* of 1 of nitrogen + 2 of oxygen; *Nitrous Acid,* of 1 of nitrogen + 4 of oxygen, and *Nitric Acid,* of 1 of nitrogen + 5 of oxygen. To complete this series a compound was required of 1 of nitrogen + 3 of oxygen, and such a body has accordingly been discovered, and is known by the name of *Per-nitrous,* or *Hypo-nitrous* acid. Numerous examples of a similar nature will hereafter present themselves, demonstrating in the most satisfactory manner the universality, as well as the uniformity, of this law of combination.

259. In order to understand the Atomic, or Daltonian theory, it must be remembered that every particle of a true chemical compound, however minutely divided, will contain a portion of each ingredient. It is true that mechanical division cannot be carried practically beyond certain limits (23), but since the most minute particle which it is capable of presenting to us, shews no difference in composition from that of the entire mass, the inference is plain that, could we proceed to the ultimate term of material divisibility, we should still have the same result; thus, chalk is a compound of lime and carbonic acid; now, however minute a portion of chalk we take, we shall find it to contain both of these ingredients. It is evident that this could not occur unless the atoms of the combining bodies united with each other; and, as all chemical compounds contain the same constant proportion of their constituents (257), it is equally evident that a union must take place of a certain determinate number of the atoms of one constituent with a certain number of the atoms of the other.

260. Assuming therefore that chemical combination takes place only between the atoms of bodies, Mr. Dalton has deduced, from the relative weights in which bodies unite, the relative weights of their ultimate particles, or atoms, and he has accordingly formed a series of such representative numbers.

261. As an illustration of the process by which we arrive at the weight of an atom, let us suppose that any two elementary bodies, A and B, combine; and that they have been proved experimentally to unite in the proportions, by weight, of 5 of the former to 4 of the latter; then it follows that if they unite particle to particle,* those same numbers must necessarily express the relative weights of their atoms.

262. But besides combining atom to atom, singly, 1 atom of A may combine with 2, or with 3, 4, &c. of B. * or 1 atom of B may combine with 2, 3, 4, &c. of A. When such a series of compounds exists, it is evident that chemical analysis ought to shew the relative proportions of their ingredients to be 5 of A, to 4 of B; or 5 to (4 + 4) 8; or 5 to (4 + 4 + 4) 12, &c. because these numbers are the representatives of the weights of such atoms; and it is evident that, if the atomic theory be true, no intermediate compounds can ever exist, such for instance as 5 of A to 6 of B.

---

* When only one combination of any two elementary bodies exists, Mr. Dalton assumes, unless the contrary can be proved, that its elements are united atom to atom singly, and such combinations he calls a *binary*.

1 atom of A + 1 atom of B = 1 atom of C, *binary*
1 ————— 2 ————— = ————— D, *ternary*
2 ————— 1 ————— = ————— E, *ternary*
1 ————— 3 ————— = ————— F, *quaternary*
3 ————— 1 ————— = ————— G, *quaternary*

263. In order to establish the accuracy of the numbers thus obtained, it will be necessary to examine the combinations of A and B with some third substance, say with C. Then we will, for the sake of example, suppose that A and C unite atom to atom, so as to form a *binary* combination, in the proportion, by weight, of 5 to 3. Then, if C and B are also capable of combining, it will be immediately perceived that they can only combine in the proportion of 3 to 4, or in that expressed by some simple multiple of these numbers, which denote the relative weights of their atoms. Such was the principle of research adopted by Mr. Dalton for estimating the weights of the atoms of oxygen, hydrogen, and nitrogen; the two first, from the known composition of water, and the two last, from the proportions in which those elements exist in ammonia.

264. But before any table, expressive of the relative weights of the atoms of different bodies, can be constructed, it is evident that we must fix upon some one substance, whose atom shall be considered as the unit of comparison; and it is to be deeply regretted that philosophers should not have conventionally adopted the same general standard. Mr. Dalton has made election of *Hydrogen* for this purpose, because it is the lightest of all known bodies, and unites with others in the smallest proportion; on the other hand, Wollaston, Thomson, and Berzelius, have assumed *Oxygen*, from its almost universal relations to chemical matter, as the decimal unit, (the first making it 10, the second 1, and the third 100). Sir H. Davy has assumed with Dalton the atom of Hydrogen as the radix; but he has modified the theory of supposing water to be a compound of one proportion (atom) of oxygen, and two proportions (atoms) of hydrogen, since two measures

of the latter and one of the former are necessary to form water; and on the supposition, that equal measures of different gases contain equal numbers of atoms, whereas Mr. Dalton assumes that two volumes of hydrogen contain the same number of atoms as one volume of oxygen. It is, however, quite indifferent which we adopt, a very simple process* reconciles them.

265. But in the doctrine of proportions arising from facts, it is not in the least necessary to consider the combining bodies as composed of indivisible particles, we may reject the atomic theory altogether, and still retain all the advantages which can result from a knowledge of the general law of the definite nature of combination; Sir Humphry Davy for instance, has not adopted the theory of Dalton, but he has embraced the doctrine of definite proportions, and what Dalton calls an *atom* he merely calls a *proportion*; while Dr. Young has given the term *combining weight* to that which Davy calls *proportion*.

266. To express the system of definite ratios, in which bodies reciprocally combine, referred to a common standard reckoned unity, Dr. Wollaston has introduced the term *Chemical Equivalent*, and in order to denote that these ratios are reduced to their lowest terms, Dr. Ure usually employs the prefix *Prime*.

267. The propriety of the term *Equivalent* will be readily acknowledged, when it is perceived that as soon

---

* Thus we may reduce the numbers of Dr. Wollaston by the simple rule of proportion; thus as 8 (Dalton's number for oxygen) is to 1 (his number for hydrogen) so is 10 (Wollaston's number for oxygen) to 1·25, the number of hydrogen. To reduce Davy's numbers to Wollaston, we have only to add one third; or to reduce Wollaston's to Davy's, to deduct one fourth. Thus Davy gives 15 as the number for sulphur, and 15 + 5 = 20 the number assigned it by Wollaston; and vice versa, 20 − 5 = 15.

as we have ascertained the proportion in which any two or more bodies A, B, C, &c. of one class, neutralize any other body X, of a different class, it will be found that the same relative proportion of A, B, C, &c. will be required to neutralize any other body of the same class as X. Thus, since 100 parts of sulphuric acid, and 68 (omitting fractions) of muriatic acid, neutralize 118 of potass, and since 100 of sulphuric acid neutralize 71 of lime, we may infer that 68 of muriatic acid will also neutralize 71 of lime. In such a case 100 parts of sulphuric acid and 68 of muriatic acid, are justly said to be *Equivalents* to each other.* By adapting a series of such numbers to a sliding rule, Dr. Wollaston has constructed a scale of Chemical Equivalents,† an instrument which has contributed more to

* A series of such equivalent or proportional numbers, will be found in the Appendix of this work.

† A description of this Scale is published in the first part of the Phil: Trans: for 1814. The different substances are arranged on one or the other side of a scale of numbers, in the order of their relative atomic or combining weights, and at such distances from each other, according to their weights, that the series of numbers placed on a sliding scale can at pleasure be moved, so that any number expressing the weight of a compound may be brought to correspond with the place of that compound in the adjacent column. The arrangement is then such, that the weight of any ingredient in its composition, of any reagent to be employed, or precipitate that might be obtained in its analysis, will be found opposite the point at which its respective name is placed. For instance, if the slider be drawn upwards, till 100 corresponds to muriate of soda, the scale will then shew how much of each substance contained in the table is equivalent to 100 of common salt. It shews, with regard to the different views of this salt, that it contains 46·6 dry muriatic acid, and 53·4 of soda, or 39.8 of sodium, and 13.6 of oxygen ; or if viewed as chloride of sodium, that it contains 60·2 chlorine and 39·8 sodium. With respect to Reagents, it may be seen, that 283 of nitrate of lead, containing 191

facilitate the general study and practice of chemistry than any other invention, by having at once subjected thousands of chemical combinations to all the dispatch and precision of logometric calculation.

268. The most interesting fact, connected with the reciprocity of saturating proportions, is that two neutral salts, in reciprocally decomposing each other give birth to two new saline compounds, always perfectly neutral, no excess of acid or of base ever occurring in the interchange of principles ; thus, if 100 parts of nitrate of baryta, which contain 41 nitric acid, and 59 baryta, be mixed with 67 of sulphate of potass, which

---

of litharge, employed to separate the muriatic acid, would yield a precipitate of 237 of muriate of lead, and that there would then remain in solution nearly 146 nitrate of soda. It may at the same time be seen, that the acid in this quantity of salt would serve to make 232 corrosive sublimate, containing 185·5 red oxide of mercury ; or make 91·5 muriate of ammonia, composed of 62 muriatic acid gas (or hydromuriatic acid) and 29·5 ammonia. The scale also shews, that for the purpose of obtaining the whole of the acid of distillation, the quantity of oil of vitriol required is nearly 84, and that the residuum of this distillation would be 122 dry sulphate of soda, from which might be obtained, by crystallization, 277 of Glauber's salt, containing 155 water of crystallization. These, and many more such answers, appear at once by bare inspection, as soon as the weight of any substance intended for examination is made, by motion of the slider, to correspond with its place in the adjacent column. With respect to the method of laying down the divisions of this scale, those who are accustomed to the use of other sliding rules, and are practically acquainted with their properties, will recognise upon the slider itself the common *Gunter's line of numbers*, and will be satisfied that the results which it gives are the same that would be obtained by arithmetical computation. The instruments, together with printed directions for its use, may be purchased at a trifling expense, of Mr. Cary, Optician, London, and it ought to be considered as indispensable in the laboratory of the Chemist as well as in the study of the Philosopher.

consist of 30 of sulphuric acid, and 37 of potass, there will be found 89 of sulphate of baryta, and 78 of nitrate of potass; so that 41 of nitric acid will combine with the 37 of potass, and 30 of sulphuric acid with the 59 of baryta.

269. When one proportion (or *atom*) of one substance is combined with more than one proportion of another, the first proportions may be separated with much more facility than the last. If, for instance, the black oxide of manganese be exposed to a strong heat, it gives off oxygen gas, and becomes brown; but no heat, as yet applied, is capable of depriving it of the whole of its oxygen; thus, again, the carbonate of soda, which contains two porportions of carbonic acid to one of soda, gives off half its carbonic acid with great readiness, by heat, (*Exp.* 38.) but obstinately retains the other half.

270. This fact will explain several phænomena in chemical affinity, which were referred by Berthollet to the influence of mere quantity (237); for it would appear that a base will have different attractive powers with respect to the first, second, and third, &c. proportions of the body with which it is combined; hence, where a *super*, or *sub* salt is readily formed, a substance less strongly attracted by another than a third, will sometimes precipitate this third from its combination with the second; in this manner the oxide of lead decomposes the muriate of soda, forming a *sub* muriate of lead; the tartaric acid, in like manner, decomposes the salts of potass, forming super-tartrate of potass; and the carbonic, the sub-acetate of lead, leaving the acetate. A saline draught also, consisting of the acetate of ammonia, is decomposed, and made pungent by the addition of pure magnesia, which latter substance stands below ammonia in the order of elective attrac-

tions; the magnesia, probably, in this case, forming a triple acetate with one part of the ammonia, and setting the rest at liberty.*

271. When one proportion of a body is combined with two or more proportions of another, it would seem to enter with more difficulty and reluctance into new combinations, than when it is combined with one proportion only. Thus, iron combined with two proportions of sulphur, as in the *golden pyrites*, is not acted upon by diluted sulphuric acid; but when combined only with one proportion of sulphur, as in the common artificial sulphuret of that metal, it is very readily acted upon.

272. It seems from these facts that two or more proportions of one body attract a single proportion of another body with more energy than one proportion, and that two proportions or more adhere to a single proportion with less energy than one proportion; or at least that a second or a third proportion adheres with less energy than the first.

273. It may possibly be said, observes Sir H. Davy, from whose work† the above statement has been quoted, that the effect of two or three proportions, in defending one proportion from the action of a new substance, may depend upon *mechanical* causes, from their more completely enveloping its parts; but-the other solution of the effect seems to be most probable.

274. Amongst the numerous practical uses to which the doctrine of equivalents may be applied, that which enables the pharmaceutist to estimate the value and strength of certain preparations is not the least important. By such means, for instance, he may easily

---

* Pharmacologia, vol. 1.
† Elements of Chemical Philosophy.

ascertain the quantity of real hydro-cyanic acid in any
dilute solution; thus, since the equivalent number of
hydro-cyanic acid is exactly one-eighth of that of the
peroxide of mercury, and as this acid combines in the
proportion of two of the former to one of the latter, we
have, at once, the relation of one to four in the formation of this compound.   Hence it is evident, that if to
a hundred grains of diluted hydro-cyanic acid, we add
in succession small quantities of the peroxide (*red precipitate*) until it ceases to be dissolved on agitation,
the weight of the oxide so dissolved, being divided by
four, will give a quotient representing the quantity of
real hydro-cyanic acid present.   In like manner we may
estimate the quantity of real acetic acid present in any
given sample of distilled vinegar, for, since carbonate
of lime and the last mentioned acid, have the same
representative number, it is plain that the quantity of
the former which is dissolved by a hundred grains of
the sample, will represent the percentage of real acid.

ON HEAT, OR CALORIC.

275. The terms *heat* and *cold*, as denoting certain
sensations excited by bodies applied to our organs,
although incapable of strict definition, are too well
known to require explanation. Of the nature, however, of the cause which excites heat we have no satisfactory evidence. Some philosophers have regarded it
as an independent fluid, of inappreciable tenuity, capable of insinuating itself between the particles of the
most dense matter; while others have denied the separate entity of a calorific element, and have referred
all the phænomena exhibited by heated bodies to a
vibratory, or intestinal motion of the particles of common matter. The former opinion appears to be the
one best calculated for explaining the phænomena;
but it is quite unnecessary for the student to enter upon
the investigation; for whatever may be the nature of
this power, its existence, as the cause of certain effects,
is clearly demonstrated; and these effects, their relations to each other, and the general laws according to
which they are produced, may be investigated with
sufficient precision, and equal advantage, although the
nature of the cause * should remain wholly unknown.

---

* In an elementary work it will be convenient to make use of
the popular language, and to speak of heat as *existing* in bodies
in greater or smaller proportions, without meaning thereby to
decide on the question of its materiality.

276. To this power, whatever it may be, as well as to the sensation which it excites, the term *Heat* was formerly applied. But, since there was an obvious impropriety in confounding under one common name, two things so distinct as cause and effect, the terms *mat-ter of heat, igneous fluid,* or *fire,* have been employed to designate the former, but which are now universally superseded by the less exceptionable appellation of *Caloric,* or *Calorific Repulsion.* Nevertheless, in order to avoid the too frequent repetition of the same word, the term *heat* is still occasionally employed in works on Chemistry, not only to express the sensation which that word popularly denotes, but also some of the modifications of the unknown cause by which it is produced.

277. Besides the power which caloric possesses of exciting the sensation of heat, philosophers have long observed that all bodies, submitted to its influence, expand, or undergo an augmentation in volume in every direction, * and that when they are again cooled

---

* We therefore perceive the importance of attending to the temperature of a body, when we wish to estimate its specific gra-vity. (81.) The property which caloric possesses of expanding bodies is attended with advantage in some of the branches of art, and with extreme inconvenience in others; the wheelwright and the cooper are thus enabled, by applying the iron hoops hot, to make them grasp much tighter; a curious application of the same principle was made, some years ago, for the purpose of drawing a spire into the perpendicular ; iron bars attached to clamps were fixed to the inclining object, and after effecting all that could be done by mechanical force, the bars were heated, and then allowed to cool, when by their contraction they exerted so considerable a force as to accomplish the object. On the other hand, it is well known what inconvenience and inaccuracy result from the operation of this force in machinery. The chemical student will require no experiment to illustrate the subject: however great

their volumes are diminished to their original dimensions; hence it is evident that caloric has a constant
tendency to separate the particles of bodies, and is accordingly to be regarded as the *repulsive power* which
is constantly opposed to the force of *attraction.*

278. There is one important exception * to this law
of expansion by heat, in the refrigeration of water,
whose density is uniformly increased by the abstraction
of heat until it reaches the temperature of 40°, when it
not only ceases to increase in density, but, on the contrary, actually becomes of less and less specific gravity,

---

may be his caution, he will necessarily experience the frequent
destruction of his glass apparatus from the expansive force of
caloric. It may be instructive, as well as practically useful, to
inform him in what manner such accidents happen. Glass is a
bad conductor of heat, (Exp. 39), when therefore any hot fluid is
introduced into a vessel of this kind, of moderate thickness, the interior surface expands, while from the exterior remaining unaltered,
a fracture of the parts necessarily ensues. To prevent such an
accident the glass should be very thin, so as to ensure the simultaneous expansion of the whole mass, or the vessel should be gradually heated in order to allow the same effect to take place more
slowly. A similar explanation will apply to the accidents which
occur on pouring a cold fluid into a heated glass vessel.

* Other exceptions have been enumerated, but they are only
apparent ones, depending upon some chemical change in the constitution of the bodies, or on their crystalline arrangements. Thus
clay contracts considerably in dimension by a very intense heat,
and on the measure of its contractions the Pyrometer of Wedgewood is founded; but in this case the clay first gives off water,
which was united to its parts, and afterwards these parts cohere
together with greater force, and, from being in a state of loose
aggregation, become strongly united. Certain saline solutions
also expand at the moment they become solid, but this is to be
explained by the fact that the same weight of matter will occupy
much more space, by leaving greater interstices, when its particles
are arranged in certain forms. The fact is the same with regard
to cast iron, bismuth, and antimony.

as the caloric is further abstracted by the cold atmos-phere. The student may satisfy himself of the truth of this fact by examining water in a deep vessel, at the time ice is forming, when it will be found to be a little warmer at the bottom than at the surface. It will be worth while to inquire why this general law of nature has been departed from in this instance, and to con-sider what would have been the result, had water been subjected to the same law with other bodies. It has been just remarked, that as water loses caloric it be-comes more and more dense until it is cooled down to about 40°; but if no check had been given to this pro-cess, and the density of water had continued to increase until it came down to 32°, the superficial stratum would have continued to descend, followed by others in quick succession, and a body of ice would have first formed at the *bottom* instead of at the *surface* of our lakes and rivers. This would have accumulated with such ra-pidity, that one severe winter might have converted the whole of the rivers of Europe into as many solid masses of ice. †   There is another obvious advan-tage attending this important contrivance. The sheet of ice which often covers the small seas, as well as the rivers and the lakes, not only preserves a vast body of caloric in the subjacent water, but when it thaws, the fish are not destroyed by the cold; for not a particle of the cold surface-water can descend, until a change in the temperature of the atmosphere has taken place, so as to raise the temperature of the whole of the water at least ten degrees.

279. By the term *Temperature* we understand the state of a body, in relation to its power of exciting the sensation of heat, and of occasioning expansion; but, as our sensations, from causes to be hereafter explained,

---

† Parke's Chemical Essays, vol. 1. on Temperature.

are too fallacious to furnish a correct estimate of the former, we are compelled, in order to derive any precise notion of temperature, to rely upon the degree of expansion produced in some one body, which has been previously fixed upon as a standard of comparison; and it is upon this principle that the instrument invented by Sanctorius, and properly termed the *thermometer*, is constructed, and used as a common measure of temperature.

280. It would appear that the phænomena of *Caloric*, in one essential circumstance, resemble the habitudes of common matter; for it may be proved to exist in two very different states,—in that of freedom, when it is capable of affecting us with the sensation of heat, and of producing expansion; and in a combined state, in which it ceases to be cognizable by our senses, or by the thermometer. In the former case it is distinguished by the terms *free*, or *uncombined caloric*, or *caloric of temperature*; in the latter, by those of *latent*, or *combined caloric*, or sometimes, for reasons that will hereafter become apparent, *caloric of fluidity*.

281. The constant tendency which caloric possesses to diffuse itself over matter, until an equilibrium of temperature is established, is a property no less striking and obvious than that of producing expansion; thus, if a number of bodies of different temperatures are placed in contact with each other, they will all, in a certain period, acquire a mean temperature; the caloric of the hottest body will diffuse itself among those which are heated in a less degree, till they have all acquired the same temperature. This law is too familiar to require illustration. If we wish to heat a body we carry it towards the fire, if to cool it, we surround it with cold substances. This propagation of caloric takes place even through the Torricellian vacuum; and

it is evidently owing to its repulsive power which constantly tends to an equilibrium.

282. Although when bodies of higher temperature than others are brought into contact with each other, the heat is propagated from the first to the second, or the colder body deprives the warmer of its excess of heat, yet some bodies do so with much greater celerity than others. Through some, caloric passes with undiminished velocity, through others, its passage is prodigiously retarded. This disposition of bodies to receive and part with caloric, is called *the power of conducting heat*, and a body is said to be a better or worse conductor of heat, in proportion to the greater or less celerity with which this process takes place.

> *Exp.* 39.—Coat two rods of equal length and thickness, the one of glass, the other of iron, with wax, at one end of each only; and then apply heat to the uncoated extremities. The wax will be melted much sooner on the end of the iron rod, than on the glass one; because the heat will travel along the iron with much greater rapidity than along the glass, the former being a better conductor than the latter. This experiment may be varied in a very pleasing manner, by substituting a small piece of phosphorus for the wax, which will be found to inflame at different periods on the ends of the different rods. Or,—

> *Exp.* 40.—Procure slender cylinders of silver, of glass, and of charcoal, of equal dimensions, and hold their extremities in the central part of the flame of a candle, the silver will rapidly become heated throughout, and cannot be held in the hand; through the glass the heat will be more

slowly communicated; but the charcoal will become red hot at the one extremity long before any
heat is felt at the other.

283. To this difference in the conducting power of
bodies is to be attributed the different intensity with
which different bodies, at the same temperature, affect
us; a fact which, in itself, must be sufficient to render
us wholly incapable of appreciating the amount of temperature by our sensations; thus, if we apply the hand,
in succession, to a number of bodies, such as wood,
iron, marble, &c. they will appear cold in very different
degrees; and as this sensation is occasioned by the
passage of caloric out of the hand into the body which
it touches, that body will feel the coldest which carries
away the heat with the greater celerity, or which, in
other words, is the best conductor. For the same
reason, if these bodies have a temperature considerably
above that of the hand, the best conductor will be the
hottest to the touch; it is thus that the money in our
pockets often feels hotter, after standing before the fire,
than the clothes which contain it. The heat of metals
at the temperature of 120° *Fah.* is scarcely supportable;
water will scald at 150°, but air may be heated to 240°
without being painful to our organs of sensation. Dr.
Fordyce remained for a considerable time, and without
great inconvenience, in a room heated by stoves to 260°,
but the lock of the door, his watch and keys lying on
the table, could not be touched without burning his
hand. It is for such a reason that we furnish metallic
vessels with wooden handles, or interpose a stratum of
ivory or wood between the hot vessel and the metallic
handle, by which means the transfer of heat is prevented.

284. The power which air possesses of abstracting
heat, is likewise comparatively very little; in the high
northern latitudes a cold has been experienced without
injury, in which mercury froze; but if in this state of
the atmosphere metallic substances were touched, the
part was immediately blistered.* This fact of the
feeble conducting power of air has been turned to a
very useful account in a variety of instances; hence
the origin and utility of double doors, and double windows, which infold sheets of air between them, and so
preserve the apartments at one uniform temperature.
Ice-houses are thus also surrounded by a stratum or
partition of air, in order to prevent the warm atmosphere from entering and melting the ice. On the same
principle Mr. Buchanan has recommended the protecting of water-pipes from freezing, by inclosing them in tin
plate tubes, leaving a space of about an inch all round
the leaden pipes. In like manner, certain culinary and
pharmaceutical utensils are fitted up with double covers
of block tin or copper, and it is found that in such
vessels any fluid will boil sooner than it would in
others, and will also be preserved at any particular
temperature by the consumption of less fuel than would
be required if they were covered only with simple
lids.

285. The conducting power of air is, however, materially increased by the presence of moisture; hence
the cold which is experienced in a humid atmosphere,
and the sensation which is so generally felt at the com-
mencement of a thaw, when, notwithstanding the absolute elevation of temperature, it appears to be colder
than during the frost.

---

* Intense cold produces upon our organs the same destructive
effect as heat, and the sensation which attends it is not to be distinguished from burning.

286. From this view of the different conducting powers of different bodies, the fitness of different kinds of clothing for their respective purposes will become apparent. Animal and vegetable substances, in general, are very bad conductors; thus the hair and wool of animals, and the feathers of birds, * are admirably adapted for protecting .them from the cold, and they,

---

* Of all these substances the one which possesses the strongest power of resisting the passage of heat is the down which covers the breast of the *Eider Duck*, a bird peculiar to cold climates. Had not the breast of this animal, which is constantly exposed to the contact of intensely cold water, been supplied with such a protection, its internal heat would have been rapidly carried off by the colder medium, and the bird must inevitably have perished. Nearly the same fact is observable in the young swan or cygnet. From the physiologist we may derive innumerable facts in further illustration of this important subject. It was, for instance, a matter of great importance that the *cicatricula* of the egg should be preserved in a uniform temperature, and that it should not be rapidly cooled, during the occasional absence of the bird; an accident against which Nature has provided by surrounding it with an albuminous fluid of remarkably slow conducting power. We find, upon the same principle, that those fish that are remarkable for retaining their vitality long after their removal from their native element, as *Eel* and *Tench*, have the power of secreting a slimy fluid with which they envelope their bodies, and thus prevent the destructive diminution of temperature which would otherwise occur; while, on the contrary, those which do not possess this power, as *Mackarel*, &c. quickly perish after they are drawn on shore. The common garden snail is indebted to its profusion of slime for a similar protection against cold. In some cases, perhaps, a prodigious accumulation of fat may answer the same end, thus the *Silurus Glanis* which is the fattest of all fresh-water fish, growing to the weight of 300 lbs. lives for a long time after it is taken out of the water. Fat, like the cells in which it is deposited, is a very bad conductor of heat. It is a well known fact that excessively corpulent persons scarcely feel the severe cold of winter. In the same manner the oil of the cetaceous tribe is no doubt intended to preserve the temperature of the animal.

moreover, inclose and retain air, which being a still worse conductor, enhances the effect. For the same purpose we wrap our bodies in woollen garments, and the air inclosed in their folds, greatly enhances their utility; hence loose clothing is generally warmer than that which is fitted to the body; the *tight* great coat may contribute to the ornament, but not to the comfort of our persons, and is in direct opposition to the effect which it was intended to produce.

287. In general, it may be stated, that the most dense bodies are the best conductors of heat, thus the metals conduct better than any other solids; gases are worse conductors than fluids, and fluids than solids; but there are many important exceptions with respect to this correspondence between conducting power and density, and a remarkable one is afforded in the densest known body in nature, *Platinum,* which is perhaps the worst conductor amongst the metals.

288. Liquid and aëriform bodies transmit heat on a different principle from that which regulates its propagation through solids, *viz.* by an actual change in the situation of their particles, they may therefore be said to *carry,* rather than to *conduct* heat. The portion of the fluid which is the nearest to the source of heat expands, and, consequently becoming specifically lighter, ascends, and is replaced by a colder portion from above. This, in its turn, becomes heated and dilated, and gives way to the second colder portion; and thus the process proceeds, as long as the fluid is capable of imbibing heat. This is illustrated by a very simple experiment.

Exp. 41.—Let a sensible thermometer (an *air* thermometer) be inverted in a vessel of water, so that
the extremity of the bulb is barely beneath the
surface; then pour a little æther upon
the water so as to form a stratum about
one-eighth of an inch above the thermometer, and let the æther be inflamed,
as represented in the annexed figure.
However delicate may be the thermometer, the air in it will not soon expand; the æther boils violently, but a
very long process of this kind is required to communicate any sensible
heat to the water.

289. From such facts Count Rumford concluded,
that water is a perfect non-conductor of caloric, pro-
pagating it in one direction only, *viz.* upwards, and
that, in consequence of the motions which it occasions
among the particles of the fluid. This inference, however, has been set aside by the inquiries of Dr. Thomson
and Dr. Murray, from which it appears that water is
a conductor, though a very slow and imperfect one.

290. Besides these modes, by which changes of temperature amongst bodies are propagated, either by caloric being communicated from particle to particle, or by
its being transported by change of place in the matter
which receives it, or from which it is abstracted, there
is still another mode,—that by RADIATION, or projec-
tion from the surface of bodies, in right lines, with extreme velocity; and it may be stated, in general terms,
that one half of the caloric, lost by a heated body,
escapes by radiation, and that the rest is carried off by
the ambient atmosphere.

291. The effect we perceive in approaching a fire chiefly results from radiation; and is little connected with the immediate conducting power of the air; and if a concave *metallic* mirror be held opposite to the fire, a heating and luminous focus will be obtained.

292. The process of radiation appears to be constantly going on from the surface of the Earth; a phænomenon which is rendered particularly obvious from its effects during the night, especially when the sky is cloudless, for a covering of clouds serves as a mantle to the Earth, and prevents the free escape of radiant heat. Under favourable circumstances, it has been shewn by Dr. Wells, that the temperature of the ground, especially when its covering is formed of some substance that radiates freely, is several degrees below that of the atmospheric stratum, a few feet above it. It is this diminished temperature of the Earth's surface, that occasions the deposition of dew and hoar frost, which are always observed to be most abundantly formed under a clear unclouded sky.

293. It has been satisfactorily shewn by Professor Leslie that the phænomena of radiation are connected with the nature of the radiating surface; and that those surfaces which radiate most, are also gifted in the greatest degree with, what has been termed, their *absorbing* power. In other words, those bodies that have their temperatures most easily raised by radiant heat are likewise those that are most easily cooled by their own radiation; unmetallic and unpolished surfaces are the best radiators, and also the best receivers of radiant heat; while polished metallic surfaces are the worst radiators, and have the lowest absorbing powers. In the ingenious experiments of Professor Leslie it was found, that a clean metallic surface produced an effect $= 12$ upon the thermometer; when covered with a thin

coat of glue its radiating power was so far increased as
to produce an effect = 80; and, on covering it with
lamp-black, it became = 100.

294. These doctrines of radiant heat admit of some
highly useful applications to the various processes of
art. It is evident, for instance, that vessels which are
intended to retain their heat, as those employed for the
purposes of infusion, &c. should be metallic and highly
polished, while those parts which are to receive heat
should be blackened, and not polished; and, independent of elegance and delicacy, there is a reason obvious,
from the preceding facts, why metallic vessels for the
purposes of the table, should be kept as bright as possible. Steam or air pipes for warming houses should
be polished in those parts where the heat is not required
to be communicated, and covered with some radiating
substance, such as lamp-black, or plumbago, in those
rooms which are to be heated by them. For the same
reason, Count Rumford has stated, that the heated
surfaces of fire places or stoves should not be metallic;
but of stony or earthy materials, for in that case much
more heat will be thrown into the apartment by radiation. The inquiry respecting radiant caloric presents
a wide and interesting field for speculation; but I have
purposely abstained from those discussions which do
not immediately tend to practical advantage.

295. It has been said (280) that caloric may exist
in two very different states, in that of freedom, and in
that of combination; we have hitherto only described
those phænomena which belong to it in the former
state, the student must now be conducted, by easy
steps, to a knowledge of those which appertain to its
latter condition.

296. It may be stated as a general proposition that, when bodies pass from a denser to a rarer state, they absorb caloric, or render it *latent,* and *vice versa.*

Exp. 42.—To a pound of water at 172° *Fah.* add a pound of ice at 32°, and the resulting temperature will be only 32°. What then has become of the excess of caloric, amounting in this case to 140°, and which has wholly disappeared? It has entered the ice, and being no longer sensible to the thermometer is said to have become *latent,* and since it is the cause of the liquefaction of the ice, it is sometimes called *caloric of fluidity.*

Exp. 43.—Dilute a portion of nitric acid with an equal weight of water; and add, as soon as the mixture has cooled, a quantity of light fresh-fallen snow. On immersing the thermometer in the mixture, a very considerable reduction of temperature will be observed, an effect which is to be attributed to the absorption, and intimate fixation, of *free* caloric by the liquefaction of the snow.

297. The same effect is observable in all cases of liquefaction, and we are thus enabled, by the rapid solution of certain saline bodies, to produce artificial cold, a principle upon which the action of freezing mixtures entirely depends.† As such agents are not

---

† The history of this discovery, and that of the progress of the art to which it has given origin, afford striking instances of a fact which was at first regarded as a mere philosophical curiosity, having furnished the chemist with one of the most important of his instruments of investigation, and the public with the most elegant luxury that can adorn their tables. The salt which was originally used, for the purpose of refrigeration, appears to have been Nitre, and the Italians were the first people by whom it was employed. About the year 1550, all the water, as well as the

only of great œconomical utility, but are subservient to chemical and medicinal purposes, I shall hereafter introduce a copious table, representing the proportions in which different salts should be dissolved, in order to produce different degrees of temperature.

---

wine, drank at the tables of the great and opulent at Rome, was cooled in this manner. It was customary to place the bottle containing the wine within a larger vessel of water, and into this they gradually put as much nitre as is equal to one-fourth, or one-fifth of the weight of the water, and during its solution the bottle was driven round by the hand with a quick motion on its axis in one direction. In the year 1621, Barclay's Argenis, an interesting romance, was published at Paris, and its author places on the table of Juba, in the middle of summer, fresh apples for Arsidas, one half of which were incrusted with transparent ice. A basin of ice filled with wine was also handed to him, and he was informed that to prepare all these things in summer was a new art. The author has told us how this basin of ice was made. " Two cups," says he, " made of copper, were placed, the one within the other, so as to leave a small space between them, which was filled with water. The cups were then put into a pail, amidst a mixture of snow and unpurified salt coarsely powdered, and the water in three hours was converted into a cup of solid ice, as well formed as if it had come from the hands of the pewterer." Towards the latter end of the 17th century, Mr. Boyle made experiments with various kinds of salts, and other substances, for reducing the temperature of water, and in 1683 published his " Experiments and Observations touching Cold." By these researches he discovered that either common salt, alum, vitriol, sal ammoniac, lump sugar, oil of vitriol, nitrous acid, caustic ammonia, or alcohol, when mixed with snow, had the power of freezing water; these experiments were subsequently extended by Dr. Blagden, Mr. Cavendish, and Mr. Lowitz, but we are indebted to Mr. Walker, apothecary to the Radcliffe Infirmary in the city of Oxford, for having first produced a cold sufficient to freeze mercury, without using a particle of ice or snow. The materials which Mr. Walker employed for this purpose were strong nitrous acid, sulphate of soda, and nitrate of ammonia.

298. Having shewn that when a body passes from a denser to a rarer state it absorbs caloric, and produces cold, the converse of this proposition remains to be considered, *viz.* that when a body undergoes condensation, as in the case of a liquid becoming solid, it evolves caloric, or, in common language, gives out heat. This fact has already been exemplified; if the student refer to *Experiment 22,* he will perceive that the solidification of Glauber's salt is attended with the evolution of caloric; so again, the separation of a salt from its solution, by alcohol, as in *Experiment 34,* is accompanied by an evolution of temperature, which is exactly the reverse of what happens during its solution (297.) It has been also stated (214) that water is capable of being cooled a number of degrees below its freezing point. If when cooled thus far, it be agitated, it immediately becomes solid, and from the evolution of caloric which accompanies the change of form, its temperature instantly rises to that of its usual freezing point. The heat given out during the slacking of quicklime is the result of the solidification of the water, and when sulphuric acid and water are mingled together, a diminution of bulk takes place, and considerable heat is evolved. In order, however, to render the theory of *latent* heat perfectly intelligible to the student, it will be necessary to examine with some attention the more recondite phænomena of Fluidity and Vaporisation.

FLUIDITY.

299. The elementary state of all bodies is solidity, (30.) When they assume the form of a fluid, or that of a vapour, it is the result of the repulsive operation of caloric.

300. All the different substances in nature, under certain circumstances, are probably capable of assuming all these forms; thus solids, by a certain increase of temperature, become fluids, and fluids gases; and, *vice versa*, by a diminution of temperature, gases become fluids, and fluids solids.

301. When a solid is heated, it is expanded; as its temperature is raised, this expansion proceeds, till its particles are separated to such distances as to be easily moveable in every direction, and, by thus sliding over each other, they impart to the mass the qualities of fluidity. By a reduction of temperature the particles again approach; when within certain distances they unite, and the fluid is brought back to the solid state.

302. These changes of form, then, depend on the actions of two opposite powers upon matter. By the mutual attraction, or force of cohesion, which subsists between the particles of bodies, they are held together so as to form solid masses; by the repulsive power of caloric they are separated to distances at which this attraction ceases, and the state of fluidity is produced; which is therefore obviously not essential to any particular species of matter, but always dependant on the presence of a quantity of caloric.

303. The particular temperature necessary for the production of this change is exceedingly various, but for the same bodies it is always the same; thus lead melts at 612°, tin at 422°, mercury at 59°, sulphur at 218°, lard at 97°, spermaceti at 112°, and ice at 32°, &c.

304. In some cases the transition of a body from a solid to a liquid state is sudden; in others, it passes through several degrees of softness before it is perfectly liquefied: the conversion of ice into water is an example of the first; the melting of wax and other

unctuous matters are instances of the second. To express the former of these operations the term *Liquefaction* has been employed; to denote the latter, that of
Fusion; although in popular language this distinction
is not always observed, and some writers have indiscriminately used both terms.

305. There are some bodies which cannot be melted,
owing to their suffering chemical decomposition at a
lower temperature than that required for their fusion;
a piece of wood, for instance, cannot be melted by the
application of any degree of heat. In come cases, however, such decomposition is prevented by great pressure,
and bodies may be thus made to undergo fusion, which
under other circumstances would be infusible; thus
marble, if heated in the open air, soon parts with its
carbonic acid, and pure lime remains, which has not
hitherto been fused; but if the same body be intensely
heated under pressure, the escape of the carbonic acid
is prevented, and the marble enters into a state of
fluidity.†

306. When a substance acquires by fusion a degree
of transparency, a dense uniform texture, and great
brittleness, and exhibits a conchoidal fracture, with a
peculiar surface, and the edges of the fragments very
sharp, it is said to be *vitrified*.

307. Certain saline bodies are occasionally added
to refractory substances to promote their fusion, and
are termed *Fluxes*.

308. Fusion is performed with the intentions,

    a. Of weakening the attraction of aggregation,
        1. To facilitate mechanical division;
        2. To promote chemical action.
    b. Of separating from each other, substances
        of different degrees of fusibility.

---

† See Sir James Hall's interesting account of these experiments
in the 6th volume of the Edinburgh Philosophical Transactions.

309. The most convenient vessels for conducting the process of liquefaction are earthenware pans. That of fusion is usually performed in pots called *Crucibles*,† which are of various sizes. The larger ones are generally conical, with a small spout for the convenience of pouring out; the smaller ones are truncated triangular pyramids,, and are commonly sold in nests.‡

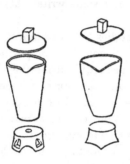

---

† The origin of this term has afforded a curious subject for etymological inquiry; see Pharmacologia, Edit. 6, Vol. 1, p. 19. (*Note.*)

‡ Those formed of common clay with calcareous or siliceous earth are easily vitrified and then melt. The Hessian crucibles are composed of better clay and sand, and when good, will support an intense heat for many hours, without softening or melting, but they are disposed to crack when suddenly heated or cooled, a circumstance, however, which may be remedied by using a double crucible, and filling the interstice with sand, or by coating the crucible with a paste of clay and sand, by which means the heat is transmitted more gradually and equally. Wedgewood's crucibles are made of clay mixed with baked clay finely pounded, and are in every respect superior to the Hessian, but they are expensive. The black lead crucibles, formed of clay and plumbago, are very durable, resist sudden changes of temperature, and may be repeatedly used; but they are destroyed when saline substances are melted in them, and suffer combustion when exposed red-hot to a current of air. Crucibles are also made of cast-iron, of fine silver, and of platinum. The first, however, are destroyed when saline substances are melted in them, and when made red-hot in

VAPORISATION.

310. After a body has been rendered fluid by heat, it is expanded in the same manner, as when it existed in the state of solidity, by the farther addition of caloric; and this expansion continues to increase as the temperature is raised, until the particles of the body are at length separated to such distances that their attractive forces cease to operate, when a change of form again takes place, and, instead of a cohesive, a repulsive or elastic power is established, and the body becomes invisible, or passes into the aeriform state. This effect is denominated *Vaporisation*; and the bodies existing in such a state are termed *Vapours*, *Airs*, or *Gases*; and sometimes, in contradistinction to Fluids properly so called, *Elastic* Fluids.

311. Some solids, by an increase of temperature, are converted into vapours, without passing through the intermediate state of fluidity; but when heated under a greater pressure than that of the atmosphere, they also may be rendered fluid

312. The point at which bodies pass into the gaseous or vaporous state is very various. Some assume it at so low a temperature, that even the most intense cold that has been produced is insufficient to reduce them to the fluid form; others are convertible into vapour at a moderately high temperature, and condense again when that temperature is reduced; while there is a third class of bodies, the metals and earths, for instance, not convertible into vapour but by the most intense heat, and there are even some of them which

a current of air are apt to suffer oxidation: but in other respects they are durable, and can sustain sudden alterations of heat and cold without cracking.

have not suffered this change, because no temperature sufficiently elevated has been hitherto produced. These latter substances are termed *Fixed*, in contradistinction to those which are *volatile*, or easily convertible into vapour, but the term is merely relative.

313. This difference in the temperature at which bodies assume the aërial form, has given origin to a distinction established in chemical language, between *Vapours* and *Gases*, but which the late important discoveries of Mr. Faraday appear to have rendered in some measure nugatory. The term vapour was exclusively applied to those elastic fluids, which by cold or pressure could be reduced to the liquid form; while that of gas was applied to those that were uncondensable. Mr. Faraday, however, has succeeded in condensing into liquids the greater number of those which were considered as permanently elastic; if therefore the above distinctive terms be retained, they must be received with such limitations as the discoveries of Mr. Faraday will suggest.

314. The transition from the liquid state into that of elastic fluidity is usually accompanied with certain explosive movements, termed *Ebullition*, or boiling. In order to understand the nature and progress of this process, we have only to examine the phænomena which attend the boiling of water in a florence flask. Upon the first impression of the heat, the liquid will be observed to expand (277); in a few seconds, very minute bubbles will be seen to ascend from the bottom of the vessel to the surface; these consist of the atmospheric air previously combined with the water, and which becoming elastic from the elevation of temperature assumes the aeriform state, and escapes; at this period an intestinal motion may be easily perceived throughout the fluid, depending upon the circulation

which is taking place between the warmer and colder strata as already explained (288), and the surface may be observed to emit vapour. In a short time bubbles of steam will be noticed rising from the bottom, and suddenly disappearing in their ascent, in consequence of being condensed in their progress through the colder medium above; this phænomenon is accompanied with a hissing or simmering noise depending upon the vibrations occasioned by the successive vaporization and condensation of the particles in immediate contact with the bottom of the vessel; the process now proceeds more rapidly, the bubbles of steam arise with greater force and frequency, and are less and less condensed; the whole liquid is in a violent state of ebullition, and a large quantity of vapour continues to escape from the surface until the water is entirely dissipated. If, during the progress of this operation, a thermometer be inserted into the liquid, it will continue to rise until it reaches the temperature of 212°, beyond which it will never pass, however greatly we may urge the heat, and if the steam be examined at the same time, it will be found to have the same temperature as the boiling liquid. Fluids in the act of ebullition in glass vessels frequently exhibit sudden starts; this sometimes occurs to such a degree in distilling sulphuric acid as to break the retort. It would seem to arise from the adhesion of the fluid to the vessel, which may be considered analogous in its action to viscidity, and occasions the steam to rise to a temperature a little above its boiling point. The evil may be effectually obviated by introducing into the retort some small pieces of platinum, which appear to operate by virtue of their greater conducting power.

315. If the steam of boiling water be made to issue from a small tube, as, for instance, the spout of a tea-

kettle, the vapour will not be visible for nearly an inch
from the orifice, because perfectly formed steam is invisible; it is only when it begins to suffer condensation
that it becomes apparent.

316. Every liquid, when of the same degree of chemical purity, and under equal circumstances of atmospheric pressure, has one peculiar point of temperature
at which it invariably boils, and this is termed its *boiling point*; thus, pure water boils at 212°; when saturated with common salt, at 225°; æther at 96°; alcohol (sp. gr. 800) at 176°; oil of turpentine at 316°;
mercury at 660°, &c.

317. As the phænomenon of Vaporization depends
upon the energy of calorific repulsion becoming superior to that of cohesion, it is evident that it will be
retarded by whatever contributes to increase this latter
force. Hence the boiling point of the same fluid varies
under different degrees of atmospheric pressure. In
general, liquids boil *in vacuo* with about 124 degress
less of heat than are required under a mean pressure of
the atmosphere. Water, therefore, in a vacuum must
boil at 88°, and alcohol at 49°, *Fah.* Even the ordinary variations in the weight of the air, as measured by
the barometer, are sufficient to make a difference in the
boiling point of water of several degrees between the
two extremes.† Thus on the summit of Mont Blanc,
the pinnacle of Europe, water was found by Saussure
to boil at 187°, and I have been informed that the

---

† On this principle the late Mr. Archdeacon Wollaston invented a thermometric barometer to measure altitudes, from
which it appears that the height of a common table is even sufficient to produce a manifest difference in the boiling point of
water, as ascertained by this sensible instrument. The whole apparatus, weighing 20 ounces, packs in a cylindrical tin case, two
inches diameter, and 10 inches long.

monks of St. Bernard complain at being unable to make good *Bouillie*; the reason is obvious, in consequence of the altitude of their monastery, the water boils before it can arrive at a temperature sufficiently high for the necessary solution of the animal matter; the evil might easily be obviated by the application of a slight pressure.

318. The influence of diminished pressure, in facilitating ebullition, may be illustrated by a very pleasing experiment.

Exp. 44.—Fill a bolt head completely with boiling water, and immediately close its mouth with wetted bladder, or a tight cork. In a few seconds, the water from loss of temperature will contract and leave a partial *vacuum* in the upper part of the tube; the water will then recommence boiling, and continue in that state for many minutes, presenting the singular appearance which is represented in the annexed figure.

319. The following apparently paradoxical experiment will also illustrate the same principle.

Exp. 45.—Introduce a little water into a common florence flask, and heat it over a lamp until it boils; then remove the vessel, and cork it securely; the water will soon cease to boil, in consequence of the pressure occasioned by the steam on its surface, but if plunged into a vessel of *cold* water ebullition instantly recommences, from the condensation of vapour thus occasioned, but it ceases again if held near the fire.

320. On the fact of liquids boiling at a lower temperature *in vacuo,* or under diminished atmospheric
pressure, have been ingeniously founded several contrivances for evaporating at comparatively low temperatures. Mr. Barry has lately obtained a patent for an
apparatus of this description, † which he uses with
great practical advantage in obtaining vegetable extracts for medicinal purposes. Mr. Pope, Chemist, of
Oxford Street, has also constructed an apparatus of
the same description, by which he prepares similar
articles of very superior excellence. The advantages
of such an arrangement depend upon the fact that
vegetable bodies, when heated at lower temperatures
than the boiling point of water, suffer less change in
their essential properties than when exposed to greater
heat.

321. The same reasoning which explains why liquids
boil at lower temperatures under diminished pressure,
will shew that, under increased pressure they must
require a higher temperature to produce their ebullition, since, in this latter case, a greater repulsive force
will be necessary to separate their particles. By means
of an instrument invented by Papin, about the beginning of the last century, termed a *Digester,* and which
is a strong tight kettle, furnished with a valve of safety,
water may be heated several hundred degrees above its
ordinary boiling point, by which its solvent power is
very greatly increased, and it has therefore furnished
the easy means of dissolving the gelatine of bones.

322. To produce the evaporation of liquids, ebullition is by no means essential; all bodies that boil at

† For an account of this apparatus, see the Medico-chirurgical
Transactions for 1819 (vol. x.), and Dr. Ures Chemical Dictionary, Art. *Evaporation.*

moderate temperatures † seem to evaporate, so as to produce a certain quantity of elastic matter, in the common state of the atmosphere; and this quantity, as well as the degree of its elasticity, is greater in proportion as the temperature is higher. In hot, dry weather, it is obvious that there must be much more vapour in the atmosphere, than in cold wet weather; and the largest quantity will exist in summer, and in the tropical climates, when moisture is most required for the purposes of life. It was formerly supposed that this *spontaneous* evaporation depended upon an affinity existing between the air and water, but Mr. Dalton has shewn, very satisfactorily, that such an hypothesis was altogether erroneous.

323. The conversion of a liquid into vapour is always attended with great loss of free, or thermometric heat; in other words, a great quantity of caloric becomes *latent* during the formation of vapour. We have seen that during the boiling of water, although the source of heat remains, and a quantity of caloric must be perpetually entering the fluid, neither the water nor the steam ever acquires a higher temperature than 212°; the heat therefore must become latent, and be expended in the formation of steam; and it has been calculated by Dr. Black that the quantity of caloric thus rendered insensible to the thermometer, would be sufficient to raise the water 810° above its boiling point, or to 1022°, had it continued in the liquid state. There are other means, besides that employed by Dr. Black, by which this fact may be ascertained, and such inquiries lead to the conclusion that the latent heat of steam is between 900 and 1000 degrees.

---

† Mercury even has been found to pass into the state of vapour at the ordinary temperature of the atmosphere; see the chapter on Aerial Poisons in my work on Medical Jurisprudence, vol. 2, p. 456.

This law may be illustrated in a very striking manner by the following experiment.

> *Exp.* 46.—Let a pound of water at 212°, and eight pounds of iron-filings at 300° be suddenly mixed together. A large quantity of vapour will be instantly generated; and the temperature of the mixture will be only 212°; but that of the vapour produced is also not more than 212°; and the steam must therefore contain, in a latent or combined form, all the caloric which raised the temperature of eight pounds of iron-filings from 212° to 300°

324. On the contrary, the following experiment will prove that vapours, during their conversion into a liquid form, evolve, or give out, much caloric.

> *Exp.* 47.—Mix 100 gallons of water at 50° with one gallon of water at 212°; the temperature of the water will be raised about 1½°. Condense by a common still-tub, one gallon of water, from the state of steam, by 100 gallons of water, at the temperature of 50°. The water will be raised 11°. Hence one gallon of water, condensed from steam, raises the temperature of 100 gallons of cold water 9½° more than one gallon of boiling water. This is owing to the much greater quantity of caloric, existing in a gallon of water when in the state of steam, than in a gallon of boiling water, though steam and boiling water affect the thermometer in precisely the same degree.

325. The processes of Nature and Art * are alike fertile in illustrations of this important law, and when we come to speak of the different artificial processes for increasing and diminishing temperature, the student will be presented with many striking examples; at present it will be sufficient to observe that in hot climates, the seas, rivers, and lakes, are prevented from acquiring any thing like the heat of the adjacent lands, by the evaporation which is constantly going on at the surface, and which carries off the greater part of the heat as fast as it is received. The same principle is also operative in regulating the temperature of the human body; for whenever it becomes oppressed by heat, if it be in a state of health, perspiration commences, and this carries off the superabundant caloric as fast as it is produced, in the same manner that steam conveys away the heat which is constantly entering into water during its ebullition.

326. To account for the absorption of caloric, and the phænomena of liquefaction and vaporization, two

---

* Many beautiful illustrations of this principle have been afforded by the late researches of Mr. Faraday on the condensation of gases. One of these I shall here describe as being remarkably striking and satisfactory. The substance termed *Chloride of Silver* has the property of absorbing a very large proportion of pure ammonia. If this body thus saturated be exposed, in the end of an hermetically sealed tube, to heat, the ammonia is driven off, but, being unable to escape, is condensed into a liquid and distills to the other extremity of the tube. In a short time, however, the ammonia again passes slowly into the state of gas, and is re-absorbed by the *Chloride*; we have thus then the two converse operations of vaporisation, and condensation proceeding at the same time, at the two extremities of the tube, and by applying the hand we shall derive from one end the sensation of cold, and from the other, that of heat. See also Mr. Leslie's beautiful contrivance of freezing water, 346.

theories have been proposed; the one by Dr. Black, the other by his ingenious pupil Dr. Irvine. Dr. Black considered the absorption of caloric as the *cause* of the change of form, and as necessary to the constitution of the fluid or vapour. Dr. Irvine regarded it as the *effect*, in consequence of the *capacity* of the body thus becoming enlarged. Dr. Black's theory is founded on the general analogy of the phænomena to those of chemical combination. We have seen that when one chemical agent is merely *mixed* with another, the properties of neither are altered; but, when chemically combined, they are more or less changed; and it has been shewn that this combination always takes place in certain definite proportions. Caloric acts upon matter in a manner similar to this; it may be diffused through bodies to a certain extent without having its properties altered; but in liquefaction and vaporisation it is united to the body, so that its properties are entirely lost. Ought not this, in conformity to general analogy, to be ascribed to its having formed a chemical union; especially since it is absorbed by different bodies in different quantities, but always in certain determinate proportions? circumstances which extend the analogy between this absorption and chemical combination. But, although the theory appears so far complete, when more minutely examined it will be found deficient. We know of no case of chemical combination in which the properties of one of the bodies combining are entirely lost, while those of the other are not altered; but in liquefaction and vaporisation this must be supposed to be the case; since, though the properties of the caloric absorbed can no longer be recognised, those of the substance fused or evaporated remain the same. Every solid too may be rendered fluid, and every fluid converted into vapour, and consequently it must be supposed, that caloric is capable of combining chemically with every other body, a property possessed by no other chemical agent. And, lastly, the contact of any body at a low temperature is sufficient to reduce a vapour to a fluid, or a fluid to the solid state; and in such cases, the caloric abstracted does not enter into combination with the body to which it is communicated, but is merely diffused through it so as to raise its temperature. But that union cannot be termed chemical which is not elective, and which can be so easily broken by the abstraction of one of its principles, without the interference of a stronger affinity.

327. The opinion of Dr. Irvine is not open to the same objection, and it is equally adapted to the explanation of the phænomena; but before we proceed to its investigation, it will be necessary to shew what is meant by the *capacity* of a body for heat.

328. Equal weights of the *same* body, at the same temperature, contain the same quantities of caloric. But equal weights of *different* bodies, at the same temperature, contain unequal quantities of caloric. The quantity of caloric which one body contains, compared with that contained in another, is called its *Specific Caloric*; and the power or property, which enables bodies to retain different quantities of caloric, has been called *capacity for caloric*. The method of determining the *specific caloric*, or *capacities* of different bodies, is shewn by the following experiment.

Exp. 48.—Mix together equal measures of water at 100° *Fah.* and at 40°, and the temperature of the resulting mixture will be found to be the arithmetical mean, viz. 70°.

Exp. 49.—Mix a pint of quicksilver at 100° *Fah.* with a pint of water at 40°, and the resulting tem-

perature will not be 70° (the arithmetical mean)
but only 60°. In this case it is evident, that the
quicksilver has lost 40° of heat, which have nevertheless raised the temperature of the water only
20°, it appears therefore that double the quantity
of caloric is required to raise the temperature of
a pint of water, than that of a similar measure of
quicksilver, through the same number of degrees.
Hence we say that water has a greater capacity
for caloric than quicksilver, in the proportion of
two to one.

329. If, instead of equal *bulks* of quicksilver and
water, we had taken equal *weights*, the disparity between the specific caloric of the mercury and water
would have been still greater. Thus a pound of water
at 100°, mixed with a pound of mercury at 40°, will
give a temperature of $97\frac{1}{2}$°, or $27\frac{1}{2}$° above the arithmetical mean. In this experiment, the water being
cooled from 100° to $97\frac{1}{2}$° has lost a quantity of caloric
reducing its temperature only $2\frac{1}{2}$°; but this caloric,
communicated to the pound of mercury, has occa-
sioned, in its temperature, a rise of no less than $57\frac{1}{2}$°
Therefore, a quantity of caloric, necessary to raise the
temperature of a pound of water $2\frac{1}{2}$°, is sufficient to
raise that of a pound of mercury $57\frac{1}{2}$°; or, by the rule
of proportion, the caloric, which raises the temperature
of a pound of water 1°, will raise that of a pound of
mercury about 23°. Hence it is inferred, that the
quantity of caloric contained in water, is to that contained in the same weight of quicksilver as 23° to 1°;
or stating the caloric of water at 1°, that of quicksilver
will be $\frac{1}{23}$ part of 1°, or 0·0435.

330. By a similar mode of inquiry may the *capacities* of other bodies be easily ascertained. If a pound

of water at 100°, and the same weight of oil at 50°, be mixed, the resulting temperature is not 75° (the mean) but $83\frac{1}{3}°$; the water, therefore, has lost only $16\frac{2}{3}°$, while the oil has gained $33\frac{1}{3}°$; or, if equal weights of water at 50°, and oil at 100° be mixed, the resulting temperature is $66\frac{2}{3}°$, so that the oil has given out $33\frac{1}{3}°$, and the water has increased only $16\frac{2}{3}°$. Hence the heat which raises a given weight of water 1°, will raise the same weight of oil 2°; and as the specific heats are inversely as the changes of temperature, the specific heat of water may be called 1, and that of oil 0·5.

331. The following is a general formula for determining the specific heat of bodies, from the temperature resulting from the mixture of two bodies at unequal temperatures, whatever be their respective quantities. *Multiply the weight of the water by the difference between its original temperature, and that of the mixture. Multiply, also, the weight of the other liquid by the difference between its temperature and that of the mixture; divide the first product by the second, and the quotient will express the specific heat of the other substance, that of water being* $= 1$. Thus 20 ounces of water at 105, mixed with 12 ounces of spermaceti oil at 40°, produce a temperature of 90°. Therefore, multiply 20 by 15 (the difference between 105 and 90) $= 300$. And multiply 12 by 50 (the difference between 40 and 90) $= 600$. Then $300 \div 600 = \frac{1}{2}$, which is the specific heat of oil; that is, water being 1, oil is 0·5.

332. When this comparison is extended to a great variety of bodies, they will be found to differ very considerably in their capacities for caloric, and as a knowledge of this difference is of great importance to the chemist, a Table, collected from the best authorities, has been inserted in the Appendix.

333. The capacities of bodies for heat have considerable influence upon the rate at which they are heated and cooled. Those bodies which are most slowly heated and cooled have generally the greatest capacity for heat. Thus, if equal quantities of water and quicksilver be placed at equal distances from the fire, the quicksilver will be more rapidly heated than the water, and the metal will cool most rapidly when carried to a cold place. Upon this principle Professor Leslie ingeniously determined the specific heat of bodies, observing their relative times of cooling a certain number of degrees, comparatively with water, under similar circumstancs.

334. Such is the simple view of the subject; and it must be remembered that the term *Capacity* is not designed to point out any particular cause, mechanical or chemical; neither does it express any vague obscure idea, as some have maintained. It is merely a general expression to denote the property which bodies have of containing *at the same temperature*, and in equal weights or volumes, certain quantities of caloric.

335. Upon what this property depends has not yet been discovered, nor is it, perhaps, a matter of material consequence, so long as the phænomena themselves are correctly observed and stated. It is clear that whenever the capacity of body is augmented there must be an absorption of caloric, and a diminution of temperature, and *vice versa*. It cannot then be doubted but the absorption and latent state of caloric which take place during the liquefaction and vaporisation are fully accounted for, on the supposition of Dr. Irvine, by the change of capacity which accompanies these phænomena. That there should be such a change is *a priori* extremely probable; for since, as there is reason to believe, the capacities of bodies depend on certain situ-

ations of their minute particles, it is a probable inference, that when the form of a body is altered, its capacity also will be altered; and as rare bodies have in
general greater capacities than those which are more
dense, there is some reason to presume that the capacity
of the fluid will be superior to that of the solid, and
that of the vapour or gas to the capacity of the fluid.

336. This conclusion is established by experiment,
for Dr. Irvine found, that the capacity of water is
greater than that of ice by one-tenth, and Dr. Crawford
states the capacity of aqueous vapour to that of water
as 1550 to 1000.

337. To the theory of Dr. Irvine it has been ob-
jected, that if the absorption of caloric be not considered as the *cause* of the change of form, from a solid
to a liquid, or a liquid to a vapour, no adequate cause
is pointed out. The reply to such an objection is obvious,—It is owing to the body being expanded to a
certain degree, by which its particles are so far separated, that the force of cohesion by which they were
held together is diminished or overcome, in consequence
of which a new arrangement takes place, it passes into
the fluid or aerial form, and in such a state as more
caloric is contained in it, at a given temperature, a
quantity must be absorbed in order to preserve that
temperature. The general question would be unequivocally decided, were it possible to determine whether
the change of form precedes the absorption or extrication of caloric, or whether the reverse be the case.
But this cannot be directly ascertained, since the two
appear to be simultaneous; it is, however, more probable that the effect of the reduction of temperature is
first to change the form, and that the extrication of
caloric is the consequence of this. The late discoveries
of Mr. Faraday add considerable support to such a

supposition, for he has shewn that the form of a body may be reduced by merely causing the particles of a body to approximate more closely. By mere pressure he has reduced gases to the fluid form. This pressure, it is evident, can have only a mechanical effect on the aeriform matter; it must merely occasion a change of form by bringing the particles into closer contact, and can have no effect in separating caloric, were it chemically combined. Several other facts come in aid of the same conclusion; by the mechanical compression of air much heat is given out; thus, if air be suddenly compressed in the ball of an air-gun, the quantity of caloric liberated by the first stroke of the piston is sufficient to set fire to a piece of tinder; a flash of light is said, also, to be perceptible at the moment of condensation. This fact has been applied to the construction of a portable instrument for lighting a candle. It consists of a common syringe, concealed in a walking stick. At the lower extremity, the syringe is furnished with a cap, which receives the substance intended to be fired, and which is attached to the instrument by a male and female screw. The rapid depression of the piston condences the air, and evolves sufficient heat to inflame the tinder. When, on the contrary, air is suddenly rarefied to many times its volume, its temperature falls sufficiently to sink a very sensible thermometer 50° of *Fah.* its sensible heat instantly passing, in this case, into a latent form. This principle is well exemplified in a machine erected at one of the mines at Chemnitz in Hungary. In this apparatus the air within a large cylinder is very forcibly compressed by a column of water 260 feet high. Therefore, whenever a stopcock which is attached to the lower part of the cylinder is opened, the compressed air rushes out with violence, and its expansion is so sudden and considerable that the moisture which was contained in the compressed air is immediately condensed, and falls in a shower of snow.

338. From the above considerations it appears more philosophical to conclude, that the absorption of caloric, which accompanies liquefaction and vaporisation, is owing not to its entering into chemical combination, but to the enlarged capacity which the body acquires by the change of form.

On the different methods of measuring Temperature.

339. On the principle of bodies expanding by heat is founded the construction of the *Thermometer*, an instrument of so much importance in every philosophical investigation, that it is absolutely necessary for the student to become acquainted with the construction of the different kinds which are employed.

340. To Sanctorius of Padua we are originally indebted for the invention. The instrument, however, employed by that philosopher was of a very simple description, and measured variations of temperature by the variable expansion of a confined portion of air. This instrument is represented in the margin. It consists of a glass tube, eighteen inches long, open at one end, and blown into a ball at the other. If a warm hand be applied to the ball, the included air will expand, and a portion be expelled through the open end of the tube. And if, in this state the aperture is quickly immersed in a cup filled with some coloured liquid, it will ascend in the tube, as the air

in the ball contracts by cooling, and its altitude will in every case depend upon the degree of expansion which the air has previously undergone. In this manner it is prepared for use, and will indicate increase of temperature by the descent of its fluid, and *vice versa.* These effects may be exhibited, alternately, by applying the hand to the ball, and then blowing on it with a pair of bellows. The amount of the expansion or contraction is measured by a graduated scale attached to the stem of the instrument. It is evident, however, that such a form of apparatus must be extremely inconvenient, and Mr. Boyle, accordingly, modified its construction in the following manner. He took a glass vessel, of a spherical shape, and introduced into it a quantity of coloured liquid sufficient to occupy about a fourth of its capacity. The glass tube, open at both ends, was then cemented into the neck, with its lower aperture beneath the surface of the liquid. In this case the expansion of the included air drives the liquid up the stem, to which a graduated scale may be attached to mark its extent. Another very simple modification of this instrument has been proposed, and consists in introducing a small column of tinged liquor, of about an inch in length, into a tube of very fine bore, and from 9 to 12 inches in length, at one end of which is blown a ball, from half an inch to an inch in diameter, and which is blackened by paint, or by the smoke of a candle. To render this instrument applicable, the coloured column ought to be stationary, about the middle of the tube, at the common temperature of the atmosphere, when its station will be altered by the slightest variation of temperature affecting the air contained in the bulb, and a scale of equal parts may be so adjusted as to measure the amount of the effect.

341. The advantages of the *Air* Thermometer consist in the great amount of the expansion of air, when compared with that of any liquid; by which it is enabled to detect minute changes of temperature, which the mercurial thermometer would searcely discover; for air is increased in volume by a given elevation of temperature, about twenty times more than quicksilver. The advantages, however, which attend this excessive delicacy are counterbalanced by several serious objections. It will be readily seen, for instance, that the air thermometer will not only be affected by changes of temperature, but by variations of atmospheric pressure.

342. Professor Leslie,* under the name of the *Differential Thermometer*, has introduced a very important modification of the air thermometer. It consists  of two glass tubes of unequal length, each terminating in a hollow ball, as represented in the annexed figure, which are united by a tube twice bent at right angles, containing coloured sulphuric acid. Whenever a hot body approaches one of the bulbs, it must necessarily drive the fluid towards the other. It is evident then that this instrument cannot be employed to measure variations in the temperature of the surrounding atmosphere, because, as long as both balls are of the same temperature, whatever this may be, the air contained in both will have the same elasticity, and, consequently,

* Experimental Inquiry into the Nature and Propagation of Heat, by John Leslie. London, 1804. p. 9, &c.

the intercluded coloured liquor, being pressed equally
in opposite directions, must remain stationary. If,
however, any change of temperature is effected in one
of the balls, the instrument will immediately indicate
this *difference* with the greatest nicety. The amount
of this effect is ascertained by a graduated scale, the interval between freezing and boiling being distinguished
into 100 equal degrees. This thermometer is peculiarly
adapted to ascertain the difference of the temperatures
of two contiguous spots in the same atmosphere, and
its application has thrown considerable light upon several of the more obscure phænomena of caloric.

343. The Thermometer in ordinary use consists of
an hermetically sealed glass tube, terminating at one
extremity with a bulb. The bulb and part of the tube
are filled with an appropriate liquid, which, when designed to measure very low temperatures, is spirit of
wine, under other circumstances quicksilver is better
adapted for the purpose. A graduated scale is attached
to the stem; and, whenever the instrument is applied
to bodies of the same temperature, the mercury, or spirit, being similarly expanded, indicates the same degree
of heat. In dividing the scale, the two fixed points
usually resorted to are the freezing and boiling of water,
which always take place at the same temperature, when
under the same atmospheric pressure; the intermediate
part of the scale is divided into any convenient number
of degrees. In the *Centigrade*, this space is divided
into 100°, as its name imports; the freezing point of
water being marked 0°, the boiling point, 100°. In
this country we use the scale of Fahrenheit, of which
the 0° is placed at 32° below the freezing of water,
which is therefore marked 32°, and the boiling point
212°, the intermediate space being divided into 180
degrees. The scale more commonly used on the con-

tinent is that of Reaumur, in which the freezing point
is 0°, the boiling point 80°. As the medical student
in his perusal of foreign works will frequently have occasion to reduce the degrees upon these scales to corresponding ones on that of Fahrenheit, it may be
useful for him to learn the following simple formula,
for that purpose. It will be seen that each degree of
Fahrenheit is equal to $\frac{4}{9}$ of a degree of Reaumur; if,
therefore, the number of degrees on the former scale,
above or below the freezing of water, be multiplied by
4 and divided by 9, the quotient will be the corresponding degree of Reaumur, thus

Fahrenheit.   *Reaumur.*
$$68° - 32° = 36 \times 4 = 144 \div 9 = 16°$$
$$212° - 32° = 180 \times 4 = 720 \div 9 = 80°$$

To reduce the degrees of Reaumur to those of Fahrenheit, they are to be multiplied by 9 and divided by 4,
when the quotient $+$ 32 will be the number required,
thus

Reaumur.   *Fahrenheit.*
$$16 \times 9 = 144 \div 4 = 36 + 32 = 68°$$
$$80 \times 9 = 720 \div 4 = 180 + 32 = 212°$$

To adapt the degrees of the Centigrade to those of
Fahrenheit, since every degree of the latter is equal
to $\frac{5}{9}$ of a degree of the former, we have only to multiply
the degrees of the centigrade by 9, and divide by 5,
when the quotient $+$ 32 will solve the problem, thus

Centigrade.   *Fahrenheit.*
$$100 \times 9 = 900 \div 5 = 180 + 32 = 212°$$

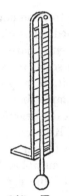

344. As the Pharmaceutist will frequently have occasion to employ the thermometer for ascertaining the temperature of corrosive liquids, the graduated scale should be provided with a hinge, so as to double back, and leave the bulb exposed, as represented in the margin.

345. For measuring high temperatures, it is evident that the Thermometer cannot be employed. An instrument, called a *Pyrometer*, has accordingly been constructed, founded on the property which metals possess of expanding by heat, an account of which may be found in most elementary works on chemistry. Sir Isaac Newton exposed a body, whose temperature he wished to discover, to the action of cold air, and observed the length of time which it required to reduce it to a temperature which fell within the range of the thermometer; he then marked the period it took to cool every degree, and from such data inferred the original temperature of the body. It is obvious that such a process is liable to incalculable fallacies. Lavoisier and La Place proposed to inclose the heated body in ice, and to estimate the quantity which it liquefied during its refrigeration, and for conducting the operation, an instrument termed a *Calorimeter* was invented, but it has been found incapable of affording results sufficiently precise.

On the different Artificial Processes by which Temperature may be increased or diminished.

1. *Methods of producing Cold Artificially.*

A. *By Liquefaction.*

346. The theory of the production of cold by the rapid conversion of solids into liquids has been already fully explained (297.) It is only necessary in this place to furnish a list of the bodies which are best adapted for such a purpose, and to offer some practical advice respecting the manner of conducting the operation. The following table shews the results of some of Mr. Walker's experiments on the subject of Frigorific mixtures.

Mixtures.	Thermometer sinks.
Muriate of ammonia .. 5 parts Nitre .... 5 Water 16	From 50° to 10°
Muriate of ammonia .. 5 Nitre .... 5 Sulphat of Soda ...... 8 Water 16	From 50 to 4
Nitrate of ammonia... 1 Water 1	From 50 to 4
Nitrate of ammonia... 1 Carbonate of soda .... 1 Water 1	From 50 to 7
Sulphate of soda...... 3 Diluted nitric acid 2	From 50 to 3

Mixtures.	Thermometer sinks.
Sulphate of soda...... 6 parts Muriate of ammonia .. 4 Nitre .... 2 Diluted nitric acid 4	From 50° to 10°
Sulphate of soda...... 6 Nitrate of ammonia... 5 Diluted nitric acid 4	From 50 to 14
Phosphate of soda..... 9 Diluted nitric acid 4	From 50 to 12
Phosphate of soda..... 9 Nitrate of ammonia... 6 Diluted nitric acid 4	From 50 to 21
Sulphate of soda...... 8 Diluted sulphuric acid. 5	From 50 to 0
Sulphate of soda...... 5 Diluted sulphuric acid. 4	From 50 to 3
Snow .... 1 Common salt 1	From 32 to 0
Muriate of lime ...... 3 Snow 2	From 32 to —50
Potass .... 4 Snow 3	From 32 to —51
Snow .... 1 Diluted sulphuric acid. 1	From 20 to —60
Snow or pounded ice.. 2 Common salt......... 1	From 0 to —5
Snow and diluted nitric acid	From 0 to —46
Muriate of lime....... 2 Snow 1	From 0 to —66
Snow or pounded ice.. 1 Common salt .... 5 Muriate of ammonia } 5 and nitre.......... }	From —5 to —18

Mixtures.	Thermometer sinks.
Snow .... 2 parts Diluted sulphuric acid. 1 Diluted nitric acid 1	From —10° to —56°
Snow or pounded ice. 12 Common salt .... 5 Nitrate of ammonia... 5	From —18 to —25
Muriate of lime........ 3 Snow 1	From —40 to —73
Diluted sulphuric acid 10 Snow 8	From —68 to —91

347. In order to ensure the success of these operations, the salts employed must be fresh crystallized, and newly reduced to a very fine powder. The vessels in which the freezing mixture is made should be very thin, and just large enough to hold it, and the materials should be mixed together as quickly as possible, for upon the rapidity with which the solution is effected the degree of refrigeration produced will in great measure depend. The vessel ought, moreover, to be enveloped in some bad conducting substance, such as flannel. As the water of crystallization seems to perform an important part of the cooling effect, efflorescent salts should be rejected. The materials to be employed, in order to produce great cold, ought to be first reduced to the temperature marked in the table, by placing them in some of the other freezing mixtures; and then they are to be mixed together in a similar freezing mixture. If, for instance, we wish to produce a cold $=$ —46°, the snow and diluted nitric acid ought to be cooled down to 0°, by putting the vessel which contains each of them into the 12th freezing mixture in the above Table, before they are mixed together.

It may be observed that the expense of making frigorific mixtures is trifling, because, in many cases, the
salts may be recovered by evaporating the water in
which they were dissolved. Mr. Walker, who was in
the habit of recovering the salt, remarks that no diminution was observed in its effect, after many repeated
evaporations. But the cheapest and most efficacious
salt which has been employed for the reduction of temperature is the muriate of lime, first used by Mr. Lowitz;
and which, when mixed with snow, is extremely active
in the production of cold. Mr. Lowitz directs, not
only that this salt should be perfectly well dried, but
that it should be in that state in which it is crystallized
with the largest possible quantity of water of crystallization, which is effected by putting the solution aside
to cool when of the specific gravity 1·5 or 1·53. When,
however, a reduction of 15° or 20° of temperature is
only required, equal parts of nitre and sal ammoniac
are found most advantageous and œconomical, as the
salts are easily recovered by evaporation.

348. The Physician or Surgeon in directing the use
of saline solutions for the purpose of refrigeration,
unless guided by the principles here explained, will
fail in producing the effects he might anticipate. A solution of nitre, for instance, is often prescribed with a
view of producing cold in the stomach; in such a case,
the salt should be swallowed the instant it is dissolved,
or no advantage can possibly result; (*see Pharmacologia, Edit.* 6, *vol.* 1, *p.* 212.)

B.   *By Evaporation.*

349. This is perhaps the more universal and efficacious mode of reducing temperature, and is practised
different ways, in various parts of the globe; thus, in

India, where the apartments are separated from their courts by curtains instead of walls, slaves are employed in perpetually sprinkling these curtains with water, the evaporation of which, when constantly kept up, will reduce the temperature of the rooms ten or fifteen degrees. The principle of cooling by evaporation is well understood by the caravans who cross the great desert of Arabia. These people have occasion for a large quantity of water, which they carry with them on camels, in bottles of earthen-ware, and which in passing over the burning sands of that country, would become very disagreeably hot, were it not for the following expedient, which is universally adopted by them. When they lay in their stock of water, each bottle is enfolded in a linen cloth, and some of the company are appointed to keep these cloths constantly wet during the journey; by which means a perpetual evaporation is produced, and the contents of the bottles are preserved at a cool and refreshing temperature. In like manner, in the nights in Bengal, when the temperature is not below 50°, by the exposure of water in earthen-pans upon moistened bamboos, thin cakes of ice are formed, which are heaped together and preserved under ground by being kept in contact with bad conductors of heat. At sea, wine and other liquors may be cooled by enveloping the bottles in wetted linen, and exposing them to a current of air in the sails. The *Alcarazas* of Spain for cooling wine act on the same principle of evaporation. These, which are very porous earthen vessels, are prepared for use by soaking them in water for a considerable time, so as to saturate them with that fluid. Within these jars, vessels containing the wine are introduced, when the perpetual oozing and evaporation of the water cools the interior of the vessel, and consequently reduces, in some measure, the temperature of the wine or other liquor placed in it.*

350. Upon the principle of the sudden rarefaction of air producing cold, as so beautifully illustrated by the hydraulic machine at Schemnitz, M. Gay Lussac has lately proposed a refrigerating apparatus, which is, in fact, a miniature representation of the machine above mentioned. He exposes the small body to be cooled to a stream of air escaping by a small orifice, from a box in which it had been strongly condensed. But, of all the expedients devised for this purpose, none equals in ingenuity or effect that invented by Mr. Professor Leslie. This apparatus is constructed upon the well known fact, that *water evaporates with much more rapidity under a diminished pressure, especially if the vapour, which is formed, be condensed as soon as it is produced, so as to maintain the vacuum.*

The annexed sketch exhibits the most approved form of the apparatus. The water to be congealed is contained in a shallow vessel supported on a stand, which is placed in another containing concentrated sulphuric acid, or dry muriate of lime, or even dried garden mould, or parched oatmeal, or indeed any other substance that has the property of attracting moisture powerfully. The whole is then covered by the receiver of an air pump,

---

* The wine coolers which are manufactured in this country for the same purpose are ill calculated for the accomplishment of their object. They are, in general, much larger than is neces-

which is rapidly exhausted; and, as soon as this
is effected, crystals of ice begin to shoot in the
water, and a considerable quantity of air makes its
escape, after which the whole of the ice becomes solid.
The rarefaction required is about 100 times; but to
support congelation, after it has taken place, 20 or
even 10 times are sufficient. The acid continues to
act until it has absorbed an equal volume of water.
If the vessel of water is covered with a plate of metal
or glass fixed to the end of a sliding wire, and which
passes through the neck of the receiver so as to be, at
the same time, air tight and moveable, as shewn in the
annexed figure, the water will continue fluid, after the
exhaustion of the receiver, until the cover is removed,
when in less than five minutes, needle-shaped crystals
of ice will shoot through it, and the whole will shortly
become frozen. In this process, it is evident that were
not sulphuric acid, or some other absorbent, present,
an atmosphere of aqueous vapour would soon fill the
receiver, and, by pressing on the surface of the water,
prevent the further production of vapour. But the
steam which arises, being condensed the moment it is
formed, the evaporation goes on very rapidly, and has
no limits but the quantity of the water, and the diminishing concentration of the acid. If in this experiment the temperature of the sulphuric acid be examined, it will be found to have been raised considerably, as we should, *a priori* have conjectured. So that
the two converse operations of vaporization and condensation have on this occasion been accompanied

---

sary, in consequence of which a sheet of air of the temperature
of the room envelopes the bottle, and prevents the proper action
of the Alcaraza upon it; whereas, if they were made perfectly
cylindrical, instead of being bulged, and no larger than just
sufficient to admit the bottle within them, they would be much
better refrigerators than those which are usually sold.

with corresponding changes of temperature, as already explained by the interesting experiment of Mr. Faraday (325, *note.*)

351. In our operations to produce cold by evaporation the effect will be more striking and rapid, if we employ fluids that evaporate at a lower temperature than water; thus æther may be made subservient, on many occasions, to the purposes of refrigeration, and this fluid as well as alcohol accordingly affords the physician a valuable ingredient for the formation of cooling lotions. The frequent abuse of such applications will afford a striking illustration of the necessity of chemical knowledge for the preparation and direction of remedies. I have known a lotion of this kind applied to the head, when the patient has immediately covered it with a flannel cap, and thus converted into a rubefacient that which was intended to act as a refrigerant. As an instance, the converse of this, we have heard of the application of Brandy to the feet, with the view of preventing the ill effects of previous cold, having occasioned, by its evaporation, such a diminution of temperature as to have aggravated the evil it was intended to counteract. If we would prevent the mischief that usually accrues from wet clothes, we have only to prevent the evaporation that is thus occasioned. If a person in such a situation covers himself with a dry great coat, he will have little occasion to fear the effects of his accident.

352. Liquid *Sulphurous acid,* in consequence of the rapidity with which it evaporates under the ordinary pressure of the atmosphere, has been recently employed with great success for the reduction of temperature.

353. A convenient mode of keeping up a uniform evaporation, for reducing the temperature of any part.

of the body, as in the case of fractured limbs, &c. is to allow the gradual distillation of water upon it through the medium of skeins of cotton as before described. (120.)

354. The Pharmaceutist frequently avails himself of the cooling property of vaporization, for regulating the heat employed in his various processes. It is upon the same principle that the Pharmacopœia directs the ingredients of several plasters to be heated with water, which, during its passage into vapour, carries off a considerable portion of caloric, and thereby prevents the temperature from becoming sufficiently high to enable the fatty matter to decompose the metallic oxides, or to undergo a chemical change by which the adhesive power of the plaster would be diminished.

355. Besides the modes above enumerated, the temperature of bodies may be lowered by connecting them with good radiators and conductors of caloric, a principle which has been beautifully illustrated by the splendid invention of Sir H. Davy's Safety Lamp.

2. Methods of producing Heat Artificially.

356. The means which have been employed for this purpose in different parts of the world have been so various, that I can only attempt to enumerate a small proportion of them; but I shall select those which may be applied to some useful purpose connected with Medicine, or which may serve to throw some light upon the production of the temperature of animals.

A. By Percussion.

357. If a piece of iron be smartly and quickly struck with a hammer it becomes red hot, and nothing is more

o

common in some countries, where fuel is dear and fires
are not kept up during the night, than for a smith to
procure a light in the morning by such an expedient.
It deserves notice, however, that heat has never been
observed to follow the percussion of liquids, nor of soft
bodies which easily yield to the stroke. The importance of this fact to the Physiologist will become apparent when we treat of the source of animal heat.
The evolution of caloric, by percussion, seems to be
the result of a permanent or temporary condensation
of the body struck, by which its capacity is diminished.
Part of the caloric, however, which is thus evolved, in
some cases, originates in another manner; for, by condensation, as much caloric is evolved as is sufficient to
raise the temperature of some of the particles of the
body high enough to enable it to combine with the
oxygen of the atmosphere.

B. *By Friction.*

358. Friction seems to be only a succession of per-
cussions, and yet the density of bodies rubbed against
each other is not increased, as happens in cases of percussion, for heat is produced by rubbing soft bodies
against each other; the density of these cannot therefore be increased by such means, as any one may convince himself by rubbing his hand smartly against his
coat. It is indeed true that heat cannot be produced
by the friction of liquids; but then they are too yielding to be subjected to strong friction. It is not owing
to the specific caloric of the rubbed bodies decreasing;
for Count Rumford has shewn that there was no sensible decrease; nor if there were a decrease, would it
be sufficient to account for the vast quantity of heat

which is sometimes produced by friction. Nor is it owing to the combustion of oxygen with the bodies themselves, or with any part of them. Are we then to conclude, with Count Rumford, that there is no such substance as *Caloric*, but that it is merely a *peculiar kind of motion?* by no means; we are not sufficiently acquainted with the laws of the motion of caloric, to be able to affirm that friction cannot cause it to accumulate in the bodies rubbed. At present, we must rest upon it as an ultimate fact, and be satisfied with availing ourselves of the advantages to which a knowledge of it may conduce. The original inhabitants of the New World, throughout the whole extent from Patagonia to Greenland, procured fire by rubbing pieces of hard and dry wood against each other, until they emitted sparks, or kindled into flame; some of the people to the north of California produced the same effect by inserting a pivot in the hole of a very thick plank, and causing it to revolve with extreme rapidity. This fact explains how immense forests have been consumed, from the violent friction of the branches against each other by the wind.

C. *By the Condensation of Steam.*

359. That vapours, in passing to the state of liquids, part with a considerable proportion of caloric, has been sufficiently demonstrated in the preceding section, we have now merely to consider the modes by which the fact may be made subservient to œconomical purposes.

360. Large quantities of water, or other liquids, may be heated by steam in three ways;—the steam pipe may be plunged with an open end into the water cistern; or the steam may traverse the fluid in pipes, as

in the case of the condensing tub of the still; or the vessel may be provided with double sides, and the steam may be diffused around in the interstice. This latter mode is of universal applicability, and is that which has been adopted with particular advantage in the laboratories at Apothecaries' Hall. * For operations on a smaller scale vapour baths may be employed with much advantage. Dr. Ure has given us directions for constructing a simple apparatus which is well designed for the use of the medical practitioner; it is œconomical, efficacious, and extremely easy of application. A square tin box, about 18 inches long, 12 broad, and 6 deep, has its bottom hollowed a little by the hammer towards its centre, in which a round hole is cut of five or six inches in diameter. Into this, a tin tube, three or four inches long, is soldered. This tube is made to fit tightly into the mouth of a common tea kettle, which has a folding handle. The top of the box has a number of circular holes cut into it, of different diameters, into which, evaporating capsules of platinum, glass, or porcelain, are placed. When the kettle, filled with water, and with its nozzle corked, is set on a stove, the vapour, playing on the bottoms of the capsules, heats them to any required temperature; and being itself continually condensed, it runs back into the kettle, to be raised again, in ceaseless cohobation. With a shade above, to screen the vapour chest from soot, the kettle may be placed over a common fire. The orifices not in use are closed with tin lids. In drying precipitates, the tube of a glass funnel may be corked, and placed with its filtre, directly into the proper sized opening.

---

* An account of this apparatus, with an illustrative plate, may be found in Mr. Thomson's London Dispensatory. A gallon of water in the form of steam will be adequate to heat 18 gallons of water to 100.

For drying red cabbage, violet petals, &c., a tin tray is provided, which fits close to the top of the box, within the rim which goes about it. The round orifices are left open when this tray is applied. Such a form of apparatus is well calculated to inspissate the pasty mass from which lozenges and troches are to be made.

361. The heating of apartments is another valuable application of steam. Safety, cleanliness, and comfort, are combined in this mode of giving genial warmth for every purpose of private accommodation, or public manufacture. It has been ascertained that one square foot of surface of steam pipe, is adequate to the warming of two hundred cubic feet of space.

362. Chambers filled with steam, heated to about 125° *Fah.* have been introduced with advantage into medicine, under the name of *Vapour Baths.* Dry air has been also used for this purpose, and the latter can be tolerated at a much greater heat than moist air. Dr. Ure observes that a large cradle, containing sawdust heated by steam, should be kept in readiness at the houses erected by the Humane Society for the recovery of drowned persons; or a steam chamber might be attached to them, for this, as well as other medical purposes.

D. *By Combustion.*

363. As this is not only the most powerful, but the most certain and manageable source of heat, for chemical and pharmaceutical purposes, it is necessary for the student to become acquainted with the means of employing it in the most œconomical and efficient manner. Although the nature of combustion has not, for obvious reasons, been yet considered, we may nevertheless entertain the subject of its practical application in this place, without any reference to its theory.

364. The fluid combustibles, such as Alcohol, Oil, and melted Tallow, are burnt, for this purpose, in lamps of various construction. Wood, turf, coal, charcoal, and coke, are consumed in grates and furnaces. The subject therefore naturally resolves itself into two divisions, the one comprehending the choice, management, and application of LAMPS; the other, the construction and regulation of FURNACES.

365. All the necessary manipulations of Chemistry, as far as they relate to instruction, may be conveniently and satisfactorily performed by means of a lamp. The one most simple in its construction consists of a vessel, of almost any shape, containing oil or alcohol, with a tube projecting a little above the surface of the liquid, and containing any fibrous substance capable of raising the liquid to the top of the tube, by capillary attraction. The oil, thus raised and diffused through the fibrous substance, is so detached from the main body of the liquid as to admit of being heated to a temperature sufficiently high to volatilize it, the vapour of which in a state of combustion constitutes the flame of the lamp. Its perfection consists in its power of sustaining a combustion sufficiently intense to burn every particle of

matter as it is volatilized, so that none shall escape undecomposed. If the inflammable vapour should occupy a considerable area, as must always occur where a solid wick is employed, it is evident that the atmospheric air cannot combine with the interior part of the ascending column; the exterior alone will enter into combustion, while the interior must pass off in the state of smoke. In which case, besides the offensive smell and appearance of the unburnt matter, there is evidently a great waste of combustible materials. Various attempts have been made to unite the advantages, and, at the same time, to remedy this defect of a thick wick; in some instances a number of smaller ones have been substituted, in others, it has been made flat instead of cylindrical; no contrivance, however, is so effectual as that known by the name of the *Argand Lamp.* In this lamp, the wick forms a hollow cylinder or tube, which slides over another tube composed of metal, so as to form an adjustment with regard to its length. When it is lighted, the flame itself has the figure of a thin tube, to the inner as well as outer surface of which, the air has access from below; and a cylindrical shade of glass serves to keep the flame steady, and in a certain degree to accelerate the current of air. The combustion of the Argand lamp, therefore, may be said to bear the same analogy to that of the candle or common oil lamp, as the fire in a furnace, to one kindled on the ground.

366. To the Chemist who requires a steady and manageable heat, without smoke, this lamp is highly serviceable, and if a copper chimney be substituted for the glass shade, we shall obtain greater heat, while it will enable us to introduce, through holes perforated in its sides, tubes, and other bodies, into the centre of the flame itself. By affixing such a lamp to an upright

pillar, to which moveable rings of various dimensions are adapted, we shall be enabled to raise or lower it at pleasure, and to adjust retorts, flasks, evaporating basins, &c. at any distance from the flame which the operation may require. The cut introduced under the head of Nitric acid will serve to explain more clearly the utility of such an apparatus.

367. For experiments in the small way, spirit of wine is upon the whole the neatest and most convenient fuel, and hence the spirit lamp may be employed for most operations of analysis as a very convenient furnace. The annexed cut re-

presents the more ordinary form of this apparatus; should this, however, not be at hand, the student may easily provide himself with a convenient substitute by the following process, as recommended by Mr. R. Phillips. I et a piece of tin plate, about an inch in length, be coiled up into a cylinder of about 3-8ths of an inch in diameter, and if the edges be well hammered, it is not necessary to use solder.

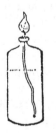

Perforate a cork previously fitted to a phial, and put a cotton wick through the short tin tube, and the tube through the cork; the lamp is then complete, and will afford a strong flame, taking care of course not to prevent the rise of the spirit by fitting the cork too closely.

368. In order to increase the heat of a candle or lamp, a stream of air is driven through a tube called the *Blow-pipe.** In its most simple form it consists merely of a brass pipe about one-eighth of an inch in diameter at one end, and the other tapering to a much less size, with a very small perforation for the wind to escape. The smaller end is bent on one side. For philosophical, or nicer purposes, the blow-pipe is provided with a bowl or enlargement (as represented in the annexed figure *a*), in which the vapours of the breath are condensed and detained; and also with three or four small nozzles *b*, with different apertures to be slipped on the smaller extremity. There is an address in the blowing through this pipe, which it is perhaps impossible to describe by words; a little practice, however, will soon render the operation easy. A continued stream of air is absolutely essential, and to produce this, without fatigue to the lungs, an equable and uninterrupted inspiration must be maintained by inhaling air through the nostrils, while that in the mouth is forced through the tube by the compression of the cheeks. After habit has rendered the operation familiar, a current may be kept up for ten or fifteen minutes, without inconvenience. A large wax or tallow candle supplies the best flame, and it should be snuffed rather short, and its wick turned on one side towards the ob-

---

* It is not known at what time, or by whom, this very useful instrument was invented, but it appears to have been used by glass-workers, enamellers, and jewellers, long before it was adopted as an article of chemical apparatus. The first intimation of its value to the Chemist is to be found in Kunkel's treatise on glass-making.

ject, so that a part of it may lie horizontally. The
stream of air must be blown along this horizontal part,
as near as may be without striking the wick. In this
way the flame will be made to exhibit two distinct
figures; that which is internal being conical, blue, and
well defined, at the end of which the most violent degree of heat is excited; the external flame is red, vague,
and undetermined, and of very inferior temperature to
the former. The body intended to be acted upon by
the blow-pipe ought not to exceed the size of a pepper-
corn. It may be laid upon a piece of close-grained,
well-burned charcoal; unless it be of such a nature as
to sink into the pores of this substance, or to have its
properties affected by its inflammable quality; such
bodies may be placed in a small spoon made of pure
gold, or silver; or on platinum foil.

369. Many advantages may be derived from the
use of this simple and valuable instrument. Its small-
ness, which renders it suitable to the pocket, is no
inconsiderable recommendation. The most expensive
materials, and the minutest specimens of bodies may
be used in these experiments; and the whole process,
instead of being carried on in an opaque vessel, is under
the eye of the observer from beginning to end. It is
true that very little can be determined in this way
concerning the quantities of product; but the first inquiry to be made is *what* a substance contains, not *how
much*, and these trials in small, suggest the proper
method of instituting experiments at large.

370. To supersede the labour and difficulty of
urging the blow-pipe by means of the breath, various
ingenious contrivances have been suggested. For this
purpose, M. Paul, and Professor Pictet of Geneva, first
employed the vapour of alcohol; and an instrument
constructed by Mr. Hooke, will afford the most conve-

nient mode of applying this principle. In some cases, double bellows have been made to communicate with the blow-pipe; such an apparatus will be found extremely effective for blowing glass. Where we wish to urge the flame by means of oxygen gas, instead of atmospheric air, or to produce heat by the combustion of a mixture of hydrogen and oxygen gases, * gasholders, and the various instruments employed for the management of gases, will afford adequate means.

371. The great heat produced by the common blowpipe would seem to depend upon two causes, *viz.* to the concentration of the flame into a small focus, and to the more rapid combustion excited by the current of air.

372. The medical student will have but little occasion to trouble himself with the construction of furnaces; all his ordinary manipulations may be accomplished by the skilful application of a lamp, aided occasionally by the blow-pipe; but it is necessary that he should become acquainted with the general principles of their construction, especially, if he descends into the details of Pharmaceutic Chemistry, and for these reasons the following observations are introduced to his notice.

373. The essential parts of a furnace are the *body*, or fire place, in which the fuel, and the substance to be operated on, are placed; the *chimney* by which the smoke and heated air are carried off; and the *ash-pit*, into which the ashes fall, and through which air is admitted to the burning fuel. The principles on which the production of heat in furnaces depends, are, that inflam-

---

* The Blow-pipe invented by Mr. Brooke is far superior to any other contrivance for this purpose, a description and figure of which may be found in *Henry's Chemistry.*

mable matter cannot burn without the access of air,
and that the rapidity of the combustion, and consequently the quantity of heat produced in a given time,
are proportioned to the quantity of air transmitted over
the burning matter. When fuel is placed in a closed
cavity, like that of a furnace, connected with a chimney, as soon as it is kindled, the air in the upper part of
the furnace is necessarily rarefied, and ascends by the
chimney; the pressure of the external atmosphere forces
a quantity of fresh air through the openings below,
which, rising through the fuel, occasions a strong combustion. In furnaces, therefore, the strength of the
combustion depends on two circumstances; on the
access of the atmospheric air from below, and on the
height of the column of heated air. When the tube or
chimney is lengthened, the difference between the specific gravity of the column of heated air which it contains, and of the column of external air being greater,
a larger quantity of fresh air is constantly forced
through the fuel, and a strong draught, as it is termed,
is formed. This is proportional to the height of the
vent, to a certain extent; for beyond a certain point,
the air in the vent being cooled, no addition to the
draught is gained. What may be the limits for the
height of a chimney has not been ascertained from any
precise trials, but Dr. Ure thinks that thirty times its
diameter would not be too high. It is also obvious,
that the draught of air may be diminished or increased
by lessening or enlarging the access of air from below;
and by closing up that access, the combustion will be
totally stopt. Registers consisting either of a number
of holes fitted with brass plugs, or, what is more convenient, of a moveable semicircular plate, are employed
to regulate the admission of air. In the construction
of furnaces there is also another important object to be

attained, the confining the heat, or preventing it from being carried off by the surrounding air This is accomplished by coating the internal surface with some substance which transmits the heat very slowly; a lute of clay and sand is commonly used for this purpose, which farther serves the important purpose of defending the substance, of which the furnace is made, from the action of the fire. Furnaces may be formed either of solid brick work, or of such materials as admit of their removal from place to place. The following are those of most general utility.

374. *The Air, or Wind Furnace*, which acts by the draught of the chimney, as above described; and in which the body submitted to the action of heat, or the vessel containing it, is placed in contact with the burning fuel.

375. *The Blast Furnace* differs from the preceding in having the air forced through the fuel by means of bellows, instead of entering by the mere draught of the chimney.

376. *The Reverberatory Furnace.* In which the fuel is contained in an anterior fire place; and the substance, to be submitted to the action of heat, is placed on the floor of another chamber, situated between the front one and the chimney. The flame of the fuel passes into the second compartment, and is *reverberated* upon the substance by the form of its roof.

377. A portable furnace, however, is amply sufficient for all the purposes of the student, and should he conduct his experiments on a scale that may require more heat than the lamp will afford, I should recommend the small, but powerful furnace invented and sold by Mr. Knight, of which he is here presented with a sketch.

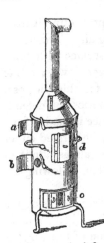

This furnace is made of wrought iron and lined with fine brick; the inside diameter being six inches; *a* is a door for the passage of the neck of a retort, when distillation is performing in the open fire; *b* is an aperture to which there is a corresponding one on the opposite side for the admission of a tube to pass through the furnace; *c*, the ash-pit door, acting as a register; *d*, the door of the fire place when used as a sand heat. For this furnace the proper fuel, when it is used as a wind furnace, is wood charcoal, either alone, or with the admixture of a small proportion of coak. For distillation with a sand heat, charcoal, with a little pit coal, may be employed.

378. The Heat communicated from bodies in combustion must necessarily vary according to circumstances; whenever, therefore, an equable, rather than a high temperature is required, it will be expedient to interpose a quantity of sand, or other matter, between the fire and the vessel intended to be heated. The *sand bath,* and the *water bath* * are most commonly used for such a purpose; it has been already stated that a *bath of steam* may in some instances be found preferable (360). A considerably greater heat may be given to the water bath by dissolving various salts in it; thus a saturated solution of common salt boils at 225° 3 *Fah*; by using a solution of muriate of lime, a bath of any temperature from 212° to 252° may be conveniently obtained.

---

* The water bath was called *Balneum Mariæ* by the older chemists.

E. *By Chemical Action.*

379. This source of heat, is scarcely applicable to any practical purpose, except by furnishing a ready method of kindling fire; thus if matches impregnated with sugar and *chlorate* of potass, be dipped into sulphuric acid, they burst into flame. A lamp has been lately presented to the public for procuring fire, or an instantaneous light, which owes its operation to the singular action of Hydrogen upon a particular preparation of Platinum.

F. *By the Sun's Rays.*

380. The rays of the sun are applied for the drying of many vegetable substances, as well as for promoting spontaneous evaporation. By means of a double convex lens, a concave mirror, or by a combination of plain mirrors, intense heat may be obtained from this source. There are difficulties, however, in applying such instruments to purposes of practical utility. A cloud passing over the sun, during the process might prove fatal to our experiments; and when the sun does shine in its full splendour, the motion of the earth would prevent the focus from ever being kept for a minute at a time on one spot.

G. *By Electricity.*

381. When electricity is transmitted in considerable quantities through bodies, it raises their temperature; and its heating power is so great, that it is able to melt even those metals that are most difficult of fusion. The electrical discharge, therefore, affords a powerful

mean of promoting chemical combination. The electric
spark is also effectual in producing the combination of
several of the gases, which, when merely mixed together, do not unite. It is equally capable, in many
cases, of effecting decomposition, apparently from its
power of exciting a high temperature in the point at
which it is received.

LIGHT.

382. Light is usually regarded as a substance, or
emanation of particles of inconceivable rarity, projected
in right lines from the sun, and from all luminous
bodies, and moving with extreme velocity.* The
optical affections of light are foreign to this work, its
chemical relations are to be alone considered.

383. When a ray of light passes obliquely from one
medium into another of greater density, it is bent *towards* the perpendicular; but if the second medium
be of less density, it is bent *from* the perpendicular.
The light in both cases is said to be *Refracted.* It
was however, discovered by Newton, that unctuous
or inflammable bodies occasioned a greater deviation
in the luminous rays than was consistent with their
known density, whence he was led to the inference
that both the diamond and water contained combustible matter. We perceive therefore that from the refractive power of bodies we may in many cases infer
their chemical constitution; and the principle has been
happily applied for discovering the purity of essential
oils. For these researches an instrument invented by

---

† Its velocity is estimated at 200,000 miles in a second.

Dr. Wollaston * may be conveniently employed. I have inserted in the Appendix a table of the refractive powers of different substances employed in medicine.

384. Light is separable by a prism into seven primary rays or colours, as well as into others, which appear to be distinguished by certain chemical powers; but it is not necessary, on this occasion, to inquire to what part of the prismatic spectrum the heating, illuminating, and chemical properties belong. As Physiologists and Chemists we have to consider the nature of light in its common character, the effects of which upon natural bodies we shall find to be very compounded. Healthy vegetation depends upon the presence of the solar beams, or of light; and whilst the heat gives fluidity and mobility to the vegetable juices, chemical effects are likewise occasioned, oxygen is separated from them, and inflammable compounds are formed. Plants deprived of light become white, and are said to be *etiolated*, as shewn in the common process of blanching celery; they acquire at the same time an excess of saccharine and aqueous particles. The Indian finds his way through the uncultivated forests of America, with no other guide than that of the colour produced by the light of the sun on the sides of those trees which are more directly exposed to its action. Flowers owe the variety of their hues to the same cause; and even animals require the

---

* Dr. Wollaston has invented a very ingenious apparatus, in which, by means of a rectangular prism of flint glass, the index of refraction of each substance is read off at once by a vernier, the three sides of a moveable triangle performing the operations of reduction in a very compendious manner. (*Phil. Trans.* 1802, or *Nicholson's Journal*, 8vo. Vol. IV. p. 89.) This instrument is capable of some practical applications highly interesting to the chemist.

rays of the sun, and their colours seem materially to
depend upon their chemical influence: a comparison
between the polar and tropical animals, and between
the parts of their bodies exposed, and not exposed to
light, demonstrates the truth of this statement. The
plumage of birds affords, perhaps, the most striking
illustration; those, for instance, which are natives of
the torrid zone are generally rich and gaudy, while
those in the frigid zone are principally whiter. The
colour of fish is also materially affected by the agency
of light, for we invariably find that the backs, or upper
parts, which are exposed to the direct influence of
light, are dark when compared with the under parts,
which are excluded from it. Nor is man himself insensible to the same action, hence the pale, sallow, and
sickly appearance of those persons who are excluded
from the " light of heaven."

385. The effects of light upon inorganic bodies have
been casually adverted to under the consideration of
crystallization (215). Its influence in de-oxidizing
bodies will be illustrated hereafter. In this place it
may be sufficient to adduce a single example which
will be afforded by the following interesting experiment.

Exp. 50.—Prepare a solution of Nitrate of Silver,
in the proportion of one part of the nitrate to ten
of water; with this solution let a piece of white
paper or leather be soaked, and placed behind a
painting on glass, and be then exposed to the
action of the sun; the rays of light transmitted
through the differently painted surfaces will produce tints of brown or black, sensibly differing in
intensity, according to the shades of the picture;
and where the light is unaltered, the colour of the

nitrate becomes deeper, or approaches more nearly to the metallic state.*

386. Light appears to enter into combination with various bodies, and to be again separated from them, and some philosophers have supposed that the light produced by combustion is to be thus explained.

387. Certain bodies, termed *Solar Phosphori*,† have undoubtedly the property of aborbing light, and of retaining it for some time; and of again evolving it unchanged, and unaccompanied by sensible heat; such are certain marine animals, both in a living and dead state, and peculiar preparations formed by art.

---

* Sir H. Davy found that the images of small objects, produced by means of the solar microscope, may be copied without difficulty on paper thus prepared. In this experiment, however, it will be necessary to place the paper at a small distance from the lens.

† Mr. Skrimshire has given an extensive catalogue of such substances in Nicholson's Journal, 8vo. vols. 15, 16, and 19.

ELECTRICITY,

AND ITS CHEMICAL AGENCIES.

388. The earth, and all bodies with which we are acquainted, are supposed to contain a certain quantity of an exceedingly subtle fluid, which is called the *Electric fluid.* This certain quantity, belonging to all bodies, may be called their natural share; and as long as each body contains neither more nor less than that quantity, it seems to lie dormant, and to produce no effect; but when any body becomes possessed of more, or less than its natural quantity, it is said to be *electrified;* and is capable of exhibiting appearances which are ascribed to *Electricity.* In the former case the phænomena depend upon the body yielding its excess; in the latter, to the surrounding bodies transferring to it a portion of theirs, to supply the deficiency; for the electric, like the calorific fluid, tends to a universal equilibrium. This equilibrium, however, could never be disturbed, or, if it were disturbed, would be immediately restored, and therefore be insensible were it not for the fact that some bodies admit the passage of the electric fluid through their pores, whilst others refuse its entrance; the former are termed *Conductors,* the latter *Nonconductors.* The metals are all conductors; glass, sulphur, and resins are nonconductors.

Conductors, again, are divided into those which are *perfect,* and those which are *imperfect;* conductors do not, in general, become electric by friction, and are therefore called *non-electrics;* non-conductors, on the contrary, are *electrics,* as they acquire electricity by being rubbed.

389. It has been stated that some bodies may contain an excess, and others, a deficiency of the electric fluid, but that both, under such circumstances, will exhibit the phœnomena of Electricity; in the former case, however, it is said to be *positive;* in the latter, *negative.*

390. When two bodies differently electrified are made to communicate, they attract each other, a discharge takes place, and an equilibrium is restored; when, however, they are similarly electrified, they repel each other.

    Exp. 51.—Rub a piece of sealing wax with dry flannel, and bring each of these substances in contact with two pith balls suspended by a line of silk. They will be found capable of attracting each other. This arises in consequence of the wax and flannel having communicated different states of electricity.

    Exp. 52.—Touch both of the pith balls with the wax, or with the flannel, and they will be found to repel each other, because they have been similarly electrified.

391. When the sealing wax is rubbed, it loses a portion of its electricity, which the flannel gains; the former is therefore *negative,* the latter *positive.*

392. When the natural electricity of a body is thus changed, if it be surrounded by bodies that are non-conductors, it must remain in that state, for it is evident that in such a condition, it can neither receive

electricity from surrounding substances, nor impartit to them, in order to regain an equilibrium. A body so circumstanced is said to be *insulated.*

393. A knowledge of these simple propositions will enable the student to comprehend the nature of the common electrical machine, an account and engraving of which may be found in Mr. Brande's excellent *" Manual of Chemistry."* When friction is applied to the glass cylinder of this apparatus, that part of the glass, which is in contact with the rubber, attracts the electric fluid from it, as well as from every conducting body which may happen to be connected with it. The glass, instantly regaining its natural state, repels the electric fluid, which is received by the prime conductor placed for that purpose. All then that is effected by this machine, is a disturbance of the natural quantity of electricity in bodies, or a transfer of it from some to others; in consequence of which, while the latter acquire a redundance, the former become proportionally deficient in their quantity of electricity.

394. The electrical equilibrium of bodies is disturbed by many other operations than that of friction. Whenever bodies change their form, their electrical states are also changed. Fusion, evaporation, and change of temperature, are also attended with electrical phænomena.

395. It appears, therefore, that for the excitation of ordinary electricity, substances called *Electrics* (388) are required; by the friction of which the electric fluid is accumulated, and from which it may be collected by a different class of bodies termed *non-electrics*, or *conductors.* The student is now to be introduced to a species of Electricity, which is excited by very different means, and which is termed, in honour to its discoverer Galvani, an Italian Philosopher, GALVANISM, or GAL-

vanic ELECTRICITY. Unlike ordinary electricity, it does not require for its excitement any electric; the simple contact of *different* conducting bodies with each other is sufficient.

396. The most simple arrangement which can be formed for exciting this electricity, is that termed a SIMPLE GALVANIC CIRCLE. It consists of three conductors, two of which may be perfect, and the other imperfect; or two *imperfect*, and one *perfect*.

> *Exp.* 53.—Place a piece of zinc upon the tongue, and a piece of silver under it, no sensation is excited so long as the metals are kept apart; but, on bringing them into contact, a metallic taste is distinctly perceived, and if a connection be made with these metals and the globe of the eye, by means of a metallic wire or rod, a flash of light will be perceived at the same instant.

397. In the above experiment we have an example of the arrangement of two perfect conductors *(silver* and *zinc)* with one imperfect one *(the saliva)*. The metallic taste is the effect of a small quantity of electricity thus excited, and its action on the nerves of the tongue.

> *Exp.* 54.—Take the lower extremities of a recently killed frog, and arm the crural nerve with a piece of zinc, or tin foil; at the same time bring the muscular parts of the animal in contact with a piece of silver. As soon as the two metals are made to touch each other, or a communication is established between them by means of a metallic conductor, violent convulsions are produced.

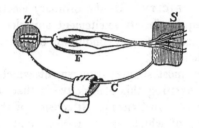

398. Upon the discovery of this singular fact, Physiologists hastily adopted the opinion that the source of nervous influence was electricity, and the most sanguine expectations were entertained, that the nature and operation of this mysterious power would be at length developed by this new train of research. It is a curious fact in the history of science, that an experiment so full of promise to the physiologist should have hitherto afforded him little or no assistance, while the chemist, to whom it appeared to offer no point of interest, has derived from it a new and important instrument of research, which has already multiplied discoveries with such rapidity, and to such an extent, that it is impossible to anticipate the limits of its power.

399. The applications, however, of Galvanic electricity to physiology, must not be abandoned as chimerical. The late researches of Dr. W. Philip, and other English physiologists, are well calculated to revive our confidence, and to stimulate us to farther inquiry; from the experiments of this respectable physician it would appear that when a nerve is divided, so as entirely to intercept the transmission of its action, the place of the nerve may be supplied by a galvanic apparatus. He divided the eighth pair of nerves, which are distributed to the stomach, and are subservient to

digestion, by incisions in the necks of several living rabbits. After the operation, the parsley which they ate remained without alteration in their stomachs; and the animals, after evincing much difficulty of breathing, seemed to die of suffocation. But when, in other rabbits, similarly treated, the galvanic power was transmitted along the nerve below its section, to a disc of silver, placed closely in contact with the skin of the animal opposite to its stomach, no difficulty of breathing occurred. The electrical action having been kept up for twenty-six hours, the animals were killed, and the parsley was found in as perfectly digested a state, as that in healthy rabbits fed at the same time; and their stomachs evolved the smell peculiar to that of a rabbit during digestion. These experiments were several times repeated with similar results. Hence it would appear that the galvanic energy is capable of supplying the place of the nervous influence, so that, while under it, the stomach, otherwise inactive, digests food as usual. These experiments are very striking, but they cannot be admitted as affording conclusive evidence of the identity of galvanic electricity and nervous influence. The electricity may merely act as a powerful stimulant upon the twigs which proceed from other nerves to the stomach, and communicate under the place of section of the *par vagum*, by which they may be made to perform such an increase of action, as may compensate for the want of the principal nerve. The respiration is also materially connected with the digestive process, and the increased action thus communicated to the respiratory organs may offer another mode of explanation.

400. The powers of a simple galvanic circle are, however, extremely feeble, and little progress was made in the investigation of this singular modification

of Electricity, until a method was discovered of multiplying those arrangements which compose simple circles. This was atchieved by Signor Volta, early in 1800, in the construction of a *Galvanic Pile.* Before we enter, however, upon the subject of compound Galvanic circles, or *Batteries,* it may be necessary to introduce some observations on the simple circle.

401. Sir H. Davy has presented us with the following Tables in which the different simple circles are arranged in the order of their powers, the most energetic occupying the highest place.

Table of some Electrical Arrangements which, by combination, form Batteries, composed of two Conductors, and one imperfect Conductor.

Zinc. Iron. Tin. Lead. Copper. Silver. Gold. Platinum. Charcoal.	Each of these is *positive* to all the metals below it, and *negative* with respect to the metals above it in the column.	Solutions of Nitric acid. of Muriatic acid. of Sulphuric acid. of Sal Ammoniac. of Nitre. of other neutral salts.

Table of some Electrical Arrangements consisting of one perfect Conductor and two imperfect Conductors.

Solution of Sulphur and Potass. of Potass. of Soda.	Copper. Silver. Lead. Tin. Zinc. Other Metals. Charcoal.	Nitric Acid. Sulphuric Acid. Muriatic Acid. Any Solutions containing Acid.

402. The *Voltaic Pile*, or *Battery*, is constructed by alternately arranging discs of copper and zinc, with pieces of woollen cloth, moistened with acid, or some solution which is an imperfect conductor, as shewn in

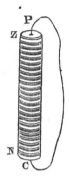

the annexed figure, in the order of zinc, copper, cloth; zinc, copper, cloth, and so on, for thirty or more alternations.

On bringing a wire communicating with the last copper disc into contact with the first zinc disc, a spark will be perceptible, and also a slight shock, provided the number of alternations be sufficiently numeous. The two ends of the apparatus are termed its poles, and the zinc pole will be found to exhibit *positive*, the copper extreme *negative* electricity.

403. The pile of Volta is now very generally superseded by a more convenient, and efficient apparatus; in which the metals are arranged in the form of a trough; it was the invention of Mr. Cruickshank, and is represented by the following cut.

Double plates of copper and zinc soldered together, are cemented into wooden troughs in regular order, and the intervening cells are filled with some appropriate fluid, for the purpose of rendering the combination active. Such an apparatus, it will be perceived, affords an example of galvanic arrangement of the *first kind,* formed by two *perfect,* and one *imperfect* conductor. (396, 401) but it admits of being so modified as to furnish a

battery of the *second kind, viz.* with one *perfect* and
two *imperfect* conductors. In this case plates of one
metal only are to be cemented in the grooves, and the
cells are then filled, alternately, with two different
liquids. The former scheme of apparatus is, however,
the one universally preferred for all purposes of experiment.

404. As the use of the trough is attended with great
inconvenience in filling and emptying it; and as the
metals become rapidly corroded without such a precaution is observed, the plates are now usually so constructed that they may be immediately removed from
the liquid, and as speedily restored to it. The annexed
sketch represents the most approved apparatus for this
purpose.

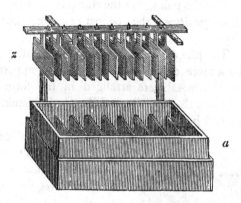

The trough *a*, with its partitions, is made of earthenware; the metallic plates *z*, are attached to a bar of
wood, so that they can be immersed and removed at
one operation. The troughs are filled with dilute
acid, and by uniting them in regular order the apparatus may be enlarged to any extent. This is the form

adopted for that splendid apparatus in the Royal Institution, which, in the hands of its illustrious Professor, Sir Humphry Davy, decomposed the fixed alkalies, and developed the composition of other bodies which was previously unknown.

405. Philosophers make an important distinction between *intensity* and *quantity* of electricity. By the former is meant its power of passing through a certain stratum of air, or other ill conducting medium ; by the latter the absolute quantity of electric power in any body. In the Voltaic pile, the *intensity* of the electricity increases with the number of alternations, but the *quantity* is increased by extending the surface of the plates. Thus if a battery, composed of thirty pairs of plates two inches square, be compared with another battery of thirty pairs of twelve inches square, charged in the same manner, no difference will be perceived in their effects upon bad or imperfect conductors; their power of giving shocks will be similar; but upon good conductors the effects of the large plates will be considerably greater than those of the small; they will ignite and fuse large quantities of Platinum wire, and produce a very brilliant spark between points of charcoal.

406. This distinction between *quantity* and *intensity* is important, and will suggest the mode best adapted for the construction of a galvanic battery, according to the objects for which it is designed.

Electro-Chemistry.

407. So important and extensive is the agency of electricity in producing chemical decomposition, that its investigation forms, as it were, a distinct branch of

science, and has been denominated *Electro-Chemis-*
try.

408. It has been considered probable, by Sir H.
Davy, that the power of electrical attraction and repulsion is identical with chemical affinity. If this be
true, we obtain at once a solution of the problem, and
can explain the action of the electric and galvanic
fluids, in disuniting the elements of chemical combinations; for it is evident that if two bodies be held
together by virtue of their electrical states, by changing their electricity we shall disunite them.

409. In this view of the subject, every substance,
it is supposed, has its own inherent electricity, some
being positive, others negative.   When, therefore,
bodies in such opposite states are presented to each
other, they combine.   Upon this fact an arrangement
has been founded, and bodies are said to be *Electro-*
Negative, or *Electro-Positive*, according as they are
attracted to the Positive or Negative pole of the battery, for it is a common law of electrical attraction,
that bodies will be repelled by surfaces which are in
the same state of electricity as themselves, and attracted
by those which are in opposite states, (390).   Upon
this principle it would appear that the inherent or
natural electrical state of the inflammable substances
is *positive*, for they are attracted by the negative or
oppositely electrified pole; while the bodies called supporters of combustion, or acidifying principles, are attracted by the positive pole, and, therefore, may be
considered as possessed of the *negative* power.

410. One of the first discoveries of the chemical
agency of the pile was its power of decomposing water.
To exhibit this phænomenon various forms of apparatus have been contrived, but the most simple is that

represented in Dr. Henry's Chemistry, and which is here introduced.

n p is a glass syphon, containing water, into which are introduced two gold or platinum wires, through corks inserted into the open mouths, which are made to terminate before they reach the bend, in which a small hole is ground, so as to form a communication with the water in which it is placed. If the parts of the wire which project from without the tube at *n* and *p*, be made to communicate, the one with the zinc or positive end, and the other with the copper or negative end, of a galvanic battery, minute bubbles of air will be seen proceeding from the extremities of each wire, and after a short time oxygen gas will be found to have collected in the leg connected with the positive end, and hydrogen gas in that connected with the negative end; and in the exact proportions which, by their union, compose water.

411. But it is a most extraordinary and important fact, that these gases may be thus obtained from two quantities of water, not immediately in contact with each other.

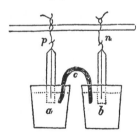

If two glass tubes, *p* and *n*, about one-third of an inch in diameter, and four inches long, having each a piece of gold wire sealed hermetically into one end, and the other end open, be filled with distilled water, and placed inverted in separate glasses filled, also with the same fluid, and the two glasses *a*, *b*, are made to communicate by the interposition of moistened thread as shewn at *c*, as soon

as the wires, projecting from the sealed ends of the
tubes be connected, the one with the *positive*, the other
with the *negative* end of the trough, gas will be immediately evolved, as in the preceding experiment. Now,
since these gases must necessarily arise from the decomposition of one and the same particle of water, and
since that particle must have been contained either in
the tube *p*, or in the tube *n*, it is clear that either the
oxygen, or the hydrogen, must have passed invisibly
from *p* to *n*, through the intervening substance *c*.
Facts of this kind, evincing the transference of the
elements of a combination, to a considerable distance,
and in a form that escapes the cognizance of our senses,
however astonishing, are sufficiently numerous and well
established.

412. If instead of simple water, the galvanic energy
be made to act upon the solution of a neutral salt, as
that of sulphate of potass (a salt composed of sulphuric
acid and potass) the same transference of its elements
will take place, and we shall obtain pure potass in the
negative vessel, and sulphuric acid in the positive one.
For performing this experiment the following arrangement will be necessary.

p, and *n*, are two vessels, containing the solution to be decomposed, and connected together with moistened thread;
w w, are two platinum wires to
be connected with the two ends
of the galvanic trough. In this
case the acid and alkali will
traverse the thread *a*, without uniting; in consequence
of being under the influence of electrical action; so
completely, indeed, is chemical affinity suspended upon
these occasions, that if an intermediate vessel, contain-

ing nitric acid, be placed between the positive and negative ones, the potass will traverse it without any combination being produced.

413. The fact of the transference of the elements of a combination to a considerable distance, through intervening substances, has been very ingeniously supposed capable of affording the means of eliminating calculi from the bladder.* Could the functions of the part be protected against the influence of so powerful an agent, it is evident that, by a galvanic battery of sufficient intensity, a calculus composed of alkaline or earthy salts might be transferred from the bladder by the simple introduction of a double sound, communicating on one hand with the calculus, and, on the other, with two vessels filled with water, in which are plunged the opposite poles of a galvanic apparatus. This arrangement would transfer the acid constituents into the vessel connected with the *positive* end, and the bases into that of the *negative* end.†

414. Nor are the effects produced by the action of a *simple* galvanic circle less interesting to the medical practitioner, especially in their connection with some curious and important facts in Toxicology. It has been long known that the poisonous effects of copper utensils might be prevented by a film of tin; but it was supposed that this protection ceased as soon as any portion of the surface was abraded. Proust, however, was the first to announce the fallacy of such an opinion, but his explanation of the fact was purely chemical;—from the superior affinity of the tin for

---

* Journal de Physiologie; Juillet, 1823.

† For a further account of this interesting subject, and of the experiments which have been instituted for its illustration, see PHARMACOLOGIA, Vol. 1. p. 231. Edit. 6.

oxygen; upon the same principle he stated that no harm could arise from the alloy of tin with lead, since this latter metal cannot be dissolved by any acid as long as an atom of tin exists. The true explanation of the fact, however, is to be derived from a knowledge of the electric relations of these metals, which the student will at once discover by referring to Table 1. (401), where it will be seen that tin is positive with respect to lead, and that copper is negative in relation to both. For the same reason, if acid matter be contained in a copper vessel it will become poisonous if a silver spoon be used in stirring it, because this metal is negative with respect to copper, whereas, if a leaden or tin spoon be substituted, we shall derive a protecting influence from its action. It was by reasoning upon this principle of Electro-chemical agency that Sir H. Davy arrived at the important fact that the copper of ships might be protected from the corrosive action of the sea water * by the juxta position of discs of zinc, iron, or tin; in which case the muriatic acid, instead of acting on the *negative* copper, is transferred to the metal which is *positive*. The pursuit of this enquiry has not only confirmed the justness of the views which suggested it, but it has furnished the Toxicologist with a striking fact, in support of the assertion of Proust, for it has been found on trial, that, when the copper is thus protected, marine insects attach themselves to its surface with impunity.†

---

* The muriate of magnesia contained in the water of the ocean is the salt which is the most destructive to the copper.

† At the moment this sheet was going to press, I read an account of Dr. Bostock's Experiments on the applicability of Sir H. Davy's discovery to copper vessels employed in cookery, (*Annals of Philosophy for September*, 1824.) by which he has shewn that, although Copper is preserved by Tin from the action of

415. Nor will the Surgeon be excluded from his share of benefit from the discovery. Mr. Pepys has proposed to preserve steel instruments from rust by a simple application of the same principle; it will be seen in Table 1, that zinc is *positive* with respect to iron; if therefore a portion of this metal be inserted in a sheath, composed of some imperfect conductor, the introduction of the steel instrument will complete a simple galvanic circle of the first order, and the iron will be protected from oxidation.

416. There is yet another application of galvanism to a purpose of medical utility, which deserves notice,— the detection of minute quantities of *Corrosive Sublimate*. For this object we have merely to let fall a drop of the solution upon the surface of a piece of gold, as for instance, on a *Sovereign*, and then to bring a piece of iron, as a key, in contact with both; a galvanic circle is immediately formed, and as the iron is positive the acid will be transferred to it, and the quicksilver will be deposited on the gold.

417. It also appears from the experiments of Mr. Brande,* that galvanism may be applied to the discovery of very minute quantities of albumen, which are not rendered sensible by any other test. In this way he produced a rapid coagulation, at the negative pole, in several animal fluids, in which albumen had not been supposed to exist.

418. How far electrical action may be concerned in some of the more recondite phænomena of animal life

---

acetic acid, in the same manner as it is from that of sea water, yet that the principle cannot be applied with safety in culinary operations, in consequence of the volatility of the acetic acid, by which it is enabled to act upon the copper, without constituting any part of the galvanic circle.

* Philosophical Transactions, 1809.

we have already offered a conjecture (399). It must
be confessed that the principles of Electro-chemistry
offer, on many occasions, a very close analogy to those
of vitality. The phænomena of secretion may receive
a more plausible explanation from this than from any
other source of action. We may imagine, says Dr.
Young, that at the subdivision of a minute artery, a
nervous filament pierces it on one side, and affords a
pole positively electrical, and another opposite filament
a negative; then the particles of oxygen and nitrogen
contained in the blood, being most attracted by the
positive point, tend towards the branch which is nearest
to it, while those of the hydrogen and carbon take the
opposite channel; and that both these portions may
again be subdivided, if it be required, and the fluid
thus analysed may be recombined into new forms, by
the re-union of a certain number of each of the kinds
of minute ramifications. In some cases the apparatus
may be somewhat more simple than this, in others,
perhaps, much more complicated; but we cannot expect to trace the processes of Nature through every
particular step; we only enquire into the general direction of the path that she follows, as much in order
to avoid being led away by false opinions, as for the
sake of any direct advantage that can be gained from
our partial views of the true state of the operations.

PART II.

ON ELEMENTARY BODIES, AND THE COMPOUNDS WHICH RESULT FROM THEIR COMBINATION WITH EACH OTHER.

419. Having investigated the general forces, which, by their action on matter, give rise to the principal phænomena of Chemistry, we have next to consider the chemical properties and actions of the individual substances on which they operate.

420. In the language of modern Chemistry, the term *Elementary*, or *Simple*, has a signification very different from that attached to it by the ancient philosophers, who considered it as expressing substances possessing absolute simplicity, which, by modifications of form, or combination with each other, form the numerous substances which compose the material world ; whereas the *Elements* of the moderns are considered as simple, merely in relation to the present state of the art of analysis, for it is assumed as a general principle, that every substance is to be regarded as simple which has not been resolved into two or more constituent parts. The number and nature of our elements will accordingly be constantly liable to vary during the pro-

gress of science. Within the last few years, for instance, how many bodies have been decomposed, by
the powers of electricity, which had been long considered as simple? It is, however, probable, that our
experimental elements are still very remote from those
of Nature, and although future discoveries may enable
us, by producing farther decomposition, to approach
more nearly to them, it may nevertheless be doubted
whether the true elements of matter are not so recondite as for ever to elude the sagacity of human research.

421. In the earlier ages of the world, Philosophers
acknowledged but four elements, three of which are
now demonstrated to be compound bodies. Paracelsus
maintained the existence of three principles, which he
called the *tria prima*, viz. Salt, Sulphur, and Mercury.
The Chemists of the present day consider the primary
substances, which can be subjected to measurement and
weight, to be fifty-two in number. To which some add
the imponderable elements,—light, heat, electricity,
and magnetism; but their separate identity, although
probable, is not clearly ascertained.

422. The arrangement of the different simple and
compound bodies, like the number of the elements, has
varied with every prevailing doctrine; and the propriety of the more popular and generally received classification which distributes substances according to
their power of undergoing combustion, or of supporting it, has become questionable from late researches
into the nature of combustion. That which is founded
upon the electrical relations of bodies appears to be,
by far, the most philosophical, and is the one which
should be adopted in all elaborate systems. In composing, however, a chemical work, the arrangement of
the subjects should be conformable with the objects for
which it is designed. In a work intended for those al-

ready instructed in the first principles of Chemistry, great advantage will arise from presenting the different substances in the order of their relations to each other, by which all the advantages of analogy will be ensured; but in a work strictly elementary, intended for the instruction of the mere student, the great object is to lead him from known to unknown propositions by the least abrupt steps; and, above all, to avoid in our explanations of the simpler subjects the anticipation of the more recondite phænomena. It is true that, in many cases, such a difficulty cannot be avoided, for the objects of science are connected in a circle, and from whatever point we start, some previous knowledge will be required. Thus, for instance, if we commence with the subject of Attraction, its laws cannot be explained without a reference to the composition of substances, the names even of which may be supposed to be unknown. If, on the contrary, we commence with a history of such substances, their nature and habitudes cannot be understood without a knowledge of the attractions by which they are influenced; the same difficulty occurs in treating the subject of Electricity. Without an acquaintance with the composition of several bodies how can we exemplify the decomposing powers of this wonderful influence; and without a knowledge of this agent how are we able to explain the history of those substances whose composition has been discovered through its application? In many cases, however, these difficulties are but apparent, and are not to be placed in competition with the numerous advantages which attend such an arrangement. These observations, it is hoped, will reconcile the reader to the method which has been pursued in the following pages, and will explain my reasons for having deviated from the ordinary routine of elementary works.

423. The following is an outline of the arrangement which I have usually followed in my Lectures, in treating of the different simple and compound substances of Nature and Art. WATER is presented as the first object of examination, not only as being a fluid universally known, but as entering very generally into the composition of other bodies, and as influencing, by its presence, almost all the phænomena of chemistry. Its decomposition, moreover, brings the student at once acquainted with two leading elements,—*oxygen* and *hydrogen,*—the nature of which, from the important part which they perform cannot be too early understood, while their gaseous form will afford us an opportunity of illustrating the methods to be adopted in collecting, transferring, and examining gases. CARBON, which may be said to constitute the basis of organized matter, is considered next, and by its union with oxygen and hydrogen, we shall learn the nature and properties of *Carbonic oxide*; *Carbonic acid;* and *Carburetted Hydrogen.* In this part of our arrangement, the consideration of the ATMOSPHERE may be entertained, since, with the exception of Azote, it includes no principles whose nature has not been already explained. Under this head the theories of combustion, of animal heat, and of oxygenation, may be conveniently introduced. The history of the Simple Supporters of Combustion, CHLORINE, and IODINE, and their combinations, will follow. The compounds produced by the union of OXYGEN and AZOTE; of HYDROGEN and AZOTE, and of CARBON and AZOTE, will succeed. The histories of SULPHUR; PHOSPHORUS; the METALS; and NEUTRAL SALTS, will conclude the first part. The second and third divisions will comprise the subject OF ORGANIC CHEMISTRY.

ON CHEMICAL NOMENCLATURE.

424. The brilliant idea of constructing a nomenclature that might convey, with the name, a knowledge of the prominent character of each elementary body, as well as the constitution of every compound produced by their union, originated with Fourcroy, Morveau, Berthollet, and Lavoisier, who, in the year 1787, delivered a plan to the chemical world in as perfect a form as the then existing state of science would allow. The extraordinary discoveries which have taken place since that period have necessarily rendered some parts of this system objectionable, and led to the modification of others, but the basis remains entire and undisturbed, as an imperishable monument of the genius and industry' of its illustrious founders. In giving names to the undecompounded, or elementary bodies, they endeavoured to preserve the ancient denomination, whenever it was free from any of those vices which they were anxious to eradicate; the metals, earths, alkalies, and several other bodies, were for this reason allowed to retain the names by which they had been long known. But the discovery of the constituent parts of water, and of the atmosphere, rendered the adoption of certain new titles indispensable, and, in constructing these, they were guided by some striking peculiarity presented by each of the elementary bodies. The two ingredients, for example, of which water is composed, received the appellations of *Oxygen* and *Hydrogen*,* in consequence of the former

---

* The roots in these cases, from which the derivation is effected, are the Greek words Οξυς, *acidum*, and ὑδωρ, *aqua*, the verb γεινομαι, *gignor*, being added to each.

having been, at that period, regarded as the exclusive
principle of acidity, and from the latter being the ge-
nerator of water. *Azote* † received its name from its
constituting that portion of the atmosphere which is
incapable of supporting life. In all the additions
which recent discoveries have made to the catalogue
of elements, the principle thus established has been,
either in letter or spirit, uniformly acknowledged, and
names perfectly arbitrary have been rejected; thus,
Chlorine received its appellation from its characteristic
green, and *Iodine* from the violet colour of its vapour.
In like manner a person, at all acquainted with the
principles of chemical language, cannot fail to recognise, in the terms *Barium* and *Strontium,* the bases of
the earths Baryta and Strontia, or in those of *Potassium* and *Sodium,* the bases of the two fixed alkalies.
The termination also of the names of the simple bodies,
as well as their derivation, is calculated to convey useful information, when regulated by some established
rule, and will at once apprize us to what class of
bodies any substance belongs. Thus, for the metals,
the termination *um* has been selected; for the alkalies
and earths, that of *a*; and for those radicals which are
neither alkaline nor metallic, but supporters of combustion, that of *ine.* We have, accordingly, Potass-
ium for the metallic basis of the alkali Potassa, Cinchonia and Quina for the alkaline principles of the
Bark, and Iod*ine,* Fluor*ine,* and Chlor*ine* for the last
description of elements. In giving names, however, to
new principles we must necessarily allow a certain
latitude for the caprice of the discoverer, whence it
must ever constitute the least perfect part of the system. In giving appellations to the various compounds,

---

* From α *non*, and Ζῶη, *vita*.

a more definite and invariable principle has been assumed; and, while the nomenclature is compatible
with any future additions, it conveys immediately to
the mind a knowledge of the composition of the substance which it is employed to represent. The compounds which arise from the union of the metals with
each other are termed *Alloys*, except in those cases
in which Mercury is an ingredient, when the resulting
compound is denominated an *Amalgam*. Where any
other inflammable element combines with a base, a
different form of expression is provided, the termination *uret* being given to the generic term; the combinations of *Sulphur*, for instance, are termed Sulphur*ets;* those of *Phosphorus*, Phosphur*ets;* and of
Carbon, Carbur*ets*. The name of a compound into
which oxygen enters, varies according to the nature
of the properties which it exhibits; when it is acid
it receives a title determined by that of the acidified
substance; but in every other case the product is
merely termed an *oxide*, and where more than one
compound of this description is produced, the terminations *ous* and *ic* are used to designate the relative
proportions of the oxygen; thus, azote forms two
oxides; that containing the smallest proportion of
oxygen is the Nitr*ous* oxide, that containing the large-
est, the Nitr*ic* oxide. The acid compounds are similarly distinguished, as Nitr*ous* and Nitr*ic* acid; Sulphur*ous* and Sulphur*ic* acids; the former termination
denoting the *minimum*, the latter the *maximum* of
oxygenation. The same principle of expression has
been extended to the other supporters of combustion,
which have been discovered since the construction of
the original nomenclature, *viz. Chlorine* and *Iodine*,
and the combinations which are formed by these elements are accordingly called *Chlorides* and *Iodides*.

In some cases the different proportions of oxygen or chlorine, which unite with the same base, are distinguished by prefixing derivatives from the Greek, as *Prot*oxide, *Proto*-chloride, *deut*oxide, *trit*oxide, &c. and when the base is combined with the largest possible quantity, the compound, if not acid, is called *Per*oxide, or a *Per*chloride. The compounds which the acids form with the alkalies, earths, and metallic oxides, are termed *Neutral salts*, and the composite nomenclature, which has been contrived to express their composition, constitutes, at once, the most precise, and the most useful part of the system. When the acid in combination with the base is in the lowest state of acidification, as expressed by the termination *ous*, the resulting salt is designated by the termination *ite*, added to the first syllable of the acid; thus Sulphur*ous* acid forms Sulph*ites*, Phosphor*ous* acid, Phosph*ites*, &c. whereas when the acid is at its maximum of acidification, as announced by its termination in *ic*, the salts are made to end in *ate;* thus the Sulphur*ic*, Phosphor*ic*, and Nitr*ic* acids form Sulph*ates*, Phosph*ates*, and Nitr*ates*. Since, however, salts may be composed of different proportions of the same acid and base, a farther provision was required to denote this difference. In the earlier period of Chemistry the prefixing the term *Super*, if the acid predominated, and *Sub*, if the base prevailed, constituted a sufficient mark of distinction; but, since the recognition of the Atomic theory (256) which teaches us that where two bodies unite in more than one proportion, the second, third, &c. are multiples of the first, these terms have received a modified meaning; and as the simplest and most regular form of combination is when the acid and base unite atom to atom, (261, *note*) the generic name is assigned to it without any additional distinction;

thus when an atom of lime unites with an atom of sulphuric acid, the product is a *Sulphate* of lime; but if two or more atoms of base be attached to one of acid, it is, in such a case, proposed that the syllable *Sub* should be prefixed; so that the term which originally served merely to express a certain quality, arising from the predominance of the base, is now used to denote a definite proportion of the ingredients. It must be confessed that such changes are attended with great inconvenience, and we shall frequently find that the new application of the term will be in variance with its former meaning; and of this, the name usually attached to the *Salt of Tartar* of the shops, and which is still retained in the Pharmacopœia, suggests itself as an immediate and striking example. This substance has been long known under the title of *Sub*-carbonate of Potass, but, as it is composed of one atom of carbonic acid, and one of potass, notwithstanding its alkaline character, it is now, in conformity with the above rule, denominated a *Carbonate,* and since no salt exists with a smaller proportion of carbonic acid, there can be no such body as a *Sub*-carbonate of this alkali. But the reform introduced by the Atomic theory does not rest here; the number of atoms of acid with which the base is combined is now designated by prefixing to the generic name of the salt a numerical syllable; thus, where two atoms of acid are combined with one of base, the term *bi* is prefixed; where three, *tri*, and so on; we thus speak of the *bi*-sulphate, and the *bin*oxalate and *quadr*oxalate of potass. In the same way, the salt of potass, still termed in our Pharmacopœia, the *Carbonate,* is, chemically speaking, the *Bi*-carbonate. The same nomenclature is extended to the compounds formed by the simple Inflammables, we have therefore *Bi*-sulphurets;—*Bi*-carburets, and *Bi*-phosphurets. There yet remains to be mentioned another form of saline combination, for which it was necessary to provide a nomenclature; I allude to what are called *Triple* salts, in which the acid unites with two bases; in this case the names of both bases enter into that of the salt, as in *Tartrate of Potass and Soda.* In some cases the latter name has been prefixed as an adjective, and we thus speak of the *Ammoniaco-muriate* of Platinum;—the *Ammoniaco-phosphate* of Magnesia, &c.

425. In concluding the history of the Composite Nomenclature of Chemistry, I may be allowed to express my strong doubts with respect to the propriety of introducing it into medical practice. Its principal value in science depends upon the assistance which it gives to the memory in distinguishing, and remembering, the multiplied combinations of nature and art; in Pharmacy such assistance is scarcely necessary; our medicines are few; and dull indeed must be that student who stands in aid of an artificial memory. At the same time, it is a matter of great consequence that the names of our remedies should be unchangeable, that the physician should be able to avail himself of the experience of his predecessors with facility, and that he should read their prescriptions without the aid of a glossary. This can only be accomplished by adopting terms that are perfectly arbitrary, and not liable to fluctuate with the tide of chemical theory.*

---

* For farther remarks upon this subject the Student may refer to my PHARMACOLOGIA. Vol. 1. p. 73. Ed. 6.

WATER.*

426. It was an ancient and very universal opinion that water constituted the first principle of almost all matter; a belief that received even some support from the experiments of the earlier philosophers, for Van Helmont shewed that plants would grow for a very long time in pure water, whence it was concluded that it was capable of being changed into all the substances found in vegetables. Mr. Boyle† supposed, that by long digestion and boiling in glass vessels, he had converted water partly into an earth, but the solid matter was afterwards shewn to be derived from the apparatus employed in the experiments.‡

427. Although the researches of modern Chemists

---

* Boerhaave says, that it was in consequence of Moses having delivered a tradition that the Spirit of God, brooding upon the face of the *waters*, had communicated to them a prolific virtue, that the ancient Persians looked upon water as the principle of all bodies. Milton favours the same idea, (*Paradise Lost*, book vii. line 234.) The same doctrine is also taught in the Koran, " Do not the unbelievers know that the heavens and the earth were solid, and I clave the same in sunder, and made every living thing of water." (*Sale's Koran, vol.* 2. p. 155.) The Latin word *Aqua* is supposed to have been derived from the same prevalent belief, viz. *a qua* omnia.

† Boyle's works, Fol. vol. 2, p. 519.

‡ The reader may find a long and curious paper upon this subject in the 4th volume of Dr. Watson's Chemical Essays.

have proved that water itself is a compound, they have, at the same time, furnished fresh evidence of its almost universal presence in natural as well as artificial substances ; of the bodies of animals it constitutes, at least, three-fourths of their weight;* and it has been supposed by many eminent physiologists, that upon the relative quantity of the water and the solid matter, depend many morbid changes of the body, as well as the natural varieties in the constitution and temperament of different individuals; we shall find hereafter that water is still farther important in furnishing *oxygen* and *hydrogen* to the different parts of the animal body.   The atmosphere always contains a notable quantity ; and various substances owe many of their properties to its presence ; acids, for instance, when deprived of water lose the characters by which they are distinguished.

428.   Water enters into combination with solid bodies in two states.   In the first, the proportion of solid matter exceeds that of water, and the liquid becomes a part of the solid body without rendering it fluid ; in the second, the solid is much exceeded by the quantity of liquid, which therefore imposes its peculiar form upon the compound.   The products of the first state are termed *Hydrates* ; the second constitutes *Solutions*.   Thus, the *Precipitated Sulphur* of our Pharmacopœia, is a true hydrate.   The whole class of saline preparations, whether assuming the form of crystals, powders, or solid masses, fall under the same denomination.   The well known operation of slacking

---

* Professor Chaussier put a dead body of 120 pounds into an oven, and found it, after several days successive desiccation, reduced to 12 pounds.   Bodies, after being buried for a long time in the burning sands of the Arabian deserts, present an extraordinary diminution of weight.

lime is also calculated to afford a striking instance of the solidification of water, and the consequent conversion of the earth into a *hydrate.* (298.)

429. The water upon these occasions is so intimately combined, that it frequently resists the decomposing power of a high temperature. Saussure, who made experiments on the hydrate of alumina, declares that this earth has so powerful an attraction for water, that it will retain a tenth of its weight of that fluid, even though it be submitted to a heat that will fuse iron. In like manner, potass, after having been subjected to a red heat, is found to retain more than 13 per cent.; and soda, under similar circumstances, nearly 19 per cent. of water.

430. In these compounds the proportion of combined water generally appears to be definite in each, although to this law there is one or two exceptions. Soap, for instance, is a *hydrate*, but the proportion of water is found to be variable. The same observation will perhaps apply also to many vegetable and animal solids.

431. For depriving airs and vapours of their moisture they are generally exposed to some absorbent salt, as muriate of lime; for other purposes of desiccation, it is most elegantly accomplished by means of the air-pump and sulphuric acid, as already explained. (350.)

432. It is scarcely necessary to give any definition or description of this universally known fluid. Every person is acquainted with its external characters, and must be aware of the moderate degree of activity which it possesses with regard to organized substances, rendering it friendly to animal and vegetable life, for both which it is indeed indispensably necessary. Hence it acts but slightly on the organs of sense, and is therefore said to have neither taste nor smell. It appears to

possess elasticity, and yields in a perceptible degree to pressure. Canton long ago proved its expansibility, but Mr. Jacob Perkins has lately contrived an instrument, which he calls a *Piezometer*, in which he has subjected water to a pressure of 326 atmospheres, and has succeeded in increasing its density 3·5 per cent. Under ordinary circumstances it appears to be at its maximum of density at the temperature of 40° *Fah.* since it expands both with heat and cold beyond this point. (278.)

433. The most simple form, in which it is probable that water will ever be exhibited, is that of ice;* for, by the mere combination of ice with caloric, fluid water will be formed, and a further portion of caloric will convert this fluid into steam, the most attenuated aqueous vapours being nothing more than ice dissolved and rarefied by the solvent and expansive power of caloric.

434. Water, as it occurs from the hand of Nature, is seldom, if ever, presented to us in a perfectly pure state. That which flows within, or upon the surface of the earth, contains various earthy, saline, metallic, vegetable, or animal particles, according to the substances over or through which it passes. Rain and snow water are much purer, although these also contain whatever floats in the air, or has been exhaled along with the watery vapours.

435. When the presence of foreign matter imparts to water any peculiarity in colour, taste, smell, or medicinal effect, it is termed a *mineral water*, the history and investigation of which constitutes an important branch of Medical Chemistry, but can scarcely be con-

---

* If it be true that ice is only water in a state of solidity, it is evident, say the sophists, that we may speak of *red hot ice* without incurring the charge of absurdity.

sidered as an object of elementary instruction.* It may, however, be stated that there is a rule, first pointed out by Mr. Kirwan, by which a knowledge of the specific gravity of a mineral water will, at once, lead us to an estimate of its saline contents. The method is this: *Subtract the specific gravity of pure water from the specific gravity of the mineral water under examination* (both expressed in whole numbers), *and multiply the remainder by* 1·4. The product is the saline contents, in a quantity of the water, denoted by the number employed to indicate the specific gravity of distilled water. Thus, let the water be of the specific gravity 1·079, or in whole numbers 1079. Then the specific gravity of distilled water will be 1000. And 1079—1000 × 1·4 = 110·6, which represent the saline contents in 1000 parts of the water in question. Dr. Ure observes that this rule may be simplified in the following manner, *Multiply by* 140 *the decimal part of the number representing the specific gravity of the saline solution, and the product is the dry salt in* 100 *grains.* This formula, it must be confessed, is one of great practical value to the Chemist and Pharmaceutist, for it may be extended to the examination of the various solutions employed in medicine. In enumerating the practical advantages to be derived from a knowledge of the specific gravities of bodies (50), this application ought not to have been overlooked.

436. It is evident, therefore, that the specific gravity of water affords a certain test of its purity;† but it fre-

---

* The reader will find some practical observations upon the subject in my Pharmacologia, Vol. 2. Art. *Aqua Marina*, and *Aquæ Minerales.*

† There is a letter still extant from Synesius, a Christian Bishop of the fifth century, to the female philosopher Hypatia, under whom he had formerly studied, in which he complains of being ill; and says that he wishes to use a *Hydroscopium*, and re-

quently happens that the sum of all the saline substances dissolved in water does not exceed a six thousandth part of its weight, and yet may be composed of
six or eight different substances. Mr. Dalton, indeed,
has asserted, from the result of his own experiments,
that the hardest† spring water seldom contains so much
as one-thousandth part of its weight of any foreign body
in solution. Such water would require for its examination with the Hydrometer, or gravity bottle (73), a
delicacy of manipulation which would be inconsistent
with practical utility; in such cases, therefore, we must
rely upon the indications of our chemical reagents.
(252, *note.*)

437. The means to be adopted for the purification
of water may be said to have been suggested to us by
Nature herself, viz. *Distillation* and *Filtration.* Mr.
Parkes has observed that the most contaminated waters
on the face of the earth are daily purified by the action
of the sun's beams, which separate the limpid particles
from the polluted mass, and elevate them into clouds,
from whence they are distilled in showers of rain, hail,
or snow. The hills and mountains of the globe also
perform a similar office, by allowing the waters in their
vicinity to percolate through them, whence they are presented to us in various degrees of purity, according to
the nature of the different strata through which they
have filtrated.

---

quests that she would cause one to be constructed for him. " It is
a cylindrical tube," he adds, " of the size of a reed or pipe. A line
is drawn upon it lengthwise, which is intersected by others, and
these point out the weight of the water." It is probable, as the
bishop was in an infirm state of health, that he was ordered by his
physicians to drink none but pure water, and that the instrument
he required was one similar in principle to our hydrometer, in
order to test the purity of his beverage.

† The property termed *hardness* depends upon *Sulphate of Lime.*

438. Water has been discovered* to consist of two elementary principles, to which the names of *Oxygen* and of *Hydrogen,* for the reasons already stated, have been assigned. The simplest form in which these bodies can be presented to us is in that of *gas,*† which,

---

* This important discovery, which may be said to have contributed more largely than any other, to the extension of our chemical knowledge, was effected by the unrivalled genius of Mr, Cavendish in the year 1784.

† As we are about to enter upon the consideration of gaseous bodies, it will be necessary to explain the apparatus, and methods of manipulation, to be adopted for collecting, preserving, transferring, and examining them. The principle is extremely simple; by filling a glass jar with water, and inverting its mouth in a tub containing the same fluid, the jar is kept full, in consequence of the pressure of the atmosphere upon the surface. In this state the jar is ready to receive any species of gas, not soluble in water; which, by its lightness, will rise through the water, and gradually displacing it from the jar, will occupy its place. For con-

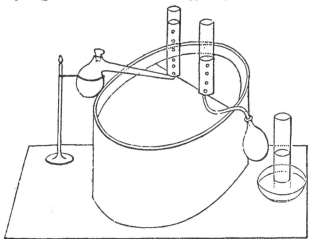

ducting this operation an apparatus termed a *Pneumatic trough,* or sometimes the *Hydro-pneumatic trough,* has been invented. It consists of a tub, which may be made of wood, or japanned

as the student has already learnt (310), depends upon the presence of a quantity of combined caloric. It

iron, as represented in the annexed sketch, and which should be twelve or fourteen inches deep, in order to allow the glass receivers to be conveniently filled. A shelf, containing several holes, is so placed as to be about an inch under the surface of the water, when the trough is filled. When any species of gas is to be collected, a jar is to be filled with water, and carefully placed, with its mouth downwards, over one of the holes. The beak of the retort from which the gas is proceeding is then brought under it, when it will bubble upwards, and displace the water; and in this manner any number of jars may be successively filled with the required gas. If our object is to preserve it for future examination, each jar may be removed from the trough, without any loss of its contents, by plunging a saucer into the water, and sliding the jar into it, taking care that its mouth is never raised above the surface of the fluid. Nor is it less difficult to transfer gas from one vessel to another. For this purpose, we have only to fill the jar, into which we wish to introduce it, with water, and to invert it in the trough; then, by bringing the mouth of the jar containing the gas, under that which is filled with water, and gradually depressing the top of the former, the gas will escape, and enter the latter. This operation may be said to be merely the converse of that of pouring water into any empty vessel, that is, into one containing only air, for in this case, the water being the heavier fluid is poured from *above*, enters the vessel, and expells the air *upwards*; whereas in the other case, the gas being the lighter fluid, it is poured from *below*, ascends, and expells the water *downwards*. When the receiving vessel has a narrow neck, as a common phial, the air may be poured through a glass funnel. If our object is to introduce the gas into a bladder, we must be provided with a jar, having a stop-cock in its upper orifice, to which the bladder is to be attached. Then by opening it, and at the same time lowering the jar, perpendicularly, into the water, the air will rush into the bladder, which may be afterwards detached from the jar, and preserved for use.

must, therefore, be distinctly understood that water is not the product of the union of these *gases*, but of the

---

For measuring gases, cylindrical vessels are employed, some of which are divided into 100 equal parts, others into tenths, and hundredths of a cubic inch. For this purpose, also, graduated tubes, sufficiently small to allow their mouths to be closed with the thumb, will be found extremely convenient.

Where large quantities of gases are required to be collected and preserved, we employ *Gas-holders*, and *Gazometers*. The most useful instrument of this description is that invented by Mr. Pepys, and known by the name of the *Improved Gas-holder*, which

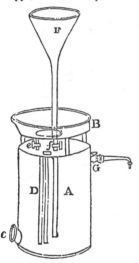

is made either of japanned iron, or copper. It consists of a body, or reservoir A, holding from six to eight gallons; a cistern B, from which issue two tubes supplied with stop-cocks, *e, f,* one entering the reservoir, the other continued, as shewn by the dotted lines, to near the bottom. C is a short oblique tube, issuing from the bottom of the reservoir, and capable of being accurately closed by a screw. D is a glass communicating at both ends with the body of the gas-holder. F is a funnel, communicating with the tube *f,* and which may be used, or not, according to circumstances, its object merely being to increase the force of the water by the height of its column. When it is intended to fill this apparatus with gas, the first step is to fill it with water, which may be effected in the following manner. Close the tube C, and open the stop-cocks *e, f,* water is then poured into the cistern, or funnel, in case the latter should be used, which running down the long tube, forces the air up the shorter one. The reservoir having been thus filled with water, the stop-cocks are to be closed, and the aperture C, opened, through which the water cannot run, in consequence of the pressure of the atmosphere. Into this orifice the beak of the retort, or tube, whence the gas issues, is to

ponderable bases which they contain. The water, therefore, resulting from their combination, bears no proportion *in volume* to that of the gases, but its weight accurately coincides with that of its constituent parts.

    Exp. 55.—The decomposition of water may be very satisfactorily exhibited by the following arrangement.

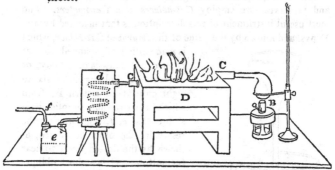

    B is a glass retort containing a given weight of water, and connected with an earthen tube C, C,

---

be introduced, which, bubbling up, displaces the water through the same opening. When it is seen by inspecting the tube D, that the cylinder is nearly filled with gas, the aperture is closed. When we wish to draw off the gas, by opening the stop-cocks *e, f,* we may receive it in any vessel placed over *e;* or by opening *f,* and *g,* it may be drawn through the latter, either for the purpose of filling a bladder, or urging a blow-pipe. In this latter case, the funnel F, should be always used to increase the force of the current.

    Those gases which are absorbed by water cannot be received in the Hydro-pneumatic trough, but must be collected over mercury. For this purpose various forms of apparatus have been contrived, with a view of furnishing the greatest accommodation with the least possible quantity of quicksilver. The most convenient and œconomical of these is that invented by Mr. Newman, an engraving of which may be seen in *Brande's Manual of Chemistry.*

which traverses the small furnace D, and terminates in the spiral pewter tube *d, d,* immersed in
water. A given weight of pure iron wire, coiled
up, is introduced into the tube C, and the whole
made red hot; the water in B is then made to
boil, and the vapour, on coming in contact with
the red hot iron, is in part decomposed, the oxygen being retained by the iron, and the hydrogen
escaping through the tube *f,* may be collected as
usual, any undecomposed portion of water is condensed in the worm pipe *d,* and drops into the
vessel *e.* After this experiment, the iron will be
found to have increased in weight; and if attention be paid to the quantity of water which has
been collected in *e,* and to the weight of the hydrogen gas evolved, it will be found that the weight
gained by the iron, added to that of the hydrogen,
will be equal to the weight of the water which has
disappeared. We shall thus be led to the inference
that nine parts of water consist of eight of oxygen
and one of hydrogen, by weight.

439. The decomposition, however, of water is more
strikingly effected by the agency of electricity, as already explained, for in this case the oxygen and hydrogen are both exhibited in their gaseous forms, and
in their exact proportions, viz. one volume of oxygen
to two volumes of hydrogen; and since, according to
the latest experiments, the specific gravity of the latter
gas, when compared to the former, is as 16 to 1, it follows that the estimate above given, with respect to the
combining *weights* of these elements, is correct.

440. It is not, however, sufficient to decompose a
body into its constituent principles; to produce com-

plete conviction we must also re-compose it with the
elements which resulted from its decomposition; happily, modern Chemistry has furnished us both with an
analytic and *synthetic* proof of the composition of
Water, but the latter will be more conveniently considered under the history of Hydrogen.

OXYGEN.

441. The most simple form in which this body can
be obtained is in that of gas, when it is at least combined with caloric, if not also with light and electricity.

442. *Oxygen Gas* was discovered by Dr. Priestley,
in August, 1774, who gave it the name of *dephlogisti-
cated air.* In the following year Scheele also discovered it, without any knowledge of the previous experiments of the English philosopher, and called it
Empyreal air. It has been also termed *Vital air.*

443. To procure it, a quantity of *Manganese,* a
mineral substance found in abundance, is introduced
into a glass retort furnished with a ground stopper,
over which is poured a quantity of oil of vitriol (sulphuric acid) sufficient to convert the powder into a
thin paste; the bottom of the retort is then gently
heated by means of a lamp, and the extremity of its
neck introduced under an inverted cylinder filled with
water in the hydro-pneumatic apparatus, as represented
in the wood-cut at page 261. Globules of gas will
soon rise through the water; the first portions collected
must be thrown away, being principally the common
air contained in the retort; the remainder may be preserved for use.

444. There are many other methods by which this gas may be obtained; the same manganese* heated to redness in an iron tube, such as a gun-barrel, the touch-hole of which is closed, or in an iron bottle usually sold for the purpose, will afford, for every pound, from 40 to 50 wine pints of gas. Nitre (common *Salt Petre*) heated strongly in a porcelain retort, will also give off oxygen gas. Red oxide of Lead offers a similar result; and from any of the salts called *Oxy-muriates*, or *Chlorates*, it may be obtained by a dull red heat. The gas obtained by this latter process is much purer than that procured by any other method.

445. Oxygen gas possesses all the physical properties of common air; it is invisible, permanently elastic, and capable of indefinite expansion and compression. It has neither taste or odour. It is rather heavier than common air, its specific gravity, according to the best authorities, being 1·1088. It is not absorbed by water, or at least, in so small a degree that when agitated in contact with that fluid, no perceptible diminution in its bulk takes place.

446. This gas is distinguished from all other gaseous matter by the following properties.

1. *All inflammable bodies burn in it with greatly increased splendour.*

Exp. 56.—Plunge a lighted wax taper, fixed to an iron wire, into a vessel of this gas, and the bril-

---

* As manganese is frequently contaminated with carbonate of lime, it will be necessary to wash it with dilute muriatic acid, before it is submitted to heat, where it is an object to obtain pure oxygen gas; and the gas should moreover be allowed to remain for some hours over water. This precaution is essential when it is intended to be respired.

liancy of its combustion will be greatly increased.
Or, if the taper be blown out, and let down into
it, while the snuff remains red hot, it will instantly
rekindle with a slight explosion, and burn vividly
In like manner, phosphorus, iron wire, charcoal,
sulphur, &c. by combustion in this gas, will exhibit the most splendid and beautiful phænomena,
during which a very large quantity of caloric, as
well as light, will be liberated.

2. *It supports animal life in a more eminent degree
than common air.* A small animal confined in a jar
filled with this gas, lives four or five times as long as
in an equal quantity of common air ;—hence it has been
called vital air.

3. *It rapidly converts dark blood to the colour of
rich vermillion.*

Exp. 57.—Introduce some venous blood into a common phial filled with oxygen gas, and shake it;
a change of colour will be instantly produced.

4. *During every combustion in oxygen gas, the
gas suffers a considerable diminution in volume.*—
This fact may be easily illustrated by burning phosphorus, or any inflammable body, in a jar of oxygen
gas. The first effect of the combustion will be a depression of the water within the jar; but when it is
finished, and the vessel has cooled, a considerable absorption will be found to have ensued.

5. *All bodies, by combustion, in oxygen gas, acquire an addition to their weight; and the increase*

is in proportion to the quantity of gas absorbed, viz. about one-third of a grain for every cubic inch of gas.

Exp. 58.—Fill the bowl of a tobacco pipe with iron wire coiled spirally, and of known weight; let the end of the pipe be slipped into a brass tube, which is screwed to a bladder filled with oxygen gas: heat the bowl of the pipe, and its contents, to redness in the fire, and then force through it a stream of oxygen gas from the bladder. The iron wire will burn; will be rapidly oxidized; and will be found, when weighed, to be considerably heavier than before. When completely oxidized in this mode, 100 parts of iron wire will gain an addition of about 30.

6. *The substances capable of combining with Oxygen, afford one or other of the following products:* 1. An ACID. 2, an ALKALI or EARTH. 3, an OXIDE. It is not easy to offer an unexceptionable definition of these three classes of compounds. This difficulty arises from the fact, that several bodies of each class are deficient in some of the common characters which distinguish it. By the term *Acid,* we necessarily associate the idea of *sourness,* but the Chemist also understands by this name, a body which reddens vegetable blue colours,*

---

† The test best adapted for this purpose is the blue infusion of the leaves of red cabbage, which becomes *red* by the action of acids, and *green* by that of alkalies. An infusion of *Litmus* is also commonly used as a test for the former of these bodies, but it is not affected by alkalies; the test usually employed for the latter is *Turmeric,* the yellow colour of which is thus converted into a brownish-red. Paper tinged with these substances should be kept in readiness by every Chemist. Litmus paper, also,

and combines with the alkalies, the earths, and metallic oxides, forming compounds in which the properties of the acid, or of the substance with which it is united, are no longer to be recognised. For these reasons, we include several bodies in the class of acids that are not distinguished by the quality of sourness, such are *White Arsenic, Prussic Acid,* &c.; nay, even *Sugar,* inconsistent as the fact may be with our popular ideas, will be found to possess some of the characters belonging to acids.

447. Nor is oxygen essential to the acidity of a compound, as Lavoisier supposed, for it has been since discovered that some bodies are rendered acid by union with an element termed *Chlorine*; and others, by combination with *Hydrogen.* The theory of Lavoisier, therefore, which considered oxygen as the essential principle of acidity, can no longer be received as correct.

448. Alkalies and Earths are chiefly distinguished by acting as bases, with which the acids combine, with the loss generally of the separate characters of each. The alkalies are soluble in water, and change some vegetable blues to green; the *Earths* are either not soluble at all, or sparingly soluble in that fluid; some of them affect vegetable colours like alkalies, whence they have been distinguished by the name of *Alkaline*

---

slightly reddened by an acid, affords a convenient test for an alkali, as its presence immediately restores the original blue colour. The changes thus produced by acids and alkalies on vegetable colours were long regarded as amply sufficient to indicate the presence of these bodies, but some late experiments by Mr. South, and Mr. Faraday, have shewn that changes of colour, inconsistent with this general opinion, are produced by a variety of different substances.

Earths. Oxides, especially those derived from the metals, agree with the earths in the quality of insolubility. They also serve as bases to the acids. We have also examples, in which the same body, combined with a small proportion of oxygen, gives an oxide that is capable of uniting with acids, and of composing salts; and, again, when united with more oxygen, of yielding an acid which is susceptible of forming saline compounds with alkaline and earthy bases.

OXYGENATED WATER.

449. It has been stated that oxygen gas has not hitherto been condensed into a liquid form, but in July 1818, M. Thenard succeeded in forming what has been termed a *Deutoxide,* or *Peroxide of Hydrogen.* It has been supposed to contain double the quantity of oxygen to that which enters into the composition of water; that is to say, if we admit water to be a compound of one atom of hydrogen and one of oxygen, the peroxide must consist of one atom of hydrogen and two of oxygen. The process by which it is obtained is complicated and difficult; very minute instructions, however, for its preparation are given by its discoverer.†

450. It is liquid, and colourless, like water. It has scarcely any smell, but when applied to the tongue whitens, and thickens the saliva, and produces a taste like that of some strong metallic solutions. It attacks the skin with considerable energy, bleaches it, and occa-

---

† Ann. de Chim. et Phys, viii, ix; Ann. of Philos. xiii, xiv, xv: and Quarterly Journal of Science, vi, 150, 379. viii, 114, 154.

sions a smarting, the duration of which differs in different persons, and in the same person according to the quantity applied. Its specific gravity is 1·452, and when poured into water it descends through it, like syrup, though easily dissolved by agitation. In its most concentrated form it has not been congealed by any degree of artificial cold yet applied to it. It is, however, decomposed at the temperature of 55°, when oxygen gas is abundantly liberated from it. By the application of heat it explodes; the addition of certain bodies also, as oxide of silver, peroxide of lead, &c. produces the same effect.

451. I am not aware that any trials have been made to ascertain its medicinal properties, but it seems probable that it may possess some virtues that would render its external as well as internal administration efficacious.

452. It is probably a solution of liquid oxygen in water.

HYDROGEN.

453. Like oxygen, this elementary body can be obtained only in the form of gas. Of all gaseous substances it is most distinctly characterised as an element; and in its relations it is opposed to oxygen. Sir H. Davy concludes from its extreme lightness, and from the small quantities in which it enters into combination, that it is unlikely it should be resolved into other forms of ponderable matter, by any instruments or processes at present within our power. It was first examined in its pure form by Mr. Cavendish in 1766.

454. To procure *Hydrogen gas*, let oil of vitriol (Sulphuric acid) previously diluted with six or eight

times its weight of water, be poured on iron filings, or pieces of zinc, in a gas-bottle, known by the name

of a *proof* and *tube*, as here represented. An effervescence will immediately ensue, and the evolved gas may be received, in the ordinary manner, in jars placed in the hydro-pneumatic trough.

455. Iron or zinc, without the intervention of the acid, are incapable of decomposing water, at least with any degree of rapidity. The addition of the acid would appear to act by forming a simple galvanic circle.

456. It may likewise be procured by passing steam over turnings of iron heated to redness in a gun-barrel, as already explained. (*Exp.* 55.)

457. This gas is characterised by the following properties:

1. *As commonly procured it has a disagreeable odour*; but this has been lately shewn to depend upon the presence of a peculiar volatile oil, for if the gas be passed through pure alcohol it becomes inodorous.

2. *It remains permanent over water*, or is not absorbed in a proportion exceeding $\frac{1}{50}$th the bulk of the water.

3. *It is considerably lighter than common air*, and is indeed the lightest of all elastic fluids. This may be easily shewn by adapting a bladder filled with the gas to a common tobacco-pipe, and blowing up soap bubbles, which instead of falling to the ground, like those commonly blown by children, will rise rapidly in the air. On this property is founded its application to the raising of balloons. According to the latest and

s

most accurate experiments of Berzelius and Dulong, its specific gravity is only 0·0688, from whence, taking 100 cubic inches of atmospheric air at 31 grains, we find the same volume of hydrogen gas to weigh 2·13 grains.

4. *It is inflammable,* and hence called *Inflammable Air;* it is the body which gives the power of burning with flame to all the substances used for the œconomical production of heat and light.

> *Exp.* 59.—Fill a small jar with the gas, and holding it with the mouth downwards, bring the flame of a candle in contact with it; the gas will take fire, and will burn away silently.

> *Exp.* 60.—In a strong bottle, capable of holding about four ounces of water, mix two parts of common air and one of hydrogen gas. On applying a lighted candle, or a red hot wire, the mixture will burn, not silently, as in the former experiment, but with a sudden and loud explosion. The same experiment may be repeated with oxygen gas, instead of atmospherical air, changing, however, the proportions, and mixing only one part of oxygen gas with two of hydrogen. The report in this case will be considerably louder. The bottle should be a very strong one, and should be wrapped round with several folds of cloth to prevent an accident.

5. *Although inflammable itself, it extinguishes burning bodies.*

> *Exp.* 61.—Bring an inverted jar, filled with this gas, over the flame of a candle, and suddenly depress the jar, so that the lighted wick may be entirely surrounded by the gas; it will be extinguished, although the gas will take fire and burn in contact with the atmosphere.

6. *It is fatal to animals.* On account of the extreme levity of this gas it may be apparently breathed for some time without inconvenience, provided that the lungs at the outset are filled with common air; but if a forcible expiration be made, before drawing in the hydrogen gas, only two or three inspirations of the latter can be made, and even these produce great feebleness and oppression about the chest. It has been also found to change the tone of the voice; this effect is observed, on the person speaking immediately after ceasing to breathe it; but it soon goes off.

458. It has been shewn (*Exp.* 60) that hydrogen gas combines with oxygen. If the two gases be pure, *water* is the only result, and the proportions are one of the former to eight of the latter in weight, or two to one in volume. We thus receive a *synthetic* proof of the composition of water. The experiment for its illustration may be variously conducted. The union of the gases may be at once effected in strong dry vessels by the electric spark; or the hydrogen may be introduced into a vessel full of oxygen through a narrow tube, by means of pressure, and inflamed by electricity, or the oxygen may be made to burn in the hydrogen in a similar manner. In these cases a sensible quantity of moisture will have condensed on the inner surface of the vessel, and by repeating the operation, a sufficient quantity of fluid may be collected to shew that water is the only product. For performing these experiments with accuracy, a delicacy of apparatus, as well as of manipulation, will be required which the student can scarcely be expected to command. He may, however, satisfy himself of the truth of the doctrine by the following simple experiment.

Hyd.	Oxy.
	7·5
1	

s 2

Exp. 62.—Into a glass bottle, supplied with a cork through which a tube passes, as here represented, 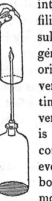 introduce a small quantity of iron filings, and pour upon them diluted sulphuric acid. Inflame the hydrogen gas which will issue from the orifice of the tube, and hold an inverted jar over the flame. In a short time its interior surface will be covered with a very fine dew, which is pure water, produced from the combustion of the hydrogen gas, evolved from the materials in the bottle, and the oxygen gas of the atmosphere. The hydrogen gas should be allowed to pass off, for some little time, before it is inflamed, in order to drive the atmospheric air from the bottle, the presence of which might otherwise occasion explosion.

459. By the combustion of these two gases, in the proportions necessary for the production of water, the heat produced is very intense, and far exceeds the highest heat of our furnaces, and may be used to fuse bodies, intractable by any other fire raised by combustion. The easiest and most efficient mode of exciting and applying this heat is by the *Oxy-hydrogen* blowpipe, in which these gases, after undergoing compression in a mixed state, are propelled through a capillary tube, and exposed to combustion.

460. Hydrogen enters largely into the composition of animal and vegetable bodies. In the human body it is found to exist in a gaseous state in the alimentary canal; to a small extent only in the stomach, but in larger proportions in the great, and in very considerable quantities in the small intestines, while oxygen has

never been found in any part of the primæ viæ except the stomach.

CARBON.

461. If vegetable matter, especially the wood of plants, be exposed to heat in close vessels, the more volatile parts are expelled or decomposed, and there remains a black shining porous body, termed CHARCOAL. This body, however, always contains several foreign ingredients, as water, air, and saline and earthy matter. It is to the peculiar inflammable matter, divested of these impurities, that the term CARBON is applied. It is an abundant principle both in vegetable and animal substances, and may be procured from them by heat. The purest known form in which it can be obtained is that deposited by oils or spirits of wine, on passing through ignited tubes.

462. Extraordinary as the fact may appear, the experiments are too numerous and conclusive to admit of a doubt that the Diamond and Charcoal, though so widely remote from each other in external characters, are, as to their chemical nature, identically the same; and that the difference between them, in all probability, results merely from the respective states of aggregation of their particles. Sir Isaac Newton, with a sagacity almost superhuman, had inferred that the diamond was a combustible body from the great refractive powers which it displays (383), but it was first shewn to contain carbon by Guyton Morveau. This distinguished philosopher was led to conclude that the diamond is the only form of true carbon, and that charcoal is a compound of carbon and oxygen, or an *oxide* of carbon; the researches, however, of Messrs. Allen and Pepys have negatived such a supposition. The proofs of the identity of these bodies are that, by combustion,

they both afford the same product, and that each is
capable of converting iron into steel, under circumstances quite free from all sources of fallacy.

463. Charcoal is now generally prepared by the
distillation of wood in cast iron cylinders. The lop-
pings of young trees, commonly called crop-wood, are
employed for this purpose; and in addition to the
charcoal, an impure vinegar, called *Pyroligneous acid,*
is obtained by the process; a description of which,
together with a sketch of the apparatus, will be found
under the history of Acetic Acid. For accurate chemical purposes, pieces of oak, willow, hazel, or other
woods, deprived of the bark, may be buried in sand in
a crucible, and exposed to the strongest heat of a wind
furnace, when the charcoal thus produced should be
used before it has time to become cold; or if it cannot
be had fresh made, it should be heated again to redness under sand in a crucible.

464. Charcoal of wood has the following properties.
It is brittle, and easily pulverized, black, perfectly insipid, inodorous, and insoluble. It is more than twice
as heavy as water, and is a conductor of electricity.
In close vessels, and entirely secured from contact with
air, it is infusible by any heat that has hitherto been
applied. By exposure to the atmosphere, it absorbs
moisture, so as to increase in weight from 12 to 14 per
cent. and possesses moreover, the singular property,
when perfectly dry, of absorbing several times its volume of any gas, to which it may be exposed, without
alteration; for which purpose it must be employed immediately after ignition, and whilst yet warm. This
effect would appear to be entirely mechanical, for when
the charcoal is reduced to powder it is much diminished.

465. Charcoal resists the putrefaction of animal substances. A piece of flesh-meat, which has begun to be tainted, may have its sweetness restored by rubbing it daily with powdered charcoal; and may be preserved sweet for some time by being buried in powdered charcoal, which is renewed daily. Putrid water is also restored by the same application; and water may be kept unchanged at sea, by perfectly charring the inner surface of the casks which are used to contain it. It produces, also, a remarkable effect in destroying the taste, odour, and colour of many vegetable and animal substances. Common vinegar, by being boiled on it, is rendered perfectly limpid. Rum and other varieties of ardent spirit, which are distinguished by peculiar colours and flavours, lose both by maceration with powdered charcoal. The colour of litmus, indigo, and other pigments, dissolved or suspended in water, is destroyed, and, in like manner, colour may be abstracted from syrups and saline solutions. Putrid animal fluids, rancid oils, and air contaminated with offensive effluvia, are also completely deprived of their odour. Even the common salt of *Hartshorn* (Carbonate of Ammonia), which is prepared from a liquor distilled from bones, and which often requires great labour to render it of the desired purity, loses the whole of its fœtid smell by being mixed with charcoal powder, and re-sublimed. Such effects are most certainly ensured by the use of *animal* charcoal. These properties render charcoal a very important article in Pharmacy and Medicine. Its application to foul ulcers proves highly antiseptic, and it furnishes the best dentifrice with which we are acquainted. It has been also recommended in certain forms of dyspepsia, and it appears to correct the putrid eructations which so frequently attend depraved digestion.

466. Powdered charcoal possesses another curious property which has not hitherto received sufficient notice. It possesses the power of abstracting certain bodies from their solutions in water. Lime water, for instance, may be deprived of the greater portion of its lime, by the action of finely divided animal charcoal. I have lately satisfied myself that the same effect is produced on a dilute solution of white arsenic.*

467. The article known in commerce by the name of *Lamp-black,* is carbon in the state of a light impalpable powder, and is produced by burning the refuse of pitch and resin in peculiar furnaces, with long flues terminating in a close chamber, the ceiling of which is covered with porous cloth, through which the gas that is disengaged in the process may escape and the soot be left behind. In drawing spirits of Turpentine from the crude Turpentine, there is a large residuum known in commerce by the name of *Rosin;* and as a considerable portion is often found to be too impure to sell as rosin, it is profitably burnt for the production of lamp-black.

468. Charcoal may also be obtained from animal substances burnt in close vessels; but this differs from that of vegetable origin in some particulars, especially in its state of aggregation, and is better adapted for the purposes of decoloration. Mr. Parkes has given us the following directions for ascertaining whether any carbonaceous matter be produced from animal or from vegetable substances—a test which is important to the manufacturing chemist who employs large quantities of animal charcoal in the state of powder, and which is frequently adulterated with *vegetable* charcoal, an

---

* Pharmacologia, Edit. 6, Vol. 2, Art. *Arsenicum.*

article that is not so efficacious, but is always to be obtained at a much lower price. *Vegetable* charcoal will burn, on a red hot iron, into white ashes, and these will be readily dissoluble by sulphuric acid into a bitterish liquor; whilst the ashes of *animal* substances are very sparingly affected by that acid, and form with it a compound having a very different taste.

469. It has been already shewn (*Exp.* 40,) that charcoal is a bad conductor of heat; and numerous are the instances in which this property might be rendered serviceable, especially in those processes of art where it is necessary to preserve for a given time an equable temperature. It has been proposed by Mr. Parkes to construct all those vessels which are heated by steam, with treble instead of double sides (360) and to fill the intermediate space with ground charcoal, by which arrangement the heat would be so prevented from escaping, that any one particular temperature might be kept up for a great length of time, and a material saving of fuel would be accomplished. Guyton Morveau has shewn that ground charcoal will conduct heat more slowly than even dry sand, and this in the proportion of three to two.

470. To the Chemist charcoal is of great service in deoxidizing various bodies, as will be more fully explained under the history of Metals.

471. Although carbon is perfectly insoluble in every menstruum with which we are acquainted, yet there appears to be a certain modification of this body soluble in acids, with which it forms a gelatinous solution; such. is the carbonaceous matter thrown down from alcohol by the action of sulphuric acid, as seen in the distillation of æther; if this matter be collected and examined, it will be found to have all the essential habitudes of charcoal, except that it dissolves in acids,

and yields hydrogen by heat. It is this peculiar form of charcoal which gives to sulphuric acid its dark tinge. The subject is a very curious one, and does not appear to have hitherto received the attention it merits.

472. Carbon combines with oxygen in two different proportions, and gives origin to two distinct combinations,—*Carbonic acid*, and *Carbonic oxide*, which we have next to consider.

CARBONIC ACID.

473. This gaseous acid, which was the first species of air discovered, distinct from common air, is formed whenever charcoal, or carbonaceous matter, is burnt in air, or in oxygen gas. It is also evolved during fermentation; by the decomposition of animal or vegetable substances, and from limestones by ignition or the action of acids. It is also given off abundantly from the lungs during the act of respiration; and it generally exists in the alimentary canal. It appears, moreover, to be transpired from the surface of the body; and plants yield it in great abundance during the night. It exists likewise in the atmosphere, and is frequently found in walls and caverns, and is known to miners by the name of *choak damp*.† It is a constituent principle of many mineral waters, to which it communicates pungency and their sparkling quality.

474. For the purpose of experiment, carbonic acid may be procured by the following process. Into a gas-bottle, similar to that employed for the generation of Hydrogen gas, (454) introduce a little powdered marble, or chalk, and pour over this some sulphuric acid diluted with five or six times its weight of water;

---

† The word *Damp* signifies *Vapour* in the German language.

an immediate evolution of carbonic acid will take place, and may be received in jars in the usual manner. Or, where our object is to disengage the gas more slowly, it will be preferable to pour muriatic acid, diluted with eight or ten times its weight of water, over fragments of marble about the size of horse beans. Carbonic acid may, also, be separated from marble or chalk by the mere application of heat. The rationale of these processes may be easily understood; in the first case, the marble or chalk, which consists of carbonic acid and lime, is decomposed by the stronger affinity of the sulphuric or muriatic acid, for the lime, with which this latter body combines, to the exclusion of the carbonic acid for which it has a weaker attraction. In the latter case the carbonic acid is rendered gaseous by increase of temperature, and can no longer exist in union with its earthy basis. The existence of this gaseous body in a *fixed* state was discovered by Dr. Black, whence it was called *fixed air*, although it is evident that this term might be applied with equal propriety to every other species of gas, since later discoveries have shewn that they are all capable of laying aside their aeriform condition, and of becoming *fixed* by combination.

475. The following experiments are well calculated to illustrate the subject under discussion, and, from the very striking and satisfactory phænomena which they exhibit, should be performed by every student.

Exp. 63. Into a quantity of carbonic acid, contained in a jar inverted over quicksilver,† intro-

---

† For the performance of this experiment, a mercurial trough is not absolutely necessary, a phial filled with quicksilver, and inverted into a saucer, may receive the carbonic acid as it is generated from a small gas-bottle. The muriatic acid may be afterwards blown into the vessel through a glass syphon.

duce a piece of recently burnt quicklime, together
with a small proportion of water. In a few seconds,
we shall perceive the quicksilver beginning to rise,
in consequence of the absorption of the gas; and
this process will continue until the whole of the
carbonic acid has disappeared, and the vessel is
filled with mercury. This having been accomplished, we may now make the lime disgorge the
gas, by an *Emetic*, if we may be allowed the use
of so technical an expression; for this purpose
introduce some dilute muriatic acid, when the
lime will be seen to effervesce, and the quicksilver
to descend, until the jar presents the same appearance as it did previous to the experiment.

476. Carbonic acid gas is characterised by the following properties :—

1. *It is heavier than common air.* If 100 cubic
inches of atmospheric air weigh 30·5 grains, the same
bulk of carbonic acid gas would weigh 46·5. Its
superior gravity may be shewn by the following experiments.

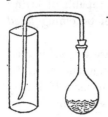

Exp. 64. Let a glass tube, proceeding from a Florence flask, containing marble and dilute sulphuric acid, be twice bent at
right angles; and let the open
end of the longer leg reach the
bottom of a glass jar, perfectly
dry within, and standing with its mouth upper-
most, as represented in the annexed cut. The
carbonic acid will expel the common air from the
jar, and take its place, because it is heavier, just
as water will, when poured into a vessel containing nothing but air. When the jar is thus filled

with the gas, (which may be known by a lighted taper being instantly extinguished in it, when plunged a little below the brim) take another jar of rather a smaller size, and place at the bottom of it a lighted taper, and pour the invisible contents of the former into the latter, as if you were pouring water, the candle will be instantly extinguished, although the eye is incapable of perceiving any thing poured upon it capable of producing such an effect.

In consequence of this superior gravity of carbonic acid gas, it is often found at the bottom of grottos, of deep wells, and of mines, the upper part of which is entirely free from it. In the *grotto del cane*, near Naples, so called from its air proving so destructive to dogs, has long been famous for the quantity of carbonic acid produced in it, which runs out at the opening like a stream of water, and kills any small quadruped that enters it, whilst man, from his greater height may pass through it with impunity.

2. *It extinguishes flame.* This has been already shewn by experiment 64.

3. *It is fatal to animals.* If we intoduce a mouse, or other small animal, into a vessel of the gas, and cover it, so as to exclude the access of common air, the animal will die in the course of a minute or two. Its fatal effects upon man has been illustrated by numerous examples; miners and well-diggers constantly fall a sacrifice to its power. According to the experiments of Sir H. Davy, it would appear to act, when undiluted with common air, by closing the glottis spasmodically, and thus preventing the ingress of the atmosphere; when diluted, however, in the proportion of three parts to seven of common air it is respirable,

and appears to produce narcotic effects. I am strongly inclined to believe that a mixture still more dilute, might be advantageously respired in many diseases; for which purpose the air of the apartment should be impregnated with it. In all those cases in which it has been accidentally breathed, the persons have expressed a delightful feeling of tranquillity, with a powerful inclination to sleep. This effect is even, in some degree, produced when the gas is taken into the stomach; the exhilarating influence of beverages containing it is well known.

4. *It is highly antiseptic, and retards the putrefaction of animal substances.* This may be easily proved, by suspending two equal pieces of fresh meat, the one in common air, the other in carbonic acid gas, or in a small vessel through which a stream of carbonic acid is constantly passing. The latter will be preserved untainted some time after the other has become putrid. This property of carbonic acid is frequently rendered serviceable in medicine; the *Yeast Poultice* owes its antiseptic property to the gas which it evolves; and it has been stated that fevers are less infectious in the vicinity of lime-kilns.

5. *It is readily absorbed by water.* This fact may be easily shewn by the following experiment.

Exp. 65. Fill a phial with water, and then displace about half its contents, by throwing up the gas; if the finger be now pressed close against the mouth, and the phial be at the same time shaken violently, so much of the gas will be absorbed as to produce nearly a vacuum within, which will be ascertained by the strong external pressure of the atmosphere on the finger that shuts the communication.

In this manner water may be charged with rather more than its own bulk of carbonic acid gas; and it acquires, when thus saturated, a very brisk and pleasant taste. This impregnation is most commodiously effected by an apparatus known the name of *Nooth's machine,* and of which the student is here presented with a sketch. It consists of three principal pieces; a lower piece *a*; a middle piece *c*; and an upper piece *d*, terminating in a curved tube. The substances from which the gas is to be extricated are introduced into the lower compartment; the middle part is filled with the fluid with which the gas is to be combined, and the upper piece is left empty. As soon as a sufficient quantity of gas is formed to overcome the pressure, it passes through the valve, and rises through the fluid to the upper part of the middle piece. At the same time it forces a quantity of fluid into the upper piece through its lower aperture. As soon as so much of the fluid has been thus forced from the middle part as to bring its surface down to the level of the lower aperture of the upper piece, a portion of gas escapes into the latter, and the fluid rises in the former. The upper piece is furnished with a conical stopper, which yields, and permits the escape of a portion of gas, as soon as its pressure becomes considerable. *b* is a glass cock for drawing off the fluid. The influence of pressure, in occasioning water to absorb a larger quantity of carbonic acid, has been very successfully investigated by Dr. Henry,[†] who has deduced as a general law, that water takes up the same volume of compressed carbonic acid gas, as of gas under ordinary pressure.

† Phil. Trans. for 1803.

And since the space occupied by any gas is inversely as the compressing force, it follows that the quantity of gas, forced into water, is directly as the pressure. Thus, if water under common circumstances take up an equal bulk of carbonic acid; under the pressure of two atmospheres† it will absorb twice its bulk; under three atmospheres three times its bulk, and so on. The carbonic acid, thus dissolved in water, is again set at liberty, on boiling the water, or by exposing it under the receiver of an air-pump, when the gas will escape with such rapidity as to present the appearance of ebullition. The same result is produced by freezing, and the ice formed under such circumstances presents the appearance of snow, in consequence of the increase of its bulk from an immense number of air bubbles.

6. *It possesses the characters and habitudes of an acid.* Its taste, for instance, is acidulous; it combines with alkaline and earthy bodies, and forms salts; and it reddens vegetable blue colours. This latter fact may be shewn by dipping into water, thus impregnated, a piece of Litmus paper; or by mixing with a portion of it, about an equal bulk of the infusion of that vegetable substance.

7. *It precipitates Lime water.* In other words, it converts the lime which is soluble, into the carbonate of lime which is insoluble in water. This fact affords a ready test of the presence of carbonic acid whenever it is suspected, as we shall hereafter have occasion to demonstrate.

---

† Chemists and mechanical philosophers always use this mode of expressing degrees of compression. If we take the pressure of the atmosphere, as unity, which is estimated as amounting to 15 pounds upon every square inch, we can easily deduce the degree of pressure of any number of such atmospheres.

Exp. 66.—Introduce a few bubbles of carbonic acid into transparent lime-water, the fluid will instantly become milky.

By an excess of carbonic acid, however, the carbonate of lime is rendered soluble, as may be readily shewn by an extension of the preceding experiment, viz.

Exp. 67.—Let a stream of carbonic acid gas be introduced into lime-water; after a short time, the milkiness which was first produced will disappear, and the liquid will recover its original transparency.

8. *By powerful compression, Carbonic acid is condensed into a Liquid.* For this interesting discovery we are indebted to Mr. Faraday, who succeeded in condensing many of those gases which had been long considered permanently elastic, by experiments so simple, that the repetition of them may be accomplished by the youngest student.

Exp. 68.—Take a glass tube of six inches in length, and bend it, at an obtuse angle, about two inches from its extremity; hermetically seal its shorter end, and pour in, through a small funnel, a portion of concentrated sulphuric acid, so as nearly to fill the short leg without soiling the long one; then introduce small fragments of carbonate of ammonia, so as nearly to fill the tube, taking care to prevent any communication between the salt and the acid. The longer end must now be carefully closed. The tube having been thus sealed, the sulphuric acid is to be made to run on to the carbonate of ammonia, when carbonic acid will be

T

immediately evolved, and after some time, the accumulating pressure will effect its condensation, and it may be seen floating upon the other contents of the tube. During the progress of this experiment the longer leg of the tube should be plunged in ice. It is a limpid colourless body, and extremely fluid. It distills readily and rapidly at the difference of temperature between 32° and 0 Fah. Its refractive power is much less than that of water. By inclosing a gage in a tube in which carbonic acid was afterwards produced, it was found that its vapour exerted a pressure of 36 atmospheres at a temperature of 32. In conducting this process great precaution is necessary; the glass tubes employed should also be strong. Mr. Faraday found that tubes which had held fluid carbonic acid for two or three weeks together, spontaneously exploded with great violence, upon some slight increase in the warmth of the weather. In endeavouring to open the tubes at one end, they uniformly burst into fragments, with powerful explosions. At present we are scarcely able to appreciate the extent and importance of the fact thus disclosed; there can, however, be little doubt but that it will lead to the explanation of many natural phænomena, as well as to the invention and improvement of several processes of art.

477. For a knowledge of the chemical nature of carbonic acid we are indebted to Lavoisier. The following experiment will afford a synthetical proof of its composition.

Exp. 69.—Into a bottle containing oxygen gas, introduce a piece of ignited charcoal; it will immediately burn with increased splendor, and throw out scintillating sparks in great abundance. After the operation is over, and the vessel has cooled, if it be opened under the surface of water, we shall find that the air has not disappeared, but that it has completely changed its nature. It has been converted into *carbonic acid.* This may be shewn by the introduction of lime-water which is instantly rendered turbid; or, by the change of colour produced by it on Litmus.

478. This experiment at once explains the cause of the deleterious nature of the fumes of charcoal.

479. In addition to the proof of the constitution of carbonic acid, derived from its synthesis, we have also equally satisfactory evidence from its analysis which may be effected by several processes. For the first of the following experiments we are indebted to Mr. Tennant, for the last, to Sir H. Davy.

Exp. 70.—Provide a tube of very thin glass, about one-third of an inch wide, and 18 or 20 inches long, sealed at one end. Coat it, within about an inch of the sealed extremity, with a lute of sand and clay; and when this is dry, put into it as much purified phosphorus, in small pieces, as will fill the uncoated part. Then cover the phosphorus with carbonate of lime. Let the part of the tube, which contains the carbonate, be made red-hot by means of a portable furnace, or chaffing dish; and, at this moment, apply heat to the part containing the phosphorus, sufficient to melt and raise it into vapour. The vapour of the phosphorus, coming into contact with the red-hot carbonate, will decompose the carbonic acid; and *charcoal* will be found in the residue of the process, in the form of a very light and black powder.

In this experiment the carbonate of lime must undergo two disuniting processes before the charcoal can be procured, viz. the carbonic acid must be separated from the lime to which it has a certain affinity, and also the oxygen of the carbonic acid must be separated from the carbon which is its base. This is effected by the affinity of one part of the phosphorus for the oxygen of the carbonic acid, and by that of the other for the lime forming the phosphuret of lime.

> *Exp.* 71.—In a glass retort, filled with carbonic acid, heat a piece of the metal called *Potassium*; it soon takes fire and burns with a red light, and charcoal in fine powder will be deposited; upon examination the gas will have disappeared, and oxygen found added to the potassium.

480. According to the latest atomic researches into the composition of carbonic acid, it appears to consist of two atoms of oxygen (8+8) and one of carbon (6), its representative number, or atomic weight will therefore be 22.

CARBONIC OXIDE.

481. By the distillation of zinc filings with chalk, we obtain a gaseous compound of carbon and oxygen, in which an atom of each exists in combination; its representative number is accordingly (8+6) 14. It possesses none of the essential characters of carbonic acid; it is lighter than common air, and its base so far predominates as to confer upon it the property of inflammability. It is extremely noxious to animals; and fatal to them if confined in it. When respired for a few minutes it produces giddiness and fainting.

CARBURETTED HYDROGEN GAS.

482. Hydrogen and carbon unite in two different proportions, giving rise to two distinct and well characterised compounds. The first of these, consisting of one atom of charcoal, and two atoms of hydrogen $(6 + 2 = 8)$, is simply called *Carburetted Hydrogen gas*. It has been also distinguished by the names of *Heavy Inflammable Air*; *Gas of Marshes*; *Hydrocarburet*; &c.

483. It may be obtained, mixed however with about $\frac{1}{20}$ of carbonic acid and $\frac{1}{15}$ or $\frac{1}{20}$ of azote, by stirring the bottom of almost any stagnant pool of water, especially if formed of clay. When this is done by an assistant, the gas is copiously disengaged in bubbles, which may be received in the usual manner. It should be washed, when collected, with lime water. This gas exists also as an ingredient in the coal gas, used for the purposes of illumination, from which it may be easily separated.

484. It burns with a bright yellowish flame. It has no taste, and but little odour. Its specific gravity, in its purest form, is to that of hydrogen as rather less than 8 to 1. 100 cubic inches weigh about 17 grains,

BI-CARBURETTED HYDROGEN GAS.

485. This gas, as its name imports, contains a double proportion of carbon. It consists of one atom of carbon and one atom of hydrogen; its atomic weight, or representative number is therefore $(6 + 1)$ 7.

486. For the purposes of experiment it may be obtained by distilling in a glass retort, with a gentle heat,

three measures of concentrated sulphuric acid, and one measure of alcohol. The mixture soon assumes a black colour from a carbonaceous deposit (471), and thick consistence, and a gas is disengaged which may be collected over water, and freed from carbonic acid by washing it with liquid potass.

487. When kindled it burns with a beautiful white flame of intense splendor. Its specific gravity is to that of hydrogen nearly as 13 to 1 ; 100 cubic inches of it weigh between 29 and 30 grains. The most remarkable character of this body is its action on a certain gas to be hereafter described, called *Chlorine.* When mixed with an equal volume of it, a mutual condensation takes place, and a peculiar fluid is formed, which has been supposed to be an *oil*; but which is a *peculiar compound*, not soluble in water, and composed of hydrogen, carbon, and chlorine.* In consequence of the effect thus produced Bi-carburetted Hydrogen has been long known by the name of *Olefiant Gas.*

488. Carburetted hydrogen gas is particularly fatal to animal life. Dr. Beddoes made many experiments upon the subject, from which it would seem to destroy life by rendering the muscular fibre inirritable without producing any previous excitement. In order to decide this question, Sir H. Davy ventured to take three inspirations of the gas produced from the decomposition of water by charcoal, and he very nearly lost his life in the attempt.

489. The infinite variety of inflammable gases produced by the exposure of moistened charcoal, of alcohol, or æther, of oil, tallow, wax, or coal, to a heat a little above ignition, and which have been regarded as

---

* In some respects this compound resembles æther, and Dr. Thomson has accordingly given it the name of *Chloric Æther.*

indefinite compounds of carbon and hydrogen, are now very satisfactorily shewn by Dr. Henry to be merely mixtures of the carburetted and bi-carburetted hydrogen, with occasionally a proportion of carbonic oxide.

THE ATMOSPHERE.

490. The Atmosphere may be defined an airy ocean surrounding the globe to an altitude of from 40 to 45 miles; producing by its pressure various mechanical, and by its chemical composition, different chemical effects. It will necessarily contain all those substances which are capable of existing in the aeriform state, at the medium temperature of the globe, and which are disengaged with greater or less abundance at its surface. These bodies, however, are to be considered as merely adventitious; they rarely exist in any considerable proportion, are only occasionally produced, and are very quickly removed by various natural processes. In some cases, they are not even discernible by the nicest chemical tests, and their presence is only inferred from their effects upon animal life. It is to the permanently elastic fluid, which constitutes the great body of the atmosphere, that we assign the name of *Atmospheric Air*, and which Chemistry has shewn to possess a uniform composition, at whatever altitude it may have been collected, or in whatever quarter of the globe, whether in cities, or in the country, on sea or on land.

491. To the Physiologist, as well as the Chemist, its *mechanical* properties are not less interesting than those which relate to its composition. Many of the functions of the living body are modified by their influence, and admit of elucidation from the doctrines of Pneumatics. If ignorant of the air's pressure and elas-

ticity, we shall be unable to comprehend the mechanism of respiration, and the effects produced on it by changes of altitude; nor could the Chemist understand the theory of those operations by which he is enabled to collect and transfer gases, without a knowledge of the mechanical properties of air.

492. Atmospheric air, although invisible, is material, and partakes of all the common properties of matter, for it occupies space, attracts and is attracted, and consequently has weight. It likewise partakes of the nature of a fluid, for it adapts itself to the form of the vessel in which it is contained, and presses equally in all directions.

493. Since air has weight, and every thing upon the Earth is surrounded and enveloped by it, it follows that all animate and inanimate bodies must be subject to its pressure, which will be exerted not only upon them, but upon itself; and since air is elastic, or capable of yielding to pressure, so of course the lower part of the atmosphere will be more dense, or in a greater state of compression, than that which is above it. To render this proposition more intelligible, let us suppose that the whole weight of the atmosphere is divided into 100 parts, and that each of these may weigh an ounce, then the Earth and all things upon its surface will be pressed with the whole 100 ounces; the lowest stratum will be pressed by the 99 ounces above it; the next by 98, and so on, until we arrive at the 99th stratum from the bottom, which will of course be subject to no more than one ounce of pressure, or the weight of the last or highest stratum. It is evident therefore that, as a body ascends, the pressure will be diminished, while the air, for the same reason, will become more and more rare, until it passes into a state of inconceivable tenuity.

This rarefaction, however, is not indefinite,† for it must cease as soon as the force of gravity downwards, upon a single particle, becomes equal to the resistance arising from the repulsive force of the medium. This simple statement will at once explain all the feelings experienced by persons who have ascended into the atmosphere; while the relations which have been already stated to exist between rarefaction and temperature will account for the cold experienced in lofty regions.

494. The pressure of the atmosphere at the level of the ocean is adequate to sustain a column of water having the altitude of 35 feet, or one of mercury of the height of 30 inches, which is about equal to 15 pounds, avoirdupoise, on every square inch of surface, so that the body of a man of ordinary stature sustains a pressure of 324,000 lbs.; yet, since the spring of the air contained within the body exactly balances, or counteracts the pressure from without, he is perfectly insensible of the existence of any pressure at all. The spring and pressure will thus in all cases balance each other, ·except the communication be cut off, and the natural equilibrium destroyed by some disturbing cause. This is effected by an instrument termed the *Air Pump*, of which we avail ourselves for the demonstration of those views, which must otherwise have remained purely theoretical. By means of this machine the air may be taken from the interior of vessels, when the effects of the external and undisturbed air will immediately display themselves.

495. The air of our atmosphere, besides small proportions of aqueous vapour, and carbonic acid, consists of two different gases, viz. *Oxygen*, which has been already considered (441), and which appears to

---

† Farther proofs of this fact have been stated at page 17, *note.*

be the principal ingredient on which the chemical effects of the air depend, and *Azote*, a gas which it will
be here necessary to introduce to the notice of the
student, before we can proceed any farther with the
history of the atmosphere.

Azote, or Nitrogen.

496. This gaseous element was discovered by Dr.
Rutherford in 1772. It constitutes four-fifths of the
atmosphere, and exists in combination with various
bodies to be hereafter described. It enters largely into
the organization of animal bodies, and is occasionally
present in vegetables, to which it imparts a peculiar
character. In consequence of its unfitness for supporting animal life, Lavoisier gave it a name, derived from
the Greek primitive *α non*, and *ζωη vita*. This, however, as being merely a negative property, has since
been deemed an improper foundation for its nomenclature; and the term *Nitrogen Gas* has been substituted,
because one of the most important properties of its base
is, that by union with oxygen it composes *Nitric Acid*.

497. It may be easily procured by extracting oxygen
from common air. If for instance mercury be heated
in contact with a certain portion of atmospheric air,
it will combine with the oxygen, and the residue will
be the gas under consideration; or, if phosphorus
be inflamed in a tube half filled with air, an elastic
fluid will remain after the combustion, which will be
found to be nearly pure *Azote*. This gas may be also
obtained by dissolving animal matters, such as glue or
muscular fibre, in diluted *aqua fortis*, or fuming nitrous
acid mixed with ten or twelve times its weight of water,
and subjecting the mixture to a heat of about 100°,

when the gas will be disengaged, and may be collected over water.

498. The properties which distinguish this gas are all of a negative kind. It will neither support combustion, nor animal life. It is rather lighter than atmospheric air; 100 cubic inches, according to Sir H. Davy, weighing 29·6 grains.

CHEMICAL COMPOSITION OF ATMOSPHERIC AIR.

499. That the air consists of oxygen and azote in the proportion of 21 of the former and 79 of the latter, by measure, admits of a synthetic as well as analytic proof, for if we mix these two gases in the proportions above stated we shall produce a mixture resembling atmospheric air in all its properties. Of this any one may be satisfied, by mixing four parts of azotic gas with one of oxygen gas, and immersing, in the mixture, a lighted taper, which will burn as in common air.

500. As the most accurate experiments have proved that no appreciable difference exists between the proportions of oxygen and azote in the atmosphere of different places, it is evident that the purity and salubrity of air must depend on some other circumstances than the proportion of these its chief ingredients. It is also evident that, as the oxygen of the air is constantly consumed by combustion, respiration, and various other operations, there must be some processes in Nature, by which a quantity of oxygen is produced equal to that consumed. One principal cause of this renovation appears to be in the process of vegetation; healthy plants exposed in the sunshine to air, containing small quantities of carbonic acid gas, destroy that elastic fluid and evolve oxygen gas; so that the two great classes of

organized beings are thus dependent upon each other.
Carbonic acid gas, which is formed in many processes
of combustion, as well as in respiration, if not removed
from air, by its excess would be deleterious to animals,
but it is a healthy food of vegetables; and vegetables
produce oxygen which is necessary to the existence of
animals, and thus is the œconomy of Nature preserved
by the very functions to which it is subservient.

501. It may seem extraordinary that the gases of
which the atmosphere is composed should be found so
uniformly intermixed. It might be supposed that they
would separate, each according to its specific gravity.
Such an effect, however, is obviated by a universal law
to which all gaseous bodies appear subservient. It has
been shewn by Mr. Dalton, and M. Berthollet, that
different elastic fluids have a tendency to rapid equable
mixture, even when at rest, and exposed to each other
on small surfaces only; and the mixture of the parts
of the atmosphere is constantly assisted by winds, by
currents of air, and by all the motions taking place on
the surface of the Earth. This explanation will supersede the necessity of regarding atmospheric air as a
chemical compound, and that its constituents are held
together by affinity, a theory which is encumbered with
many difficulties.

502. In addition to the two principal ingredients
of Atmospheric air, *Oxygen* and *Azote,* there is another, existing in a very small proportion, but which
from its constancy and uniformity cannot be regarded
as adventitious. This is *Carbonic acid,* which has not
only been discovered in air at ordinary heights, but
was ascertained by Saussure to exist in the atmosphere
of Mont Blanc, nearly 16,000 feet above the level of
the sea, and was found by Humboldt in air brought
down by Garnerin, the celebrated aeronaut, from the

height of several thousand feet. The proportion is estimated by Mr. Dalton not to exceed $\frac{1}{10000}$th, or $\frac{1}{14000}$th of its bulk, and the experiments of Saussure, junior, make it still less. It has been also discovered to be more abundant in summer than winter. Its presence in the atmosphere at all seasons and places is demonstrated by leaving a shallow vessel of lime-water exposed to the atmosphere, when its surface is soon covered with a solid pellicle, which, when removed, is succeeded by another, and so on, till the water is deprived of almost all the lime which it held in solution. From the precipitate, thus formed, carbonic acid is disengaged by dilute acids.

503. There have been several substances proposed for ascertaining with facility the quantity of oxygen in air; they have been called *Eudiometrical* substances; and the instruments in which they have been employed are named *Eudiometers*. The value of such investigations must be obvious, when it is considered how important an influence the purity of the air we breathe exerts upon the animal œconomy. Every medical practitioner ought to possess a sufficient degree of chemical knowledge to conduct such enquiries. What an accession of valuable information might we have possessed, had the Navy Surgeon included in his reports, the state of the air in different parts of the ship, during the prevalence of various epidemics?

504. For the purpose of Eudiometry various processes have been recommended. If, for instance, a stick of phosphorus be confined in a portion of atmospheric air, it will slowly absorb the oxygen present, without any visible combustion; and in six or eight hours its effect is completed. The residuary azotic gas has its bulk enlarged about $\frac{1}{40}$th, by absorbing a little phosphorus; and for this, allowance must be made in

measuring the diminution. Seguin, with the same
view, recommended the rapid combustion of phosphorus. For this purpose a small piece of phosphorus may
be introduced into the bulb of the tube
a, containing a given measure of the air
to be examined, confined over mercury,
which, to prevent loss by expansion,
should be suffered to occupy about half
the tube, or to stand at *b*. The phosphorus may then be inflamed in the
tube; and when the combustion is over,
and the tube cold, the residuary air may
be transferred for measurement into a
small cylindrical jar graduated into minute aliquot
parts. In this instance about $\frac{1}{18}$th the volume of the
residuary gas is to be deducted from the apparent
quantity of azotic gas, because, in this case also, a
small portion of phosphorus is dissolved by the latter,
and occasions a trifling expansion. Volta had recourse
to the accension of hydrogen gas as a test for the purity
of atmospheric air. For this purpose, two measures of
hydrogen are introduced into a graduated tube, with
three of the air to be examined, and fired by the electric
spark; the diminution of bulk, observed after the vessel had returned to its original temperature, divided by
three, gives the quantity of oxygen consumed. This
eudiometric test has lately received great improvement
from the discovery of the singular fact, that precipitated
platinum will, by virtue of a disposing affinity, enable
the oxygen and hydrogen gases to combine silently,
and thereby to supersede the necessity of an electric
spark; it moreover possesses a great advantage over
this latter mode of combustion, as it enables these
gases to unite, however small the proportion of hydrogen may be, whereas the electric spark will not fire a

mixture in which the oxygen is in considerable excess.
In performing this experiment we have only to mix the
hydrogen gas with the air to be examined, in the proportions above directed, and then to introduce into the
jar a pellet composed of the *spongy platinum*, and
allow them to remain in contact. Scheele in his
eudiometrical researches, employed liquid *Sulphuret
of Potass*, a substance which possesses the property of
absorbing oxygen, but not azote. It therefore acts on
atmospheric air, only as long as any oxygen gas remains,
and may be employed as the means of ascertaining the
quantity of this gas in any portion of air. For applying this test, Dr. Hope of Edinburgh invented an instrument, which is represnted in the margin (B). It

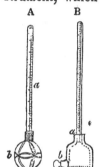

A   B   consists of a small bottle, holding
about three ounces, into which the
graduated glass tube *a* is carefully
fitted by grinding. It also has a
ground stopper at *b*. To use it, the
phial is filled with the solution, and
the tube *a*, containing the air to be
examined, fitted into its place. On
inverting the instrument, the gas ascends into the bottle, where it is to
be brought extensively into contact with the liquid by
brisk agitation;—an absorption ensues; and to supply
its place, the stopper *b* is opened under water, a quantity of which rushes into the bottle. The stopper is
replaced under water; the agitation repeated; and
these operations are renewed alternately, till no farther
diminution takes place. The tube *a* is then withdrawn,
the neck of the bottle being under water, and is held
inverted in water for a few minutes; at the close of
which the diminution will be apparent, and its amount
may be measured by the graduated scale engraved on

the tube. To this form of apparatus, however, Dr. Henry has offered the following objections. If the tube *a*, and the stopper *b*, are not both very accurately ground, the air is apt to make its way into the instrument to supply the partial vacuum, occasioned by the absorption of oxygen gas. This absorption causes a diminished pressure within the bottle; and, consequently, towards the close of each agitation, the absorption goes on very slowly. Besides, the eudiometric liquid is constantly becoming more dilute, by the admission of water through *b*. To obviate all these difficulties, Dr. Henry substitutes for the glass bottle, one of elastic gum, as shewn in the above marginal cut A.

505. There is a peculiar gas, to be hereafter described, termed *Nitrous gas*, or *Nitric oxide gas*, which has the property of combining with oxygen, and forming nitrous acid, which is immediately absorbed, if the mixture be made over water. In consequence of this effect, nitrous gas was originally applied by Dr. Priestley to the purposes of eudiometry. Many objections, however, have been urged against it, and it has been supposed that the sources of error in its employment are such as to forbid our relying implicitly on the results which it may afford. The researches of Mr. Dalton and Gay Lussac have, notwithstanding, revived our confidence in its efficacy, and it appears to be susceptible of perfect accuracy, provided certain precautions be observed. In mixing the gases a narrow tube should be avoided, and a wide vessel, such as a tumbler glass, be employed, and to 100 parts of atmospheric air, previously measured, we must add at once 100 measures of nitrous gas. A red fume will appear, which will soon be absorbed without agitation, and in half a minute, or at most a minute, the absorption will be complete. The residuum must be then passed into a gra-

duated tube, and if the air has been pure, it will be found that 84 measures have disappeared. This number must be divided by 4, which will give us the quantity of oxygen condensed, viz. 21. Sir H. Davy has proposed the use of a solution of sulphate of iron, impregnated with nitric oxide gas, for the absorption of oxygen; in which case, the apparatus of Dr. Hope will afford a convenient mode of applying it.

506. The proportion of carbonic acid present in the atmosphere may be ascertained by agitating it in contact with a solution of potass, and noting the degree of absorption.

507. In our investigations into the quality of air, we should also ascertain its power of supporting combustion.

508. We come now to consider the nature of those adventitious bodies which are present, in variable proportions, in the atmosphere, and upon which its degree of salubrity most probably depends. Water, even during the driest weather, is always present in the atmosphere, although its proportion is constantly fluctuating. Various saline bodies when exposed to the air become moist, or *deliquesce*, which arises from their attracting the water present in the atmosphere. Saussure states the quantity of water in a cubic foot of air, charged with moisture at 65° *Fah.* to be 11 grains. The quantity of water that may be extracted from 100 cubical inches of air, at 57° is 0·35 of a grain; but, according to Clement and Desormes, at 54° *Fah.* only 0·236 of a grain can be detached by exposure to muriate of lime. The experiments, both of these chemists and of Mr. Dalton, concur in proving that, at the same temperature, equal bulks of all the different gases give up the same quantity of water to deliquescent salts. The portion of water which they thus abandon, has

U

been called *Hygrometric water*. Whether they contain a still farther quantity in a state of more intimate
union, and not separable by deliquescent substances,
is still undetermined.

509. There are many bodies, which can scarcely
be considered as possessing any affinity for water, and
yet greedily absorb it from the atmosphere. Such are
almost all substances in the state of powder; porous
paper; soils which have been artificially dried; parched
oatmeal; and even the filings of metals. These bodies
are said to be *Hygrometric*.

510. Much yet remains to be discovered respecting
the manner in which water is suspended in the atmosphere. An attempt has been made to estimate the degree of moisture of the air, by ascertaining what has
been termed the *dew point*, that is, the temperature at
which moisture is precipitated.* Besides this method
a variety of instruments have been also constructed for
the same purpose. They are called *Hygrometers;*
the most common of which consist of some substance,
such as human hair, or a fine slip of whalebone, which
is elongated by a moist atmosphere, and shortened by
a dry one.† The extreme points are altered by placing

---

* For a farther explanation of this subject, as well as for other
information respecting the habitudes of the Atmosphere, the
reader is referred to Mr. Daniel's Essays on Meteorology.

† Many vegetable productions are calculated to afford the
same indications; thus, the capsule of the Geranium, and the beard
of wild oats, if fixed upon a stand, serve the purpose of an hygro-
meter, twisting itself more or less according to the moisture of
the air. The awn of Barley is furnished with stiff points, which,
like the teeth of a saw, are all turned towards one end of it; as
this long awn lies upon the ground, it extends itself in the moist air
of night, and pushes forward the barley corn, which it adheres to;
in the day it shortens as it dries; and as these points prevent it
from receding, it draws up its pointed end; and thus, creeping

it, first in air artificially dried, and afterwards in air rendered as humid as possible. The degree of expansion or contraction is rendered more sensible by connecting it with an axis, which moves a circular index, like the finger of a clock. Mr. Leslie and Mr. Daniel have proposed more refined instruments, founded upon the fact, that evaporation proceeds with a rapidity equal to the dryness of the air; and since the production of cold is proportional to the rate of evaporation, the thermometer may be thus made subservient to the purposes of Hygrometry.

511. The human body is greatly influenced by the degree of aqueous vapour present in the atmosphere, and few subjects connected with Meteorology are more interesting to the medical philosopher. How far the origin of various Epidemics may be connected with it, future observations may probably discover. Increased humidity is ever attended with the sensation of cold, because the air is thus rendered a better conductor of caloric; while, at the same time it cheeks the perspiration, since the atmosphere, when in a state of saturation with water, is incapable of carrying off the insensible

---

like a worm, will travel many feet from the parent stem. Upon this principle Mr. Edgeworth once made a wooden automaton; its back consisted of soft Fir-wood, about an inch square, and four feet long, made of pieces cut the cross-way in respect to the fibres of the wood, and glued together; it had two feet before, and two behind, which supported the back horizontaily; but were placed with their extremities, which were armed with sharp points of iron, bending backwards. Hence, in moist weather the back lengthened, and the two foremost feet were pushed forward; in dry weather the hinder feet were drawn after, as the obliquity of the points of the feet prevented it from receding. And thus in a month or two, it walked across the room which it inhabited.

perspiration as it is formed.* For the same reason, the
watery exhalation from the lungs is diminished, and
various morbid effects may be thus produced. The
subject has not hitherto received a share of attention
commensurate with its importance. The investigation
might not only lead to an explanation of many phænomena which are at present unintelligible, but to the
adaptation of an artificial atmosphere for the cure of
disease.† It, moreover, deserves notice that a humid
atmosphere becomes a more powerful solvent of vegetable and animal substances. Numerous examples
might be adduced to shew that volatile bodies are
sooner converted into a gaseous state under such circumstances. It is well known to lime-burners that the
limestone is burnt and reduced to quick-lime much
sooner in moist than in dry weather; and, indeed, in
the latter case they not unfrequently place a pan of
water in the ash-pit, the vapour of which materially
assists in carrying off the carbonic acid   In like manner, camphor is found to volatilize with much greater

---

* Under these circumstances the perspirable matter is condensed upon the surface, hence we appear upon these occasions
to perspire greatly upon the slightest exercise, whereas the cuti-
cular discharge is at such times absolutely less ; and since it must
cease to be a cooling process we experience a sensation of heat
greater than the state of the thermometer will explain. The extreme dryness of the atmosphere of Chili will furnish us with a
number of facts in farther proof of the justness of these views.
Dr. Schmidtmeyer informs us that in the climate alluded to, notwithstanding the very high temperature, the perspiration passes
off so entirely in the insensible form, that, during the most violent
exercise, it mght be doubted whether there existed any perspiration at all.

† I have offered some farther observations upon this subject in
the 6th Edition of my Pharmacologia, Vol. 1. p. 197—198.

celerity in damp situations. Every body has noticed how sensible the perfume of flowers becomes during the fall of the evening dew, or in the morning, when the dew evaporates and is dissipated by the rays of the rising sun. For the same reason the stench of putrid ditches, and common sewers, is conveyed to the organs of smell much more speedily in summer previous to rain, when the air becomes charged with moisture. We cannot therefore be surprised to find that heat and moisture have ever proved favourable to the origin and propagation of Epidemic diseases. The most subtle of all poisons,—*the matter of contagion*,—is undoubtedly modified in activity by the degree of moisture in the atmosphere, influencing its solubility and volatility. On the other hand, it may be stated that the *Harmatton*, a wind experienced on the western coast of Africa, between the Equator and 15 degrees North Latitude, blowing from north-east towards the Atlantic, and which, in consequence of its passage over a very extensive space of arid land, is necessarily characterised by excessive dryness, puts an end to all Epidemics, as the Small Pox, &c.; and it is even said that, at such a time, infection does not appear to be easily communicable by art.

512. Eudiometry, in its present state, is too imperfect to enable us to appreciate the presence of various animal and vegetable substances which are so generally present in the lower regions of our atmosphere; and by which the salubrity of the air appears to be materially affected. All living bodies, when crowded together, generate a peculiar matter which would seem to be highly destructive. No species of animal can congregate in ill ventilated apartments with impunity. Under such circumstances, the horse becomes infected with the *Glanders*, fowls with the *Pip*

or *Pep*,* and sheep with a disease peculiar to them, if they be too closely folded. The same fact occurs in the vegetable kingdom, and Nature would seem to have established this law, in order that the extent of her productions might be limited to those bounds, which were essential to the well being of the whole. The unhealthiness of crowded cities must arise from a similar cause, and, although the chemist is unable to detect the deleterious principle, it would be vain to deny its existence.

513. It has been stated that a certain portion of carbonic acid is present in the purest air; but in cities, and apartments in which persons have breathed, or combustible matter has been burnt, this proportion is exceeded, and its excess must be regarded as injurious. *Sulphurous acid* may also be occasionally present in an atmosphere which has been impregnated with the vapour of burning coal; the air of London contains it,† and to its presence we may probably attribute the well know fact of iron oxidizing with such rapidity. A quantity of carbon, in a state of extremely minute division, is also diffused through the air of our city, which cannot but prove injurious to the health of the inhabitants. The presence of such adventitious bodies is amply sufficient to explain the effects which are

---

* It is worthy of remark that these diseases, evidently engendered by congregation, become subsequently contagious.

† The ordinary mode of collecting the air of a place, for eudiometric examination, is to empty a bottle of water, and to cork the vessel carefully; but in this case, the Sulphurous acid, or any soluble matter will be washed away. It is therefore preferable to carry an empty bottle, perfectly dry, to the spot, and to dislodge the common air by the blast from a pair of bellows, and then, after the interval of a minute or two, to close it securely.

known to attend a constant residence in the metropolis,† as well as to account for the feelings of relief which its inhabitants experience from occasional migration.

514. We may now examine the function of breathing, and the changes which the air undergoes in its passage through the lungs. The construction of the human organs of respiration, and the mechanism by which the thorax is alternately enlarged and contracted, exclusively fall within the province of the Anatomist. The Chemist contributes his assistance by demonstrating the composition of the atmosphere, and the alteration which takes place in it during the act of respiration. He also shews the changes which the blood suffers from its exposure to the air's influence, and leaves the physiologist to draw his conclusions from the data with which he is thus enabled to furnish him.

---

† The unhealthy appearance of plants growing in the metropolis affords, in itself, a sufficient indication of the air's impurity. Evelyn has the following curious remark upon this subject. " That the smoake destroys our vegetation is shewn by that which was by many observed in the year 1644, when Newcastle was besieged, and blocked up in our late wars, so as through the great dearth and scarcity of coals, these fumous works were either left off or diminished, divers gardens and orchards planted even in the very heart of London, (as in particular my Lord Marquis of Hertford in the Strand; my Lord Bridgewater's, and some others about Barbican,) were observed to bear such plentiful and infinite quantities of fruits, as they never produced the like before, or since, to their great astonishment; but it was by the owners rightly imputed to the penury of coales, and the little smoake which they took notice to infest them that yeare." As the impurity of the air constitutes an interesting subject of Medical Police, I have treated it more fully in my work on MEDICAL JURISPRUDENCE, to which I must refer the reader.

515. All animals, however different in their structure and functions, appear to agree in possessing organs appropriated to the purpose of respiration, and in producing the same species of change on the air which they breathe. In man, quadrupeds, birds, and the amphibia, they consist of a cavity, with the necessary appendages, which alternately receives and emits a portion of the air of the atmosphere. Fish, in consequence of the medium in which they are immersed, perform this function by a different apparatus; they are furnished with a passage communicating with the fauces or œsophagus, and terminating in the external surface of the body, through which a part of the water received into the mouth is forcibly propelled. In this passage, the *branchiæ*, or gills, are situated, and the blood which circulates in their fringed extremities, is thus exposed to the action of a quantity of air, which the water always holds in solution. This is rendered evident from the fact that fish die in a state of suffocation, if the water in which they are contained, be deprived of its air by the action of the air pump. In many of the Insects and less perfect animals, the respiratory organs consist merely of a number of tubes or pores, provided with open mouths, and which simply admit the external air to be received into them; whence, in order to suffocate such animals, we have only to cover their bodies with oil.

516. Oxygen gas appears to be the only part of the atmosphere which is essential to the maintenance of respiration and life. No gas, with which we are acquainted, can be substituted for it, and an animal, enclosed in a definite portion of air, lives a longer or shorter period, according to the proportion of oxygen present (446). When an animal, however, has expired in a quantity of atmospheric air, in consequence

of its being no longer fit for respiration, it is still found that the whole of the oxygen is not removed from it,* but in this case the death of the animal is to be rather attributed to the presence of carbonic acid, than to a deficiency of oxygen, for it has been shewn by Lavoisier that where the expired air was exposed to a substance capable of combining with this deleterious body, the life of the animal was protracted. The azote does not appear to be absorbed in any notable degree, but remains passive, and is received into, and emitted from the lungs without undergoing any change; its only use in the atmosphere would seem to be for the dilution of the oxygen. Such, at least, is the more probable opinion, although some philosophers have concluded that there is an absorption of azote as well as of oxygen.

517. It has been a problem from the earliest ages to ascertain the capacity of the lungs, and the quantity of air received and emitted by them. Many calculations were made upon this subject by the older writers, but as they were necessarily deficient in those data which can be alone derived from the most refined experiments, we cannot be surprised that their conclutions should have been so inconsistent with each other, and so remote from truth. There are moreover, several circumstances which render the enquiry extremely difficult. Owing to a difference in stature, and to the peculiar conformation of the thorax, great varieties prevail in the quantities of air respired by different persons; and in the same person it is well known, that

---

* This proposition, however, is subject to exceptions; M. Vauquelin found that some of the species of worms possessed the power of separating the oxygen from the azote in the most perfect manner.

the respiration is materially influenced by the degree of muscular exertion, the state of the stomach, the mental impressions, and the powers of volition; all, therefore, that can be effected by experiment, is to ascertain the average quantity of air respired in that state of the body, when it is least under the influence of external agents.

518. Inspiration may be said to be performed with three different degrees of energy, 1st, *Ordinary* inspiration, which takes place by the depression of the diaphragm, and an almost insensible elevation of the thorax; 2ndly, the *Great* inspiration, in which there is an evident elevation of the thorax, and, at the same time, a depression of the diaphragm; 3rdly, *Forced* inspiration, in which the dimensions of the thorax are augmented in every direction, as far as the physical disposition of this cavity will permit the contraction of the thorax, or *Expiration*, presents also three degrees, viz. 1st, *Ordinary* expiration, 2nd, *Great* expiration, and 3rdly, *Forced* expiration.

519. According to Menzies, the mean quantity of air that enters the lungs, at each inspiration, is 40 cubic inches.—Goodwin thinks that the quantity remaning after a complete expiration is 109 cubic inches; Menzies affirms that this quantity is greater, and that it amounts to 179 cubic inches. According to Sir H. Davy, after a *forced* expiration, his lungs contained 41 cubic inches.

After a natural expiration............ 118
After a natural inspiration 135
After a forced inspiration............ 254
By a forced expiration, after a forced inspiration, there passed out of the lungs 190
Do. after a natural inspiration......... 78·5
Do. after a natural expiration.......... 67·5

Dr. Thomson thinks that we should not be far from the truth in supposing that the ordinary quantity of air contained in the lungs is 280, and that there enter or go out at each inspiration or expiration 40 inches. Thus, supposing 20 * inspirations in a minute, the quantity of air that would enter and pass out in this time would be 800 inches; which make 48·000 in the hour, and in 24 hours 1,152,000 cubic inches.

520. It will be evident from the foregoing statement that the portion of air expired is not exactly that which was inspired immediately before, but a portion of the mass which the lungs contained after expiration; and if the volume of air that the lungs usually contain is compared with that which is inspired and expired at each motion of respiration, we must suppose that inspiration and expiration are intended to renew in part the considerable mass of air contained in the lungs. This renewal will be so much the more considerable as the quantity of air expired is greater, and as the following inspiration is more complete. In cases of extreme fatigue, the ordinary inspirations are probably less ample than necessary, and hence we are under the necessity of occasionally interposing a forced inspiration to supply the deficiency. The same fact will explain the deep sighs of those affected by reverie.

521. The nature of atmospheric air having been already explained, we may now consider the physical and chemical changes which it undergoes during its passage through the lungs. In its exit from these organs its temperature is nearly the same as that of the

---

* We fairly assume this as the mean. Haller thinks there are 20 in the space of a minute. A man, upon whom Menzies made experiments, required only 14 times in that interval. Sir H. Davy informs us that he required 26 or 27 times; Dr. Thomson says that he requires generally 19 times. Majendie only 15 times.

body; and it appears from the experiments of Messrs. Allen and Pepys, that the same quantity is given out at each expiration, as was inhaled by the previous inspiration; its chemical composition, however, undergoes material change, it returns charged with aqueous vapour, called *pulmonary transpiration*, which is evidently alone derived from the secretion lining the bronchia and air vesicles, although Lavoisier imagined that it was generated by the union of the oxygen of the air with the hydrogen given off from the blood. Different authors have attempted to estimate the quantity of water thus produced; the celebrated Sanctorius, who devoted so large a portion of his life to the investigation of the quantity of matter perspired from the body, under different circumstances, estimated the pulmonary exhalation, at about half a pound in 24 hours; Dr. Hales supposes the quantity to be about 20 ounces. Dr. Menzies, about six ounces, and Mr. Abernethy, nine ounces, or three grains in a minute. The fact is, that the quantity must not only vary considerably in every individual, but with every hygrometric change in the air (510.)

522. Atmospheric air, after having been once admitted into the lungs, returns charged with 8 or 8·5 *per cent.* of carbonic acid.

> *Exp.* 72.—Blow the air from the lungs, with the aid of a quill, or the stem of a tobacco-pipe, through a glass of lime water, the liquid will become milky, and deposit *carbonate* of lime.

If the same portion of air be breathed repeatedly, considerable uneasiness is experienced; but the quantity of carbonic acid cannot be increased beyond 10 *per cent.* The proportion, however, which is produced by the same individual, is liable to some variations which have been described by Dr. Prout (Ann. of Phil. xiii. 269.) In examining the changes which the constituent parts of the air have undergone we shall find that the proportion of azote is much the same, but that a quantity of oxygen has disappeared, equal in volume to that of the carbonic acid formed; and since this acid has been shewn to contain exactly its own bulk of oxygen gas, it is very evident that all the oxygen which disappears in respiration, must have been expended in forming this acid; and that no portion of it can have united with hydrogen, as Lavoisier supposed, to form water. But in this place the important question arises as to the source of carbonic acid; whether the oxygen be absorbed through the coats of the vessels, and thus displaces an equal quantity of ready formed carbonic acid, which may be supposed to have pre-existed in the blood; or, whether this acid be not rather generated by the union of the inspired oxygen with the carbon of that fluid? Lavoisier in his first memoir proposes these two hypotheses without deciding in favour of either of them; but in his later papers he adopts the latter, and the most enlightened chemists and physiologists of the present day have sanctioned the decision. The only change, then, that has been satisfactorily proved to take place in respired atmospherical air, is the removal of a certain quantity of oxygen (its nitrogen being wholly untouched) and the substitution of a precisely equal volume of carbonic acid gas.

523. The air in the lungs then must have acquired from the venous blood a quantity of carbon, and it therefore becomes a material question, whence this inflammable matter is derived. The chemical composition of the blood cannot be well understood until the student has been made acquainted with the different

animal principles which enter into its constitution.
At present it is only necessary to state that, at the
instant in which the venous blood traverses the small
vessels of the pulmonary lobules, it assumes a scarlet
colour; its odour becomes stronger, and its taste more
distinct; its temperature rises about one degree; a part
of its serum disappears, passing off in the state of vapour; and its tendency to coagulation augments. The
venous blood, having acquired these characters, now
becomes *arterial* blood.   That these changes are produced by the operation of the oxygen of the air is
manifest from the fact, that if there is any other gas in
the lungs, or even if the air is not suitably renewed,
the change of colour does not take place. We can
easily see the colouring of the blood, even in the dead
body.   The venous blood often accumulates in the
lungs before death, and the branchial vessels being
deprived of air, it preserves the venous properties long
after death; when, if atmospheric air be injected into
the trachea, so as to distend the lungs, the *brown red*
colour of the accumulated blood will be immediately
changed into *vermillion red*.   The same phœnomena
takes place whenever the venous blood is in contact
with oxygen, or atmospheric air. (*Exp.* 57.)

524. By exposing arterial blood to carbonic acid
gas, it immediately assumes the colour of venous blood.
The immediate cause of these changes is not understood; but we shall have occasion to recur to the subject in a future part of the work.   The origin of the
carbon given off is thus explained by Dr. Crawford;—
The solids of the living body have a constant tendency
to decay; their particles are continually changing, those
which are no longer fit for performing their functions
are removed, and discharged from the body, while new
ones are deposited in their room.   The arterial blood,

which is distributed to all parts of the body in the minute capillary vessels, is the vehicle by means of which this operation is performed; it conveys the nutritious matter to the different parts of the body, and deposits it in such a manner as to repair the necessary waste, while at the same time it receives the putrescent particles, which are become useless or noxious to the system, and carries them to the lungs, where they are united to oxygen, and discharged together with the air of expiration. It is to the addition of this extraneous matter that Dr. Crawford attributed the change from arterial to venous blood, and by the removal of it in the lungs, the blood, he imagined, is brought back to the arterial state. Lavoisier considered the products of Digestion as the immediate source of carbon, but this idea is at once controverted by the fact that the blood becomes completely venalized before it receives the contents of the thoracic duct. It must be admitted that the theory of Dr. Crawford is simple and ingenious, but it is encumbered by one insurmountable difficulty; it is contrary to all analogy to suppose that the arteries are the instruments by which useless matter is ejected from the system. The body is provided with a distinct set of vessels, which from their office are called the *Absorbents*, whose appropriate function it is to remove all superfluous matter, and these vessels have no communication with the sanguiferous system, except by the intervention of the thoracic duct, which receives all the substances absorbed, and pours them into the left subclevian vein. If it were necessary to add any farther strength to this objection it might be stated, that, on some occasions, the blood is converted into the venous state; while it still continues in the great trunks of the arteries, as where the blood has been stopped by the pressure of the tourniquet. But even

supposing, for a moment, that such an objection could
be removed, is the theory of respiration thus offered
satisfactory? can we imagine that Nature would have
established so important a function, and have so carefully guarded against the possibility of its intermission
only for a few moments, for the sole purpose of removing a small portion of excrementitious matter from
the system. The ends would be wholly unworthy of
the means. It is not impossible but that this may be
one of the beneficial effects produced, but there are
obviously many others of greater importance; the chyle,
for instance, is converted into blood by its operation;*
but even this must be an effect of secondary consequence, or the act of respiration might have admitted
of temporary cessation. When it is remembered that
this is the first and last † act of life; that it cannot be
suspended, even for a few seconds, without the extinc-
tion of vitality, we are inevitably led to conclude that
it must impart some principle too subtle to be long
retained in our vessels, and too important to be dis-
pensed with for the shortest period; what this principle
or energy may be, we shall probably never learn. The
contemplation, however, of the secondary and more

---

* Hence in diseased lungs, the body is not nourished, and the patient dies from atrophy. Unless some accident happens, such as
the rupture of a blood-vessel, or the bursting of a vomica, the
phthisical patient lives as long as the body contains the least portion of fat that can be absorbed for his support.

† *Breath* and *Life* are very properly considered in the Scrip-
tures as synonymous terms; and the same synonym, as far as we
know, prevails in every language. In the Greek, the most philo-
sophically constructed language with which we are acquainted,
this *first* and *last* act of life is expressed by a word composed of
Alpha and Omega, *αω*. In Latin the connection between breath
and life is displayed in the words *Spiro* and *Spiritus*.

obvious phænomena is within our grasp; already has the Physiologist and Pathologist derived important lights from the investigation, and the progress of Chemistry will, without doubt, extend and correct the views which have been thus formed, and suggest new applications of practical utility.

525. Whatever may be the principle communicated, or the medium through which it passes, muscular energy is undoubtedly in some manner connected with the function of respiration. Those animals which are furnished with the most perfect pneumatic apparatus enjoy the highest degree of muscular power. Comparative Anatomy will furnish us with many beautiful illustrations in support of this theory; birds are enabled to sustain the exertion of flight, in consequence of their extensive organs of respiration, and many insects,* in the act of flying, disclose avenues of air, which, in their quiet state, are closed by the cases of their wings, thus procuring for themselves a larger supply of the principle of muscular energy, at a period when from their exertion, and consequent exhaustion, they most require it; flat fish, who having no swimming bladder remain at the bottom, and possess but little velocity, have gills that are quite concealed, whilst those who encounter a rude and boisterous stream, as trout, perch, or salmon, have them widely expanded.† In men and quadrupeds the same law is to be discovered, the narrow chested

---

* This is particularly evident in the different species of *Scarabæus.*

† In speaking of the respiration of fishes, it may be observed that the sum of oxygen which they receive will vary, jointly, as the momentum of the water which imparts it, and the extent of the gills; an acquaintance with this truth will at once shew that the rapid current, instead of being inimical to the strength of the animal, must prove a powerful cause of its invigoration.

individual is always averse to labour, and indeed is
incapable of sustaining it.   In animals, the *animosum
pectus* is always an indication of strength.

526.   Some Physiologists have considered muscular
energy to be the result of a nice combination of oxygen
with the animal organs; this idea, however, is far too
chemical, and rests upon a set of entirely gratuitous
positions.   It was originally suggested by Girtanner,
and gained considerable popularity, in consequence of
the zeal with which it was adopted by Dr. Beddoes,
who applied it very extensively to Pathology, and made
it the foundation of some supposed improvements in
the practice of medicine.

527. Without inquiring farther into the cause of
the phænomenon, if it be admitted that a relation subsists between the function of respiration and muscular
contractility, it will follow, as a corollary, that a greater
or less demand will be made upon respiration, according to the degree of muscular exertion which an animal
may undergo.   This fact has been established by observation and experiment.   It has been found that the
quantity of air deteriorated by respiration, in a given
time, will vary with the degree of exertion made by the
animal confined in it.   Lavoisier states that a man,
under ordinary circumstances, consumes 1300 or 1400
cubic inches of oxygen in an hour, but he found that
if he is engaged in raising weights, the consumption is
at the rate of 3200 in the hour.   The practical inference
to be deduced from this fact is obvious, that where it
is an object to œconomise the oxygen, we should remain tranquil.   It was accordingly observed in the
black hole at Calcutta,* that those who were quiet and

* For an account of this awfully interesting affair, see my work
on Medical Jurisprudence, vol. 2, p. 50.

orderly, suffered the least. In like manner, a person who falls into the water in a state of syncope will remain a much greater time submerged with impunity, than one who is in a condition to exert his muscular energies. In my work on Medical Jurisprudence (Vol. 2. p. 39.) I have related a case, from Plater, of a young woman, who having been condemned to be drowned for infanticide, fainted at the moment she was plunged in the water, and having remained for a quarter of an hour under its surface, recovered after being drawn out.

528. The proportion of oxygen consumed by respiration appears moreover to be influenced by the nature of the diet. The quantity of carbonic acid emitted at each expiration, and consequently that of oxygen inspired, varies at different periods of the day, and probably also in different individuals; it appears at its maximum during digestion, and at its minimum in the morning, when the stomach is empty, and when no chyle is flowing into the blood. Dr. Prout has shewn that fermented liquors and vegetable diet diminish the proportion of carbonic acid, and that the same thing happens when the system is affected by mercury. The influence of vegetable diet in diminishing the call for oxygen has been very satisfactorily proved by the observations of Mr. Spalding, the celebrated diver, who found that whenever he used a diet of animal food he consumed in a much shorter time the oxygen of the atmospheric air in his diving-bell, and therefore he had learned from experience to confine himself to a vegetable diet. The same effect he found to arise from the use of fermented liquors, a fact in apparent contradiction with the observation of Dr. Prout, and he therefore drank nothing but water, while he followed his profession. The same fact is well known to the Indian pearl-divers, who always abstain for many hours before

their descent, from every species of food. We perceive therefore that in certain morbid states of the pulmonary organs a spare vegetable diet should be strictly enforced.

529. It is probable that in different diseases the function of respiration receives various modifications. An investigation of this subject would unquestionably afford some fresh light, but Pathology still requires this assistance from Chemistry.

530. The quantity of pulmonary transpiration is also influenced by various circumstances, especially the liquid nature of the food, and the quantity of fluids taken into the stomach. I have paid some attention to this circumstance, for it admits of some important applications to practice. In humoral Asthma solid food, with a diminished quantity of liquids, should be prescribed. The same regimen will be useful in excessively damp weather, when the air is incapable of carrying off the *halitus.**

531. Nor does the watery part of the blood alone escape by pulmonary transpiration; various substances which find their way into the circulating current very soon pass out by the lungs. To this fact many Expectorants are indebted for their efficacy, and I have assumed it as the basis of a new classification of these remedies.† *Garlic* and various fœtid gums, when introduced into the stomach, are soon absorbed, and being transported to the lungs, they pass into the bronchial vessels, and may be recognised by their odour, in the expired air. M. Majendie has made many experiments by injecting various substances into

---

* This subject has been more fully considered in my Pharmacologia, Edit. 6. vol. 1. p. 197, &c.

† Pharmacologia, vol. 1. p. 192.

the veins; and their presence has been afterwards discovered in the pulmonary *halitus.*

532. After the ingenious researches of Lavoisier, Laplace, and Dr. Crawford, animal heat was considered, exclusively, as the result of Respiration, but this opinion has been shaken to its foundation by the experiments of Mr. Brodie. Some modification, however, may perhaps be admitted that will reconcile these conflicting opinions; I shall therefore present to the student an outline of the Crawfordian theory, and then explain the objections to which it is exposed.

533. There is nothing more wonderful in the animal œconomy than the power which it possesses of maintaining a degree of heat independent of that by which it is surrounded. It has been shewn that an inert body, when placed amongst others, very soon assumes the same temperature, on account of the tendency of caloric to an equilibrium (281.) The body of man, however, is influenced by a different law; surrounded by bodies of a lower degree of heat than itself, it nevertheless maintains its superior temperature, and when placed in a medium hotter than itself, it resists its heating power as long as life continues. There must exist then, in the animal œconomy, two different and distinct properties,'the one of generating heat, the other of producing cold.

534. It had been observed, that in different animals, the superiority of their temperature to the surrounding air is greater according to the greater size of their lungs; thus the respiratory organs of birds, compared with their size, are more extensive than those of any other animal; and they have accordingly the greatest degree of animal heat, whence it had been inferred, that respiration is the source from which it proceeds; a conclusion which Dr. Black confirmed by observing.

the analogy between this process and combustion, in the changes which are effected in the air.

535. Dr. Crawford first fully explained this doctrine, and supported it by a series of most ingenious experiments. He had previously shewn, that, in general, when inflammable bodies combine with oxygen gas, a diminution of *capacity* (328) takes place, whence there must be an evolution of caloric; and that in the combustion of charcoal, when carbonic acid is formed from the combination of the oxygen with carbon, a large quantity of caloric is rendered sensible. It is evident that, since in respiration a similar consumption of oxygen and production of carbonic acid take place, there must be a similar evolution of caloric. He further ascertained by experiment, that the capacity of the blood changes when it passes from arterial to venous; and of course from venous to arterial; that of arterial being greater than that of venous blood, in the proportion of 1·030 to 0·892. On these facts Dr. Crawford founds his theory of animal heat. In respiration a quantity of oxygen is combined with carbon, or, as he supposes with *Hydro-carbon*; and carbonic acid is formed. Caloric must be evolved in consequence of this combination. But the blood is at the same time changed from venous to arterial, and by this change acquires a greater capacity for caloric. It therefore takes up the caloric which has been extricated by the combination; so that any rise of temperature in the lungs which would be incompatible with life is prevented. The arterial blood, in its circulation through the extreme vessels, passes again into the venous state, and in this conversion, its capacity for caloric is diminished, as much as it had been before increased, in the lungs; the caloric therefore, which had been absorbed is again given out; and this slow and constant

evolution of heat in the extreme vessel over the whole body is, according to Dr. Crawford, the source of its uniform temperature. Besides the experiments from which this theory was directly deduced, it has been fortified by others, in which the quantity of caloric rendered sensible by an animal is measured. Dr. Crawford, and Lavoisier and Laplace found, that when an animal is confined in a vessel, so contrived as to measure the quantity of caloric which it gives to the surrounding medium in a certain time, and the quantity of oxygen consumed by the animal in that time, this quantity of caloric corresponds nearly to that evolved from the combustion of carbonaceous matter, such as wax or oil, in the same quantity of oxygen.

536. It must be acknowledged that this theory has the merit of being altogether independent of any particular theory of respiration. Whatever may be the nature of the differences between venous and arterial blood, whether the latter contain oxygen or not, or whether the former hold in solution hydro-carbon, carbonic acid, or any other principle; it is proved by experiment, that the blood in these two states has different capacities for caloric; and on this fact, the explanation of the origin of animal temperature is made to depend.

537. The objection to this theory may be stated in a very few words.—Mr. Brodie (*Phil. Trans.* 1812,) found that the heart was capable of retaining its functions for some hours, and of carrying on the circulation, in a decapitated animal, and consequently independent of the influence of the brain, by keeping up an artificial respiration. Under these circumstances it was observed that, although the change of blood from the venous to the arterial state was perfect, no heat was generated, and that the animal cooled regularly and

gradually down to the atmospheric standard. In more than one instance, Mr. Brande examined, at his request, the expired air, and found that it contained as much carbonic acid as was produced by the healthy animal; so that in this case, the circulation went on, there was the change of oxygen into carbonic acid, and the alteration of colour in the blood, and yet *no heat whatever appeared to be generated.* But the nervous influence was withdrawn; that mysterious power which appears to preside over all the phænomena of the animal system, and to present an insuperable barrier to the completion of all our researches.

538. There remains to be considered another subject connected with animal temperature, which cannot possibly receive an explanation from the theory of Dr. Crawford,—the increased heat in certain local diseases, in which the temperature of the part exceeds, by several degrees,* that of the blood taken at the left auricle, so that the continual renewal of the arterial blood cannot be sufficient to account for this increase of heat. M. Majendie says that this second source of heat must belong to the nutritive phænomena which take place in the diseased part. It is true that most of the chemical combinations in the body produce elevations of temperature, and it cannot be doubted that both in the secretions and in the nutrition, combinations of this nature take place in the organs. I should, however,

---

* We have no accurate experiments to shew the extent to which this increase of temperature is carried. There is, however, a curious communication upon the subject, from Dr. Thomson, in the 2nd volume of the *Annals of Philosophy,* in which he attempts to shew the quantity of heat given out by a small inflamed gland in the groin; and it appears from his statement that, in the course of four days, it was sufficient to have raised seven pints of water at 40°, to the boiling point.

be more inclined to refer the phænomenon, in cases of inflammation, to a particular state of the nervous energy of the affected part.

539. The faculty of producing cold, or, in more exact terms, of resisting heat, is certainly capable of a complete chemical explanation, from the increased evaporation which is thus produced (325).

540. Nor are the lungs the only part of the body which is acted upon by the air; the skin is unquestionably influenced by it, and has been found to exhale carbonic acid. Its effects upon ulcerated surfaces are very striking. The action of oxygen upon the matter of a common abscess renders it more acrimonious, and the benefit which is derived from the application of carbonic acid gas to cancers, and other ulcers, may in some measure be attributed to their preventing the access of oxygen, and its union with the secreted matter.

COMBUSTION.

541. The phænomena of combustion, and the distinction of bodies into combustible or inflammable, and incombustible or uninflammable, are sufficiently familiar. The latter, when exposed to heat, have their temperature raised, and that in proportion to the degree of heat applied to them; but whenever this communication of caloric, from an external source, is stopped, their temperature falls, and they return to their former state. Combustible bodies, on the other hand, when heated to a certain extent, begin to suffer a very evident change, they become much hotter than the surrounding bodies, emit light more or less copiously, and appear to be consumed, or rather are converted into substances altogether new, and which fre-

quently are not apparent to the senses. It is this rapid
emission of light and heat, and this change of properties, and apparent loss of substances, which constitute the process popularly denominated COMBUS-
TION. This phenomenon was referred by Stahl and
his disciples, to a peculiar principle which they called
Phlogiston; it was supposed to exist in all combustible bodies, and combustion was said to depend upon
its separation; but this explanation was absurdly at
variance with the well known fact, that bodies during
combustion increase in weight.

542. After the discovery of oxygen gas and the
composition of the atmosphere, it was maintained by
Lavoisier that this gas was the universal supporter of
combustion, that the process could not proceed without its presence, and that, in short, it was nothing
more than the union of the inflammable body with the
basis of the gas, in consequence of which the heat and
light, which it before contained in its aeriform state,
were evolved in the form of flame. Recent discoveries,
however, have shewn that combustion can take place
without the presence of oxygen, and it is now generally
admitted that it does not depend upon *any* peculiar
principle or form of matter, but is a general result of
intense chemical action. It may be connected with
the electrical energies of bodies; for all bodies which
act powerfully upon each other are found to be in the
opposite electrical states of positive and negative (409),
and the evolution of heat and light may depend upon
the annihilation of these opposite states, which happens
whenever they combine.

MURIATIC ACID.

543. If we pour oil of vitriol (sulphuric acid) upon
common sea salt, white fumes will be immediately disengaged, of an extremely acid and penetrating odour;
these consist of *muriatic acid gas* combined with a
small portion of atmospheric moisture.

544. If the above experiment be conducted in a
retort, the neck of which is made to terminate under
the surface of mercury, the pure gas may be collected
in jars; which will be found to possess all the mechanical properties of air. Its odour is pungent and peculiar, and when applied to the skin for some time, will
raise blisters; its specific gravity is greater than that
of common air, being about 1·278, so that 100 cubic
inches weigh about 39 grains. When brought into
contact with the atmosphere, it occasions a white
cloud, in consequence of its union with the hygrometric water. It extinguishes flame. It is very rapidly
absorbed by water; a drop or two of water admitted
to a jar full of this gas, causes the whole of it instantly
to disappear. According to Kirwan, an ounce measure troy of water absorbs 800 cubical inches (*i. e.* 421
times its bulk,) of muriatic acid gas; and the water,
by this absorption, is increased about one-third its
original volume. So eagerly does this acid gas seize
water that, if admitted to a piece of ice, it liquefies
with the rapidity of red-hot iron. It is in this state of
watery combination that muriatic acid is kept for
chemical and medicinal purposes, and a process is
accordingly given in our Pharmacopœias for its preparation.

545. By acting upon muriate of ammonia by sulphuric acid in a closed tube, as directed for the con-

densation of carbonic acid (*Exp.* 68.) muriatic acid
will be evolved, and liquefied, when it will appear as
a colourless liquid, and at the temperature of 50^0 will
require a pressure of 40 atmospheres to counterbalance
its elastic force.

546. *The watery solution of Muriatic Acid* is obtained by distilling a mixture of dilute sulphuric acid
and common salt; the most œconomical proportions being 32 parts of salts, and 20 of sulphuric acid, diluted
with one-third its weight of water. The retort containing these ingredients may be luted on to a receiver,
containing twice the quantity of water used in diluting
the sulphuric acid, and the distillation carried on in
the sand bath, the heat being gradually raised until
the retort becomes red hot.

547. The liquid muriatic acid thus obtained has
the following properties. Its specific gravity is about
$1 \cdot 160$. It emits white suffocating fumes, which consist of the muriatic acid *gas* rendered visible by contact
with the moisture of the air. When heated in a retort,
the gas is expelled and may be collected over mercury.
Upon being diluted with water, an elevation of temperature is produced, much less however, than that occasioned by diluting sulphuric acid. When perfectly
pure it is colourless, but it is rarely met with in such
a state.

548. In order to ascertain the quantity of real acid,
in liquid acid of various specific gravities, Dr. Ure
has furnished us with the following simple Formula.
*Multiply the decimal part of the number denoting
the specific gravity by* 147; *the product will be nearly
the per-centage of dry acid; or, when we wish to learn
the per-centage of acid gas, the multiplier must be*
197.

Example.—The specific gravity being 1·160, as directed in the Pharmacopœia; required the proportion of dry acid in 100 parts.

$$0·160 \times 147 = 23·520 = real \text{ acid.}$$

or

$$0·160 \times 197 = 31·520 = \text{acid } gas.$$

549. Few subjects in chemical science have given rise to a keener controversy than that which regards the chemical composition of Muriatic acid. But as the student cannot possibly understand the merits of this question, until he is made acquainted with the singular body which arises from its decomposition, or which is formed by its combination with oxygen, the history of *Chlorine* must be first investigated.

CHLORINE, or OXY-MURIATIC ACID.

550. If Muriatic acid be mixed with the black oxide of Manganese, and heated over a lamp in a glass retort, gas is copiously evolved, and may be collected over warm water. As it is absorbed by cold water, it cannot long be retained over that fluid. It may also be procured from a mixture of 8 parts of common salt, 3 of oxide of manganese, 4 of water, and 5 of sulphuric acid.

551. This gas has the following properties :

1. *It has a yellowish green colour;* a property which suggested its name, from χλωρος, *green.* Its odour is pungent and suffocating, and its action on the lungs is not only extremely painful, but highly injurious.* Its specific gravity exceeds that of common

---

* The death of the ingenious Pélletier was occasioned by his accidentally inhaling a large portion of this gas. A consumption was the consequence, which in a short time proved fatal. The vapour of Ether, or Ammonia, appears the best antidote,

air, being according to Davy as 2·5082 is to 1000, so that 100 cubic inches weigh about 77 grains.

2. *It supports Combustion.*—When a burning taper is introduced into a vessel of this gas, the flame becomes red, a dense smoke arises, and the taper is soon extinguished. Phosphorus, however, takes fire the moment it is introduced, and burns vehemently; and many of the metals, when brought in a finely divided state into contact with this gas, burn with great brilliancy. Here then are examples of combustion without the presence of oxygen; and Chlorine accordingly must be regarded as a supporter of it. Its natural or inherent electricity is also, like oxygen, *negative*, and is therefore attracted by the positive pole. The phænomena which arise from its union with combustible bodies constitute some of the most brilliant and striking experiments in chemistry, *viz.*

Exp. 73.—Into a jar of Chlorine, suspend a piece of Dutch metal, or copper-foil; it will immediately inflame, and the combustion will continue, until the whole is consumed.

Exp. 74.—Project into a jar of this gas, powdered Antimony, Arsenic, or Bismuth; the metal will inflame, and fall, in a shower of fire, to the bottom of the vessel. In order to ensure the success of such experiments the temperature of the gas should not be under 70°.

3. *It destroys Vegetable colours.* Hence its application to the purpose of bleaching. This effect may be easily witnessed by introducing any coloured liquid into a jar filled with the gas. If, however, the gas be perfectly dry, and every source of moisture is avoided, no change of colour will ensue. The decolorizing power of this body depends upon its power of trans-

ferring oxygen to the coloured body; the manner in which this is effected will be considered hereafter.

4. *Its Hydrate concretes at* 40° Fah. When a receiver, filled with this gas, not artificially dried, is surrounded by snow, or pounded ice, the gas forms on its inner surface a solid concretion, of a yellowish colour, resembling, in its ramifications, the ice which is deposited on the surface of windows during a frosty night. By a moderate increase of heat, such as to 50° *Fah.* this crust melts into a yellowish oily liquid, which, on a farther elevation of temperature, passes to the state of a gas. Before this substance was shewn by Sir H. Davy to contain water, it was considered as the gas itself reduced into a concrete form.

5. *It is condensable into a liquid, at* 60°, *by a pressure of four atmospheres.* For this fact we are indebted to Mr. Faraday. The result may be obtained either by heating the hydrate in a glass tube hermetically sealed, or, by compressing the gas itself, in a long tube by means of a condensing syringe.

6. *Chlorine Gas is absorbed by water.* This takes place but slowly, if allowed to stand over it quiescent, but rapidly, if agitation be employed; at the temperature of 60° *Fah.* water dissolves about double its volume, and acquires a strong acrid taste and a disagreeable smell; it possesses also the property of discharging vegetable colours. When exposed to the direct rays of the sun, oxygen is obtained, and the chlorine passes to the state of muriatic acid.

552. The next question for our consideration is, What is Chlorine? Is it an elementary body, or a compound? Scheele, by whom it was discovered in 1774, considered it as an element of the muriatic acid, and hence called it *dephogisticated marine acid.* Lavoisier and Berthollet, however, asserted that it was a com-

pound of muriatic acid gas and oxygen, and the parti-
sans of this theory accordingly assigned to it the name
of *Oxy-muriatic acid.* This opinion was naturally
deduced from the fact that all bodies which are capable, by their action on muriatic acid, of producing
Chlorine, contain oxygen, and that their proportion of
oxygen is diminished by the process. Sir H. Davy,
however, assisted by the researches of Gay-Lussac,
and Thenard, has formed a very different theory, instead of considering muriatic acid as the undecomposed
body, and Chlorine, or oxy-muriatic acid, as its combination with oxygen, he regards Chlorine as the elementary body, and Muriatic acid as the result of its
union with hydrogen. To convert the latter therefore
into the former, we have only, according to this view,
to abstract hydrogen from the muriatic acid; this, it
is believed, is effected by treating it with metallic
oxides, the oxygen of which combines with the hydrogen and forms water. Again, to convert chlorine into
muriatic acid, we have only to supply it with hydrogen, which may be derived from the decomposition of
water.

553. It is not a little curious to observe how satisfactorily the phænomena of chlorine may be explained
on either of these theories; and the question at last has
been only decided by some experimental refinements,
the examination of which can scarcely fall within the
compass of an elementary work.* I shall merely observe

* See a controversial paper upon this subject in the 34th
volume of Nicholson's Journal; and a paper by Sir H. Davy in
the Phil. Trans. for 1818, p. 69. The reader may also refer to
the 8th volume of the Transactions of the Royal Society of
Edinburgh; the Annals of Philosophy, xii: 379, and xiii. 26,
285, and to a paper by Mr. R. Phillips, in the new series of that
work, vol. 1. p. 27, On the Action of Chlorides on Water.

that the theory, which assigns to chlorine the characters of an element, is that which affords the most satisfactory explanation of the phænomena.

554. In all those cases in which chlorine appears to impart oxygen, as in the operation of bleaching, &c. water is decomposed, the hydrogen of which combines with the chlorine to produce muriatic acid, and the oxygen is thus evolved.

555. If the watery solution be exposed to light, oxygen will be evolved, and the chlorine will pass to the state of muriatic acid; this was explained, on the former theory, by supposing that the oxy-muriatic acid at once yielded its oxygen;—on the latter, by assuming that the water is decomposed.

556. According to the modern theory, Muriatic acid consists of equal volumes of Chlorine and Hydrogen, and has been therefore called *Hydro-Chloric acid;* accordingly, the simple mixture of one measure of each of those gases, when exposed for a short time to the sun's rays, affords two measures of muriatic acid gas. But this phænomenon admits also of explanation upon the former hypothesis, in which case we have only to suppose the hydrogen to combine with the oxygen of the oxy-muriatic acid gas, and to leave the muriatic acid.

557. Chlorine combines at once with the metals, without requiring, like the acids, that they should be in the state of an oxide; but this will be fully considered under the history of Salts. The compounds which it forms with various bases are termed *Chlorides.*

558. Chlorine and Oxygen are capable of existing in combination, and they form peculiar bodies; to the first of which the name of *Protoxide of Chlorine,* or from its brilliant yellow green colour, that of *Euchlo-*

rine has been given. It exists in a gaseous state and explodes by a gentle heat. It appears to be constituted of one atom of chlorine and one atom of oxygen. To the second compound, the name of *Deutoxide*, or *Peroxide* of Chlorine has been assigned. It consists of one atom of chlorine and four atoms of oxygen. These bodies have no peculiar points of interest to the medical student, we therefore pass to a third compound of Chlorine and oxygen, viz. *Hyperoxymuriatic*, or *Chloric Acid*.

CHLORIC ACID.

559. This acid, which consists of one atom of Chlorine and five atoms of oxygen, cannot exist, independent of water or some base. It may be prepared by passing a current of chlorine through a mixture of oxide of silver and water. Chloride of silver is produced, which is insoluble, and may be separated by filtration. The excess of Chlorine, which the filtered liquor contains, is separable by heat, and the Chloric acid dissolved in water remains. It has no sensible smell, but its taste is very acid, and it reddens litmus without destroying the colour. When concentrated it has something of an oily consistency. It combines with various bases, and forms *Chorates*, an interesting class of salts formerly known by the name of *Hyperoxygenized-muriates*, or *Oxy-muriates*.

IODINE.

560. This elementary body was accidentally discovered, in 1812, by M. de Courtois, a manufacturer of saltpetre at Paris. In his processes for procuring

soda from the ashes of sea-weeds, he found the metallic vessels much corroded; and in searching for the cause of the corrosion, he made this important discovery.

561. It may be procured by the following process. Dissolve the soluble part of incinerated sea-weed (*Kelp*) in water; evaporate the lixivium until a pellicle forms, and set aside to crystallize. Evaporate the mother liquor nearly to dryness, and pour upon the mass, half its weight of Sulphuric acid. Apply a gentle heat to this mixture in a glass alembic : fumes of a violet colour arise and condense in the form of opaque crystals, having the aspect of Plumbago.* These are solid Iodine. To purify it from the redundant acid, that comes over with it, it may be redistilled from water containing a very small quantity of potass, and afterwards dried by pressing it between folds of blotting paper.

562. Iodine is a solid at the ordinary temperature of the atmosphere; it has a metallic lustre and is soft and friable; its specific gravity is 4·946. Its smell resembles that of diluted chlorine; its taste is acrid. It stains the skin yellow. At a temperature between 60° and 80°, it produces a violet vapour, and hence its name (from ἰώδης *violaceus*). At 120° or 130° this vapour rises more rapidly. At 220° it fuses, anu produces copious violet colour fumes, which condense in brilliant plates and acute octohedrons. Like Chlorine and Oxygen it is electro-negative ; and therefore attracted by the positive surface of the Voltaic pile. It is also a supporter of combustion, which may be shewn by im-

---

* Several other processes have been proposed for its preparation. To those who are desirous of gaining farther information upon the subject I strongly recommend the perusal of the article *Iodine*, in Dr. Ure's Chemical Dictionary, 2nd Edition.

mersing a piece of the metal called *Potassium* in its vapour, when it will inflame and burn with a pale blue light. Phosphorus also burns under similar circumstances. It renders vegetable colours yellow. It is very sparingly soluble in water, that liquid not holding more than $\frac{1}{7000}$ its weight in solution. It is much more soluble in spirit. As a chemical test this solution is highly valuable to the chemist, as it forms, with a dilute solution of Starch, a very beautiful blue compound which eventually precipitates.

563. Although, according to the just logic of chemistry, a body must be regarded as elementary until it is decomposed, still when we discover the high atomic number of Iodine $= 125$, it seems reasonable to believe that future discovery will prove it to be a compound.

564. In its relations to other bodies, Iodine presents some striking analogies to Chlorine. With Hydrogen it forms an acid of some interest to the medical practitioner, as its saline combinations have been lately proposed as remedies.

HYDRIODIC ACID.

565. Under ordinary circumstances *Hydrogen* and *Iodine* exert a slow action upon each other, but when the former is presented to the latter in a *nascent** state, they readily unite, and produce a gaseous acid, the *Hydriodic Acid.*

566. It is prepared by the action of moist Iodine upon Phosphorus, and, as it is absorbed by water, it must be received over Mercury. It is colourless, very sour, and smells like muriatic acid. With bases it forms a class of Salts termed *Hydriodates.*

---

* This term is used to express the condition of a gas at the moment it is disengaged from a solid combination.

567. It is in the state of Hydriodate of potass, that Iodine exists in sea weeds, and since it is a deliquescent salt, it remains in the *mother liquor,* after separating the carbonate of soda and other salts by crystallization.

NITRIC ACID.

568. If we distill together equal weights of Nitre and Oil of Vitriol (Sulphuric acid) in a glass retort, connected with a tubulated receiver, which is made to pass into a bottle, as represented in the annexed cut, we shall obtain this acid, and be enabled to observe the phænomena which attend its production.

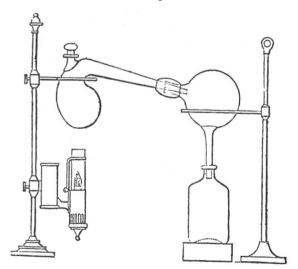

569. In this process the nitric acid is separated from its combination with potass, ·in which state it exists in the nitre, by the superior affinity of the sulphuric acid for the potass. The nitric acid thus disengaged is

volatilized by the heat and carried over into the
receiver, where it is condensed, and collected in the
bottle to which it is attached.   The phænomena which
accompany the distillation will vary in some measure
with the quantity of Sulphuric acid employed; if this
be less than that directed by the London College, as
soon as the materials are heated, orange yellow vapours
are disengaged, which in a short time, as the heat increases, become paler, and continue so until the ingredients in the retort are nearly dry and the heat is
augmented to 500°; when, owing to a partial decomposition, which will be hereafter explained (589), the red
fumes will reappear, and a quantity of oxygen gas will
be liberated.   The liquid acid in the receiver will also
deepen in colour, and present the appearance of an
orange coloured liquid.   If, however, the proportion of
Sulphuric acid be greater, the same decomposition will
not take place until the process is nearly completed,
and a paler liquid will be found in the receiver.*
After the operation the retort will contain a salt composed of the sulphuric acid, employed as the decomposing agent, and potass the base of the nitre.

570.   Nitric acid thus obtained, and purified by
cohobation (182.) is a limpid, or straw coloured fluid
having the specific gravity of 1·5, it emits white fumes
when exposed to the air; its taste is extremely sour
and corrosive, and the skin is indelibly stained of a
yellow colour by it.   When of the specific gravity 1·42

---

* The manufacturer, who prepares Nitric acid upon the large
scale, instead of the glass apparatus above described, employs
distillatory vessels of stone ware.   Nitric acid was first obtained
by Raymond Lully, in the 13th century, by distilling a mixture
of nitre and clay; a process still employed on the continent.
The name *Nitric acid* was given to it in 1787, by the French
Chemists.

it boils at 248°, and may be distilled over, without any essential change, but if it be weaker than this, it gains strength by ebullition, whereas, if it be stronger, it becomes weaker by the operation, so that all the varieties of this acid, by sufficient boiling, are brought to the specific gravity 1·42. When strong nitric acid is exposed to the air it absorbs a portion of its hygrometric water, and becomes weaker. When two parts of it are suddenly mixed with one of water, an elevation of temperature is produced to about 212°. By exposure to the light of the sun, it becomes at first of a pale straw colour, which gradually deepens until it assumes that of a deep orange. In this state it is usually, although, as we shall hereafter see, very incorrectly termed *Nitrous Acid,* or in commercial language, *Aqua Fortis.*

571. The most remarkable chemical character of this acid depends upon the readiness with which it yields a portion of the oxygen which enters into its composition. Hence it is rapidly acted upon by all combustible bodies; sugar, alcohol, charcoal, &c. excite the most violent action, accompanied with the evolution of copious red fumes. If certain essential oils, as those of Turpentine and Cloves, be suddenly poured into this acid, the mixture will instantly take fire.

572. Real or *dry* nitric acid can only exist in combination with some base; the strongest liquid acid, which has been termed *Hydro*-nitric acid,[†] that

---

† The prefix *Hydro*, to denote the presence of water, although employed by our ablest chemists, is in discordance with the just principles of chemical nomenclature. On other occasions it is used to express the presence of Hydrogen, and ought to be restricted to that purpose.

can be obtained, contains at least 25 per cent. of water.

573. Extraordinary as the fact may appear, the same elements which constitute the air we breathe, when combined in different proportions, form the highly acrid body which we are now considering. For this discovery we are indebted to the unrivalled genius of Mr. Cavendish, who furnished us with a synthetical demonstration of its truth. Having passed electric sparks through a portion of atmospheric air, or through a mixture of one part of nitrogen and two of oxygen, confined over mercury, he observed that after some time the mixture diminished in bulk, and, on admitting a little water, an acid solution was obtained which afforded crystals of nitre when saturated with potass.

574. When nitric acid is passed in vapour through an ignited tube it is resolved into oxygen gas, which may be collected in the usual manner, and nitrous acid gas, a compound which will be hereafter considered.

575. According to the most accurate experiments, nitric acid, in its dry state, may be regarded as composed of one proportional of Azot, or nitrogen, and five proportionals of oxygen, and may therefore be represented by the annexed symbol. It will be seen therefore that its representative number will be $(5 \times 8 + 14)$ 54.

Azot. 14	Oxy. 8
	Oxy. 8
	Oxy. 8
	Oxy. 8
	Oxy. 8

576. Nitrogen combines also with other proportions of oxygen than that above stated, and gives origin to a series of compounds, the history of which is so intimately connected with that of nitric acid, that it will

be impossible for the student to comprehend the changes which this latter body undergoes, without a perfect knowledge of the former; we shall therefore proceed to their investigation.

NITROUS OXIDE.

577. If we distil the salt called *Nitrate of Ammonia*, at a temperature of about 420°, this gaseous compound will pass over, and may be collected over water. Or, the same gas may be procured by exposing common *nitrous gas* for a few days to iron filings, or to various other bodies which have a strong attraction for oxygen. The theory of this change, and that which will explain the conversion of the different compounds into each other, cannot be understood, until after the nature and composition of these combinations have been described.

578. Nitrous oxide has the following properties. It is considerably heavier than common air; 100 cubic inches weighing between 48 and 49 grains. It supports combustion with considerably greater energy than atmospheric air. Animals, when wholly confined in this gas, die speedily. Its most extraordinary property, however, is evinced in its action on the human body, and we are indebted to Sir H. Davy[+] for a very curious series of experiments which were instituted with a view to ascertain its effects. When inspired from an oiled silk bag, it does not prove fatal, in consequence of the atmospheric air present in the lungs, with which it mixes and is diluted; but it occasions highly plea-

---

[+] Researches Chemical and Philosophical, chiefly concerning Nitrous Oxide. London, 1800.

surable sensations, differing in intensity according to
the temperament of the person under experiment; but,
in general, they have been compared to those attendant
on the early period of intoxication; great exhilaration,
an irresistible propensity to laughter, a rapid flow of
vivid ideas, and an unusual fitness for muscular exertion, are the ordinary feelings which it is said to produce; and it is farther stated that these sensations are
not succeeded, like those accompanying the grosser
elevation from fermented liquors, by any subsequent
exhaustion.

579. By detonating this gas with hydrogen, and
examining the products, we are enabled to infer its
composition. If, for instance, one volume of the former
be fired with one of the latter, water is formed, and one
volume of nitrogen remains. Now, since one volume
of hydrogen takes half a volume of oxygen to form
water, nitrous oxide must consist of two volumes of
nitrogen and one volume
of oxygen; these three volumes being so condensed,
in consequence of chemi-

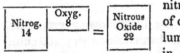

cal union, as only to fill the space of two volumes.
Its composition is justly represented by the annexed
diagram, and its representative number will accordingly
be $(14 + 8) = 22$. It is condensable into a liquid,
but requires for that purpose a pressure of 50 atmospheres.

NITRIC OXIDE, or NITROUS GAS.

580 To produce this gas we must introduce some
copper filings into a gas bottle containing nitric acid,
diluted with thrice its bulk of water; an action immediately ensues, red fumes are produced, and gas is

abundantly liberated, which may be collected in the usual manner over water.

581. The gas thus obtained, when well washed, does not exhibit any acid characters; if a portion of carbonate of ammonia be suspended in a muslin bag in the interior of a jar containing it, no action will ensue. It extinguishes most burning bodies, but phosphorus, or charcoal, readily burns in it, if introduced in a state of intense ignition. It is rather heavier than common air; 100 cubic inches weighing 32 grains.

582. At high temperatures it undergoes decomposition by the action of some of the metals. One volume of nitric oxide is by these means resolved into equal volumes of oxygen and nitrogen, the annexed

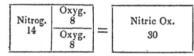

symbol will therefore afford a just representation of its composition.

583. The most remarkable property of this gas is the change which it undergoes on admixture with oxygen gas; red fumes arise; heat is evolved, and a diminution of volume takes place. The same appearances ensue, less remarkably, with atmospheric air; and the diminution in volume will correspond with the quantity of oxygen present; and hence the use of this gas in eudiometry, (505.) If this experiment be conducted in a vessel containing a bag of carbonate of ammonia, we shall perceive that an effervescence, indicating the presence of an acid, takes place on the admixture of the gases. A new compound (*Nitrous Acid*) has therefore been formed, which we have next to investigate.

Nitrous Acid.

584. When two measures of nitrous gas and one of oxygen, both freed from moisture, are mixed together in a vessel previously exhausted of air, they are condensed into half their volume, and form a deep orange coloured elastic fluid, which has been called *Nitrous acid gas*. As it is absorbed by water, it disappears on the admixture of its gaseous components over water, as we have already seen. It is also absorbed by quicksilver.

585. This gas supports the combustion of a taper, of phosphorus, and of charcoal, but extinguishes sulphur. The annexed symbol represents its composition; for we have seen that *nitric oxide* is composed of equal volumes of nitrogen and oxygen, and one additional volume of oxygen, or two proportionals by weight, are added to form nitrous acid, which therefore consists of one atom of nitrogen and four of oxygen; and consequently its representative number will be 46.

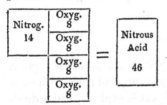

586. Amongst the variety of chemical evidence which may be adduced in proof of the composition of nitric oxide, and nitrous acid, there is none which is more simple and satisfactory than that which is furnished by the following experiment of Dr. Milner.

Exp. 75.—Into a gun-barrel, or porcelain tube, introduce a portion of black oxide of manganese, and let it traverse a furnace; to one extremity of this tube lute a small glass retort containing a solution of ammonia; to the other, a bent glass tube which is to terminate under a glass receiver, or globe. As soon as the oxygen begins to pass off, which may be known by holding a lighted taper at the extremity of the open tube, apply a gentle heat to the ammonia, which will thus be volatilized, and made to traverse the ignited tube; in a short time red nitrous vapours will fill the receiver. The result is to be explained in the following manner; one portion of the oxygen given off by the manganese unites with the nitrogen of the ammonia and forms nitric oxide, which, on coming in contact with the air, produces red fumes; another portion of the oxygen combines with the hydrogen, and forms water.

HYPO-NITROUS, or PER-NITROUS ACID.

587. We have seen that compounds of nitrogen and oxygen exist in the proportion of $1 + 1$ (*Nitrous oxide*) $1 + 2$ (*Nitric oxide*) $1 + 4$ (*Nitrous acid*) and $1 + 5$ (*Nitric acid.*) It was natural for the disciple of the atomic school to inquire whether a combination might not exist of one atom of nitrogen and three atoms of oxygen. Such a compound has been discovered and is termed *Hypo-nitrous,* or *per-nitrous acid.*

588. When 400 measures of nitrous gas and 100 measures of oxygen (in which, taken together, the nitrogen and oxygen are to each other by measure as 100 to 150) are mixed together over a solution of potass confined by mercury, we obtain the body in question, but it has hitherto never been exhibited in a separate form; for if a stronger acid be added to expel it from the potass, it is resolved into nitrous gas and nitrous

acid. The same compound appears also to be formed, during the action of nitrous gas upon common air.

589. It is obvious that the five compounds above described, since they consist of the same elements, and differ from each other only in the proportions of nitrogen and oxygen which they contain, may be converted into each other, by adding or subtracting a due proportion of oxygen; and the student is therefore now prepared to understand the different changes which these bodies undergo. In the first place, this knowledge may be usefully applied for explaining the phænomena already described as attending the distillation of nitric acid (569.) It has been stated that at different stages of the process red fumes are given off, and that oxygen gas is liberated. This arises from a portion of the nitric acid being decomposed by heat, and converted into nitrous and oxygen gases; the former of which coming into contact with the atmosphere becomes nitrous acid, and thereby occasions the appearance of the red fumes to which we have alluded. The additional proportion of sulphuric acid directed by the London College prevents this decomposition by furnishing an excess of water. We see therefore the nature of the fuming acid, which is commonly termed *nitrous*; it is merely nitric acid holding *nitric oxide* in a state of loose combination. The application of heat therefore drives off this latter compound, and leaves the acid in a colourless state. The same effect is produced by dilution with water, and in performing this operation it is curious to observe the successive changes in colour which the acid undergoes, according to the quantity of water added; thus the dark orange coloured solution passes through the shades of blue, olive, and bright green, before it becomes pellucid.

590. The same view will explain the theory of the process by which we obtain nitric oxide (580); if the student compare the symbols expressive of the composition of nitric acid and nitric oxide, it will be seen that the former is at once converted into the latter by abstracting three atoms of oxygen; which is effected by the copper filings; so again the *nitric* oxide is convertible into *nitrous* oxide by the farther abstraction of one atom of oxygen, which is accomplished by the agency of iron filings (577). The following diagram will explain the decomposition, and play of affinities, which take place during the conversion of *nitrate of ammonia* into *nitrous oxide.*

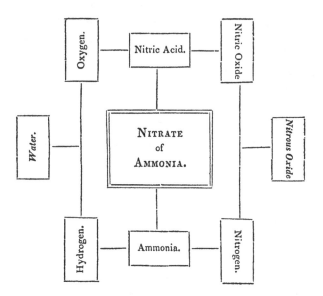

It will be seen that Nitrate of Ammonia consists of Nitric Acid, and Ammonia ; Nitric acid of Nitric oxide and oxygen; and Ammonia of Hydrogen and Nitrogen. At the temperature of about 480°, the at-

tractions of Hydrogen for Nitrogen in ammonia, and
that of nitric oxide for oxygen in nitric acid, are *diminished;* while, on the contrary, the attractions of
the hydrogen of the ammonia for the oxygen of the
nitric acid, and that of the remaining nitrogen of the
ammonia for the nitric oxide of the nitric acid, are *increased*; hence all the former affinities are broken, and
new ones produced, *viz.* the hydrogen of the ammonia
attracts the oxygen of the nitric acid, the result of
which is *Water*; the nitrogen of the ammonia combines with the liberated nitric oxide, and forms *Nitrous oxide.* If we examine the atomic composition
of the several bodies engaged, we shall find that the
proportions of each element in the new compounds
accord exactly with those which are furnished by the
decompositions, and that consequently the whole of the
nitrate of ammonia is resolved into nitrous oxide and
water, without any excess either of oxygen, hydrogen,
or nitrogen, as will appear from the following statement.

$$
\text{Nitric acid} \begin{cases} \text{Nitric oxide} \ldots\ldots = 2 \ Oxy. + 1 \ Nit. \\ \text{Oxygen} \ldots\ldots\ldots\ldots\ldots = 3 \ Oxy. \end{cases}
$$

Ammonia $\ldots\ldots\ldots\ldots\ldots = 1 \ Nit. + 3 \ Hyd.$

$$
\overline{\quad\text{2 of Nitrous oxide.}\quad\quad\text{3 of Water.}\quad}
$$

AMMONIA, or VOLATILE ALKALI.

591. If we mix one part of a salt called *Sal Ammoniac* (muriate of ammonia) with two parts of dry quick-
lime, and introduce the mixture into a small glass
retort, we shall obtain, by the application of a gentle
heat, a peculiarly pungent gas which must be collected

over mercury * as it is rapidly soluble in water. This is *Ammoniacal gas*, to which Dr. Priestley, who first obtained it, gave the name of *Alkaline air.* It exhibits the following properties.

1. *It immediately extinguishes flame, and is fatal to animal life*; it has, however, been observed that a burning candle has its flame enlarged before it is extinguished in this gas.

2. *It is lighter than atmospheric air*, 100 cubic inches weighing not quite 19 grains.

3. *It displays the characters of an alkali.* If moistened test paper be brought into contact with it, its change of colour at once indicates its alkalinity.

4. *It is condensable into a liquid.* This was effected by Mr. Faraday in the manner already described, (325, *note.*) When thus procured it is colourless, transparent, and very fluid, having a specific gravity of 0·760.

5. *It is rapidly absorbable by water.* This may be seen by introducing a portion of water into a jar of this gas over mercury, when a rapid absorption will take place. Or, by at once receiving the gas in a jar of water as fast as it is evolved from the materials, when the bubbles will be seen to undergo rapid condensation, and the water becoming charged with the ammonia, will constitute the *Liquor Ammoniæ* of our Pharmacopœia, from which by heat the ammoniacal gas may be again driven off. The pharmaceutical process for obtaining this preparation is so ably described and explained by Mr. Phillips in his translation of the

---

* If the student is not provided with this piece of apparatus, he may obtain a jar full of the gas by means of the contrivance recommended for the collection of carbonic acid gas (*Exp.* 64.), taking care, however, to use a vessel with an extremely narrow mouth.

Pharmacopœia, that I consider it unnecessary to enter upon the subject in this work.

592. The composition of this alkaline body may be proved synthetically, as well as analytically, to arise from the union of hydrogen and azote. The following instructive experiment may be performed by every student, and it will not only afford him a proof of the constitution of ammonia, but will serve to impress upon his mind the nature of the different ingredients employed in the process, while it will illustrate some of the more important laws of chemical affinity.

> *Exp.* 76.—Introduce into a capacious wine glass two drachms of the filings of tin, and pour upon. them one fluid-drachm of nitric acid, previously diluted with double its quantity of water. A vio- lent action will ensue, accompanied with the copi- ous evolution of red fumes; during this action the filings must be stirred with a glass rod, and after it has subsided, quick-lime must be added to the mixture; when the liberation of ammonia will become evident from the smell of the fumes, or it may be recognised by tests to be hereafter des- cribed.

593. Let us now examine the decompositions which have arisen during the course of this experiment; in the first place, the nitric acid, as well as the water, have been decomposed by the tin, which seizing the oxygen of both, becomes an oxide, and appears in the form of a white powder. The nitrogen from the for- mer, and the hydrogen from the latter, having in consequence been disengaged, unite and constitute ammonia; but, in consequence of the excess of nitric acid present, the alkali forms with it a nitrate of am- monia, which is decomposed in its turn by the quick-

lime; and thus is the nature of an oxide, the composition of nitric acid, of water, and of ammonia, satisfactorily exhibited by one and the same experiment.

Exp. 77.—Introduce into a jar containing nitrogen gas, iron filings moistened with water, and invert the whole over mercury. In a few days we shall find that ammonia has been generated. The rationale of this experiment will be readily understood. The iron attracting the oxygen of the water liberates its hydrogen, which combines with the nitrogen in the jar, and forms ammonia.

594. In like manner we may explain the presence of ammonia in the products of various decompositions. In the fumes, for instance, which arise from the deflagration of nitre and charcoal, an ammoniacal odour is frequently perceptible. In which case, the nitrogen of the nitric acid in the salt seizes the hydrogen of the decomposed water of its crystallization.

Its analytical proof may be conducted in several ways, *viz.*

Exp. 78.—Introduce into a gun barrel, containing some fragments of tobacco pipe,* a portion of red lead; heat the tube, and pass through it the vapour of ammonia, when azotic gas will be evolved. Here is a case of simple affinity; the oxygen given off by the red lead combines with the hydrogen of the ammonia and form water, and the nitrogen is abandoned.†

---

* This is to prevent the barrel from being choaked, by which an accident might happen, in consequence of the evolved gas not having proper vent.

† This experiment differs only from that of Milner as already related (75) by having red lead substituted for black manganese; red lead gives off no more oxygen than is sufficient for the saturation of the hydrogen, whereas the manganese affords oxygen to the

595. Ammonia may also be decomposed by passing it through a red hot iron tube, without the aid of any other body, when it will suffer expansion and be resolved into hydrogen and nitrogen gases.

596. Dr. Henry also observed that a mixture of ammonia and oxygen gas might be fired by an electric spark. In this manner we

Nitro.	Hyd.			
14	1			
	Hyd.	=	Amm.	
	1		17	
	Hyd.			
	1			

are enabled to deduce the proportion of its elements, and to learn that it consists of three volumes of Hydrogen and one of Nitrogen, condensed into two volumes; and consequently that its representative number is 17.

597. In order to detect the presence of minute protions of ammoniacal gas, a glass rod, moistened with muriatic acid, may be brought into contact with it, when dense white fumes will be produced, in consequence of the formation of *muriate of ammonia*, as shewn by *Experiment* 9. A more elegant test, however, is furnished by a mixture of nitrate of silver and white arsenic, for ·if a piece of paper moistened by a solution of this mixture be brought into contact with ammoniacal gas, an intensely yellow colour is produced.

598. Ammonia combines with the acids, and gives rise to a class of compounds, which have been termed *Ammoniacal salts*. With Sulphuretted Hydrogen, it unites in equal volumes and forms *Hydro-sulphuret of Ammonia*, a substance which affords a valuable test for the metals, and may be obtained by distilling, at nearly a red heat, a mixture of 6 parts of slacked lime, 2 of muriate of ammonia, and 1 of sulphur.

---

nitrogen. It may be necessary to observe, that red lead, as usually prepared, will give off a portion of carbonic acid upon the first impression of the heat.

PRUSSIC ACID, or HYDRO-CYANIC ACID.*

599. This compound in its more dilute form, and in various states of combination, has been long known. But we are indebted to the masterly researches of Gay-Lussac for a perfect acquaintance with its nature and characters. Macquer was the first person who discovered that the pigment known by the name of *Prussian blue* was decomposed by alkalies, and the iron separated from the unknown principle with which it existed in combination. Bergman afterwards ranked this principle amongst the acids; Sage in 1772 announced the fact of this *animal* acid, as he called it, being capable of forming neutral salts with the alkalies; and about the same time Scheele succeeded in obtaining it in a separate form, and he concluded that it consisted of ammonia and carbon.

600. The following is the process adopted by Scheele for obtaining this acid. Mix four ounces of Prussian blue with two of red oxide of mercury, prepared by nitric acid, and boil them in twelve ounces, by weight, of water, till the whole becomes colourless; filter the liquor, and add to it one ounce of clean iron filings, and six or seven drachms of sulphuric acid. Draw off by distillation about a fourth of the liquor, which will be *Prussic acid,* though, as it is liable to be contaminated with a portion of sulphuric acid, to render it pure, it may be rectified by redistilling it from carbonate of lime.

601. This acid of Scheele, which later experiments have shewn to contain six times its volume of water,

---

* It derives this title from its combination with iron, long known and used as a pigment under the name of Prussian blue.

has the following characters—It has a strong smell of peach blossoms, or bitter almonds; its taste is at first sweetish, then acrid, hot, and virulent, and excites coughing; it has a strong tendency to assume the form of gas; it does not completely neutralize alkalies, and is displaced even by the carbonic acid; it has no action upon metals, but unites with their oxides, and forms salts for the most part insoluble; it likewise unites into triple salts with these oxides and alkalies.

602. The peculiar odour of this acid could scarcely fail to suggest its analogy to the deleterious principle that rises in the distillation of the leaves of the cherry-laurel, bitter kernels of fruits, and some other vegetable productions; and M. Schrader of Berlin has ascertained the fact, that these vegetable substances do contain a principle, which like the Prussic acid, are capable of forming a blue precipitate with iron; and that with lime they afford a test of its presence equal to the prussiate of that earth. Dr. Bucholz of Weimar, and Mr. Roloff of Magdeburgh, confirm this fact. This acid appears to distill over in combination with the essential oil.

603. It has been stated (601) that the acid of Scheele was in a state of dilution. It is identical with that which is employed in the present day as a remedy. The process adopted at Apothecaries' Hall, is the following; one pound of *Prussiate* (or more correctly *Cyanide*) of mercury is put into a tubulated retort with six pints of water and one pound of muriatic acid, specific gravity 1·15; a capacious receiver is luted to the retort; and six pints are distilled over. The specific gravity of the product is 0·995; it must be preserved in bottles excluded from the light, and, being subject to decomposition, should not be long kept. The committee of the London College lately

appointed to revise the Pharmacopœia intended to have introduced a formula* for its preparation, but the resolution was over-ruled by the College, upon the ground that the medicinal efficacy ascribed to it had not been borne out by experience. Upon this subject, however, I must refer the student to my Pharmacologia. Dr. Ure considers the specific gravity of this acid as a very inadequate test of its strength; he has accordingly proposed another method of inquiry, founded upon atomic views, which has been already explained (274).

604. Prussic, or Hydro-cyanic, acid was first obtained in its concentrated form by Gay-Lussac. The following is the process: a portion of the crystals of Prussiate or Cyanide of mercury is introduced into a tubulated glass retort, to the beak of which is adapted a horizontal tube about two feet long, and fully half an inch wide at its middle part. The first third part of the tube next the retort is filled with small pieces of white marble, the two other thirds with fused muriate of lime. To the end of this tube is adapted a small receiver, which should be kept cool by a freezing mixture. Pour on the crystals muriatic acid, in rather less quantity than is sufficient to saturate the metallic

---

* The following was the proposed Formula.—℞ *Cærulei Prussici* libras duas, *Hydrargyri oxydi rubri* libram, *Acidi Muriatici* libram, *Aquæ destillatæ* octarios duodecim. Cæruleum prussicum coque cum hydrargyri oxydo rubro in aquæ destillatæ octariis sex, spatha assidue movens, donec coloris cærulei expers sit; tum cola, et liquorem paulatim consume ut fiant crystalli. Horum crystallorum libram, cum pari pondere acidi muriatici et aquæ destillatæ octariis quinque cum semisse in retorta vitreâ misce; aquæ octarium infunde in receptaculum ad gradum 32 frigefactum. Denique, retorta aptatâ, destillent acidi Hydrocyanici diluti octarii sex.

base of the salt. Apply a very gentle heat to the retort, when the hydro-cyanic acid will be evolved in
vapour, and will condense in the tube. Whatever
muriatic acid may pass over with it will be abstracted
by the marble, while the water will be absorbed by the
muriate of lime. By means of a moderate heat applied
to the tube the hydro-cyanic acid may be made to pass
successively along; and after being left some time in
contact with the muriate of lime, it may be finally
driven into the receiver.

605. On repeating this process, Vauquelin found
the product so extremely small that he was induced to
seek for a better mode of obtaining it. For this purpose he passed a current of sulphuretted hydrogen
gas through a glass tube slightly heated and filled with
cyanide of mercury, its extremity terminating in a receiver which was artificially cooled. The process was
continued until the smell of sulphuretted-hydrogen
was recognised in the receiver. By this process he
succeeded in obtaining a quantity of the acid which
amounted in weight to one-fifth the mercurial salt
employed. To prevent the inconvenience that might
arise from the presence of undecomposed sulphuretted
hydrogen, some carbonate of lead is placed at the end
of the tube next the receiver, which has the effect of
absorbing it.

606. The concentrated acid prepared by the process
of Gay-Lussac or Vauquelin, is a colourless liquid
possessing a strong odour; and the exhalation, if in-
cautiously snuffed up the nostrils, may produce sickness, or even syncope. It acts as a quick and violent
poison; an animal is instantly killed by merely drawing a feather, dipped in the acid, across the eye-ball;
in this manner M. Majendie constantly terminates the
sufferings of those animals which he has made sub-

servient to his physiological reasearches. Its taste is at first cool, but soon becomes hot and acrid, but I should not recommend the student to verify this statement by experiment. Its specific gravity at 64°, is 0·6969. It boils at 79° and congeals at about 3°, and affects a crystalline form. The cold which it produces, when raised into vapour, even at the temperature of 68°, is sufficient to congeal it. This phenomenon is easily produced by putting a small drop at the end of a slip of paper or a glass tube, when a portion of it will become instantly solid. Though repeatedly rectified on pounded marble, it retains the property of feebly reddening paper tinged blue with litmus; the red colour, however, disappears as the acid evaporates.

607. What is the composition of Prussic Acid? This problem has been solved by Gay-Lussac, who has discovered that it consists of a peculiar compound base, to which he has given the name of *Cyanogen*, acidified by Hydrogen. The name *Hydro-cyanic acid* is therefore well adapted to express its composition.

CYANOGEN*,—PRUSSINE,—OR PRUSSIC GAS.

608. This gaseous compound may be obtained by heating perfectly dry *Prussiate of Mercury* in a glass retort, or in a tube closed at one extremity. It first blackens, then liquefies, and the cyanogen comes over in the form of gas, and may be collected over mercury.

609. This gas has a strong, penetrating, and disagreeable odour. It burns with a bluish flame mixed with purple. Its specific gravity is to that of common air as 1·8064 to 1; hence 100 cubic inches at 60°

---

* This term signifies *the producer of blue.*

weigh 55 grains. Water dissolves 4·5 volumes, and
alcohol 23 volumes. By the pressure of about 3·6 atmospheres Mr. Faraday condensed it.

610. In its chemical relations Cyanogen may be
compared with Chlorine and Iodine; it is acidified by
Hydrogen, and it forms compounds with metallic bases
which are termed *Cyanides.*

611. By detonating Cyanogen with oxygen, carbonic acid and nitrogen are the results produced; and
by estimating their proportions it has been inferred,
that Cyanogen consists of two proportionals of carbon,
and one of nitrogen. Its representative number will
therefore be (6+6+14) 26.

612. Hydro-cyanic acid is composed of equal volumes of Cyanogen and Hydrogen, and hence its representative number will be (26+1) 27.

613. The student having been now made acquainted
with the composition of Cyanogen, he will easily understand the theory of the processes employed for the
production of Hydro-cyanic acid. In that directed by
Gay-Lussac, (604) Cyanide of Mercury, which is a
compound of that metal with Cyanogen, is acted upon
by muriatic acid, which by the aid of its water oxidizes
the metal and combines with it, while the Cyanogen
thus liberated unites with the disengaged hydrogen
and passes off in the state of Hydro-cyanic acid. In
the process of Vauquelin the result is brought about
by the aid of double affinity. When the sulphuretted
Hydrogen is passed through the Cyanide of mercury,
the sulphur combines with the mercury, and the Hydrogen with the Cyanogen.

614. Hydro-cyanic acid unites with salifiable bases
and forms an interesting class of salts termed *Hydro-
cyanates.* The habitudes of these bodies will form a
subject for future consideration.

SULPHUR.

615. This simple inflammable body is chiefly a mineral production, although it exists as a constituent of animal matter. The sulphur which occurs as an article of commerce is brought to this country chiefly from Sicily, although much which appears in the market is obtained from our own sulphuret of copper, which, when roasted, yields sulphureous fumes which are condensed in brick chambers, purified by fusion, and cast into sticks. It is also met with in the form of a light powder called *Flowers of Sulphur*, which is procured from the solid substance by sublimation.

616. Sulphur is a brittle substance of a pale yellow colour, insipid, and when cold inodorous; but exhaling, when heated, a peculiar smell. When a roll of sulphur is suddenly seized in a warm hand, it crackles, and sometimes falls to pieces. This is owing to the unequal action of heat, on a body which conducts that power slowly, and which has little cohesion. At about 220° it fuses, but it begins to evaporate at 180°; when kept melted in an open vessel for some time, at about 300° it becomes thick and viscid; and if in this state it be poured into water, it appears of a red colour, and is as ductile as wax, on which account it is used for taking impressions. Its specific gravity is said to be thus increased from 1·99 to 2·325. The change is not owing to oxidation, as generally supposed, for it takes place in close vessels; it would seem to depend upon a new arrangement of its parts. At 560° sulphur takes fire in the open air, and burns with a pale blue flame, emitting at the same time the most suffocating fumes. If after fusion it be slowly cooled it forms a fine crystalline mass. It is insoluble in

water; but in small quantity in alcohol and æther,
and more largely in oil, especially in that of Linseed.

617. Sulphur is classed amongst the electro-positive
bodies, that is, when separated from combinations by
voltaic electricity it is attracted to the negative pole.
It is considered as elementary in its nature, because, as
yet, nothing certain is known respecting its composition, for although Sir H. Davy by the agency of electricity separated hydrogen from it, it may be questioned
whether this might not be derived from the moisture
which it usually contains. Its atomic weight has been
fixed at 16.

618. It unites with the metals, alkalies, and earths,
and gives rise to a class of compounds termed *sulphurets*.

619. Sulphur unites with oxygen in two definite
proportions, giving origin to two different compounds
which it will be here necessary to examine.

Sulphurous Acid.

620. If Sulphur be burnt in dry oxygen gas we shall
obtain this gaseous compound; if, however, water be
present, a quantity of sulphuric acid will at the same
time be formed. The method usually employed for
obtaining this gas, is by boiling one part by weight of
mercury with six or seven of sulphuric acid in a glass
retort; sulphurous acid gas is thus evolved, and may
be collected and preserved over quicksilver.

621. The following are its properties—It has a
pungent and suffocating smell, exactly resembling that
which arises from burning sulphur. It is more than
twice as heavy as atmospheric air. It reddens vegetable blues and gradually destroys most of them; it

whitens many animal and vegetable substances, as, for
instance, silk and straw, and hence the vapours of
burning sulphur are employed in bleaching. It is absorbed by water; this fluid takes up 30 times its bulk,
gains a nauseous sub-acid taste, and according to Dr.
Thomson, becomes of specific gravity 1·0513. From
this solution, when recently prepared, the gas may be
separated by heat, but not by congelation; the solution, like the gas, discharges vegetable colour. It is
converted into sulphuric acid by all those substances
which are capable of imparting oxygen to it, but the·
presence of water seems to be essential to the effect.
A mixture of dry oxygen and sulphuric acid gases,
standing over mercury is not diminished by remaining
in contact with each other during some months; but
if a small quantity of water be admitted, the mixture
begins to diminish, and sulphuric acid is formed. The
proportions required for mutual saturation are two measures of sulphurous acid, and one of oxygen gas.

622. In like manner, perfectly dry nitrous acid gas,
produces no change on dry sulphurous acid, but when
placed in contact with a small quantity of water, these
bodies act mutually and rapidly on each other; the
nitrous acid gas yields a portion of its oxygen to the
sulphurous acid, from whence result nitrous gas and
sulphuric acid, both of which, combining with a small
proportion of water, form white crystalline flakes.
When a larger quantity of water is brought into contact with these crystals, it dissolves the sulphuric acid,
and the nitrous gas is liberated with effervescence. By
means therefore of a small quantity of nitrous gas, we
may transform a large quantity of sulphurous acid into
sulphuric acid, provided the acid gas be mingled with
half its volume of oxygen, or with an equivalent quantity of atmospheric air; but we shall have occasion to

revert to this subject, when we investigate the formation of sulphuric acid by the combustion of sulphur.

623. By a pressure of two atmospheres at 45° *Fah.*
Mr. Faraday condensed sulphurous acid into a limpid
and colourless liquid; when the tube containing it was
opened, the contents did not rush out with explosion,
as occurs in other cases of condensed gases, but a portion of the liquid evaporated rapidly, cooling another
portion so much as to leave it in the fluid state at
commmon barometric pressure. It was, however, rapidly dissipated, not producing visible fumes, but
evolving the odour of pure sulphurous acid, and leaving the tube quite dry.

624. As this gas does not require a greater pressure
than that of two atmospheres, it was justly inferred
that its liquefaction might be accomplished by distillation at a reduced temperature; this has accordingly
been effected by receiving the gas in a receiver cooled
by a mixture of pounded ice and salt.

625. Liquid sulphurous acid displays some curious
phænomena. From the rapidity with which it evaporates it produces a most intense cold, and it furnishes the means of freezing bodies which have hitherto
resisted the ordinary modes of refrigeration. If a portion of this liquid be poured into water, the latter is
immediately converted into ice; and by pouring it
over a bulb containing quicksilver, the metal is rapidly
solidified.

626. If a piece of ice be dropped into the fluid it
is instantly made to boil, from the heat thus communicated by it.

627. The salts to which this acid gives origin, by
combination with salifiable bases, are termed *sulphites.*

SULPHURIC ACID. OIL OF VITRIOL.

628. This acid has been long known * and used, although the true nature of its chemical composition is a discovery of later date. It was originally prepared by the distillation of a salt known in commerce by the name of *green vitriol* † (Sulphate of Iron) and hence it was called *vitriolic acid*, or, from the circumstance of its oil-like smoothness, as displayed upon pouring it from one vessel to another, *Oil of Vitriol.*

629. The method of forming sulphuric acid by the combustion of sulphur was first adopted in this country by Dr. Ward, ‡ who probably obtained the information respecting it during his residence on the continent, for the process is described by Lefevre and Lemery. An exclusive right to it, however, was granted to Ward by patent; and, by way of distinction, the article which

---

* The first mention of this body occurs in the writings of Basil Valentine. It was also known to Paracelsus, who died in 1541. The first person, however, who gave any thing like a correct account of it, was Gerard Doronœus, in a work which he published in 1570.

† The sulphuric acid which is prepared in Saxony, and in many other parts of Germany, is still procured from this salt. The English makers of sulphuric acid distilled the green vitriol in earthen vessels called long-necks, placed within a reverberatory furnace with a glass receiver luted to each retort. It was usual to put fifty or more of these into one furnace. The inconvenience of this process depended upon the length of time which was necessary for the operation, the small quantity of acid procured, and the intense heat required for its distillation, which always destroyed the stills with great rapidity.

‡ Dr. Ward is known to the medical profession as the vender of a noted analeptic pill, and the inventor of the *White Drop* and *Paste*, an account of which will be found in the Pharmacologia.

he prepared was long known by the name of " Oil of
Vitriol made by the bell." The glass vessels which
were employed for this purpose were made with wide
necks, and as large as they could be blown with safety,
and capable of holding forty or fifty gallons; a small
quantity of water was poured into each, and a mixture
of sulphur and nitre inflamed, the fumes from which
were thus condensed. This operation was repeated
until the water became sufficiently saturated. In the
year 1746 Dr. Roebuck of Birmingham substituted
chambers lined with lead for the glass vessels, by which
means the manufacturerer was enabled to conduct the
operation on a greater scale; and this is the method
employed at the present day for the production of this
highly important article. A quantity of sulphur mixed
with about 1-7th of its weight of nitre is placed in pans
of iron or lead, communicating with a chamber of
lead, the bottom of which is covered to the depth of
several inches of water. The mixture is then kindled,
and the combustion allowed to proceed.

630. In order to understand the theory of this process, the student must first be informed that sulphuric
acid consists of one proportional of sulphur, 16, and
three proportionals of water $(8+8+8)$ 24; its representative number will therefore be 40. And in order
to convert an atom of Sulphurous into Sulphuric acid,
we have only to add one proportional of oxygen.

631. The chemical changes by which Sulphuric acid
is produced by the combustion of sulphur and nitre have
been thus explained by Sir H. Davy.—The sulphur,
by burning, forms sulphurous acid gas, and the acid in
the nitre (nitric acid) is decomposed, giving off nitrous
gas; this coming in contact with the oxygen of the
atmosphere produces nitrous acid gas, which has no
action upon sulphurous acid, to convert it into sul-

phuric acid, unless water be present, (622) and if this substance be only in a certain proportion, the water, the nitrous acid gas, and the sulphurous acid gas combine, and form a white crystalline solid. By the large quantity of water usually employed, this compound is instantly decomposed, oil of vitriol formed, and nitrous gas given off, which by contact with air again becomes nitrous acid gas, and this process continues, according to the same principle of combination and decomposition, till the water at the bottom of the chamber has become strongly acid. It will be seen therefore that nitrous gas merely acts as a carrier of the oxygen from the air to the sulphurous acid.

632. It has been sometimes observed that a portion of sulphuric acid concretes into a white mass of radiated crystals; which has been called *Glacial* sulphuric acid. Dr. Thomson considered it as the pure acid divested of water; other chemists have regarded it as a combination of the sulphuric and sulphurous acids, but it is probably a compound of Hypo-nitrous and sulphuric acids.

633. Sulphuric acid displays the following properties.—Its consistence is thick and oily; and, when perfectly pure, is limpid and colourless; it gains, however, a dark tinge from carbonaceous matter, which would therefore appear to be so modified as to be soluble in the acid (471). Its specific gravity is 1·85, it boils at 620°, and freezes at 15°; these effects, however, are materially influenced by its degree of dilution. It is extremely acrid and caustic; straw, wood, and other vegetable substances, when immersed in this acid, without heat, are disorganized, softened, and blackened, and a certain portion of carbonaceous matter is separated from them. For water it has a great attraction, so that upon sudden mixture, a violent heat is produced,

and a considerable penetration of dimension takes place; so rapidly does it unite with the moisture of the atmosphere that, if exposed in a shallow vessel, it will soon double its weight.

634. The strongest acid that can be obtained contains 19 per cent. of water, which appears essential to its constitution, and can only be separated by combining the acid with a base. It would seem therefore that the term sulphuric acid is improperly applied to it; accordding to the principles of the French nomenclature some chemists have called it *Hydro-sulphuric acid.*

635. Those who are desirous of learning the proportion of commercial acid in diluted sulphuric acid of different specific gravities may refer to a copious Table constructed by Mr. Parkes, and published in the second volume of his Chemical Essays.

636. When this acid is perfectly pure, its transparency is not disturbed by solution; but this is never the case with the article as it occurs in commerce, for the addition of water produces a white precipitate which consists of the sulphates of lead and potass.

637. Sulphuric acid combines with various bases, and forms a class of salts, called *Sulphates.*

SULPHURETTED HYDROGEN GAS.

638. This gaseous compound of Sulphur and Hydrogen exists in many mineral waters, especially in that of Harrogate, and is found in the large intestines of animals. It is also evolved during the decomposition of animal substances, and may be artificially produced by presenting sulphur to hydrogen in its *nascent* state, that is, at the moment it is escaping from some compound in which it existed in a solid form. This happens when dilute sulphuric acid is poured upon a

substance called *Sulphuret of iron.** The following process, however, enables us to procure this gas in the greatest abundance, and in the highest state of purity.

Exp. 79.—Into a gas-bottle, or glass retort, introduce some powdered Sulphuret of Antimony (*Crude Antimony* of the shops) and pour over it five or six times its weight of the muriatic acid of the Pharmacopœia. Apply the heat of a lamp, and receive the evolved gas in the ordinary manner.

639. This gas displays the following properties:

1. *It is absorbable by water.* This fluid is found by agitation to absorb thrice its bulk. It is therefore advisable to receive it in bottles provided with glass stoppers, and after filling them entirely with the gas to introduce the stopper. Where we wish to obtain an aqueous solution, we may drive a current of the gas through a phial containing distilled water, or agitate it in contact with a certain quantity of that fluid.

2. *Its smell is extremely offensive,* resembling that of putrefying eggs, or of the washings of a gun-barrel, to which indeed it imparts their offensive odour. In the first case it is developed during decomposition; in the latter, a portion of sulphuret of potass is produced by the explosion of the powder, which, meeting with water, yields sulphuretted hydrogen.

3. *It is inflammable,* and burns either silently, or with an explosion, according as it is previously mixed, or not, with oxygen gas, or atmospheric air. The results of which are water, and sulphurous, with a little sulphuric acid.

---

* To obtain this substance, a bar of iron is to be heated to a white heat, and, in this state, to be rubbed with a roll of sulphur. The metal and sulphur unite. and form a liquid compound, which falls down in drops. These soon congeal, and must be preserved in a well-closed phial.

2 A 2

4. *It precipitates sulphur, when mixed with bodies capable of combining with its hydrogen;* thus Chlorine, when added to this gas, unites with its hydrogen and forms muriatic acid, and the sulphur is abandoned. Atmospheric air, if long kept in contact with it, produces the same effect. The watery solution undergoes the same changes; the addition of a few drops of nitric or nitrous acid instantly precipitates the sulphur.

5. *It possesses the character of an acid.** The watery solution reddens the infusion of violets, and the gas is rapidly absorbed by alkalies and by the earths, with the exception of two or three, and gives rise to a series of compounds termed *Hydro-sulphurets.*

6. *It precipitates Metals from their solutions.* This renders it a most valuable test for these bodies. If we except iron, nickel, cobalt, manganese, titanium, and molybdenum, all metallic solutions are decomposed by it. On this account it tarnishes silver, mercury, and other polished metals, and instantly blackens white paint, and solutions of acetate of lead. Dr. Henry found that one measure of this gas, mixed with 20,000 measures of common air, produced a sensible discoloration of white lead, or of oxide of Bismuth,† mixed with water and spread upon a piece of card.

7. *It is unrespirable, and highly destructive to animal life.* A small bird was found to die immediately in air containing $\frac{1}{1500}$ of its volume of sulphuret-

---

* Some of the German chemists have proposed to designate it by the term *Hydrothionic acid*; and Gay-Lussac has very improperly called it Hydro-sulphuric acid, a name which has been applied to liquid Sulphuric acid.

† Oxide of Bismuth is known as a cosmetic, under the name of Pearl White. Those, however, who used it soon found that their faces were blackened by the combustion of mineral coal, and various other processes in which Sulphuretted Hydrogen-gas is evolved.

ted hydrogen; a dog perished in air mingled with $\frac{1}{100}$; and a horse in air containing $\frac{1}{150}$th. It has long been considered a very energetic poison to man, and it would, at the same time, appear to be a very insidious one; for sensibility is quickly destroyed by it, without any previous suffering. I am acquainted with a chemist who was suddenly deprived of sense, as he stood over a pneumatic trough, in which he was collecting the gas. It would seem to act upon the nervous system through the medium of the blood, in which it is extremely soluble. It constitutes the particular gas of privies, and is the immediate cause of those accidents which so frequently befal nightmen, and of which I have given a full account in another work (*Medical Jurisprudence,* vol. 1, p. 100.) In order to detect its presence, we have only to expose a piece of card moistened with white lead, (*see above* 639, 6.)

640. The specific gravity of this gas is to that of hydrogen as 16 to 1; 100 cubic inches weighing 36 grains. By detonating it with one half of its volume of oxygen, we obtain one volume of sulphurous acid, and a portion of water; so that the sulphur is transferred to one volume of the oxygen, and the hydrogen to the half volume; it therefore consists of 16 sulphur$+$1 hydrogen, and its representative number is 17.

BI-SULPHURETTED, OR SUPER-SULPHURETTED HYDROGEN.

641. This compound is obtained by pouring *Hydro-sulphuret of potass* (formed by boiling flowers of sulphur with liquid potass,) by little and little, into muriatic acid; a very small proportion of gas escapes; and while the greater part of the sulphur separates, one portion of it combines with the sulphuretted hydrogen; assumes the appearance of an oil, and is deposited at

the bottom of the vessel. Its taste resembles that of sulphuretted hydrogen, but it is less offensive; it is heavier than water, and is inflammable. With the alkalies and earths it forms compounds called *Hydroguretted Sulphurets*, a term which might be advantageously changed for that of *Hydro-bi-sulphurets*.

642. According to Mr. Dalton it consists of two atoms of sulphur=32, with one atom of hydrogen; its representative number is therefore 33.

643. The student will therefore perceive that there are three distinct combinations of Sulphur and its compounds, with alkalies and earths. The first consist, simply of sulphur, united with an alkaline or earthy base, and are called *Sulphurets;* the second are composed of sulphuretted hydrogen, united with a base, and are termed *Hydro-sulphurets*; the third containing Bisulphuretted hydrogen and a base, and are distinguished by the appellation of *Hydroguretted Sulphurets*, which might be superseded by that of *Hydro bi-sulphurets*.

PHOSPHORUS.

644. For a knowledge of this curious substance we are indebted to the alchemist Brandt of Hamburgh, who in the year 1699, accidentally produced it during a visionary search after the *philosopher's stone*. The matter from which it was originally obtained was human urine, but in 1769, Gahn, a Swedish chemist, discovered it in bones, and a process was soon afterwards invented by Scheele for obtaining it from them. In the animal bodies above mentioned the phosphorus exists in combination with oxygen, as phosphoric *acid*, which is farther combined with lime, constituting phosphate of lime. The art therefore of preparing phosphorus consists in decomposing these compounds, for which

purpose we have to disunite the phosphoric acid from the lime, and then to detach the oxygen from the phosphoric acid. The following is the process now usually employed.

645. Having obtained Phosphoric acid, by decomposing bone earth, by means of sulphuric acid; the former of these bodies is mixed with an equal weight of charcoal, and distilled at a red heat. The apparatus required for this purpose is a coated glass or earthen retort, the body of which may be placed in a portable furnace while its extremity is immersed into a basin of water. After a short time a great quantity of gas escapes, and when the retort has obtained a bright red heat, a substance looking like wax, of a reddish colour, passes over; this is impure Phosphorus, the impurities of which are to be removed by remelting it under the surface of water, and squeezing it through a piece of fine shamoy leather. It may then be cast into cylinders.*

646. Phosphorus thus obtained is distinguished by the following properties, *viz.*—It has generally a flesh colour, but, when carefully purified, may be obtained colourless and perfectly transparent. Its specific gravity is 1·77. It is soft like wax, and readily yields to the knife. It is so inflammable that at about 100° *Fah.* it takes fire and burns with intense brilliancy, throwing off copious white fumes. At the ordinary temperature of the atmosphere it emits a white smoke, which is luminous in the dark, and has the odour of garlic. It may be set on fire by friction, for which purpose we

---

* Phosphorus may also be easily obtained by mixing a solution of Phosphate of Soda with a solution of Acetate of Lead, in the proportion of four of the former salt to one of the latter; by which a precipitate of Phosphate of Lead will occur; and this by distillation will yield Phosphorus.

have only to rub a small fragment of the substance between two pieces of brown paper. When covered with
water, so as to exclude the action of the atmosphere,
it melts at about 109 or 110°, and boils at 550°. Oils
dissolve phosphorus, provided the temperature be a
little raised, and becomes luminous; rectified æther
also takes up a certain portion, and when the phos-
phorated fluid comes into contact with water it produces a luminous appearance. Its representative number is stated to be 12. We have not hitherto been able
to reduce it into simpler forms of matter.

647. Phosphorus acts upon the animal system as an
energetic poison. It has however been employed in
medicine, in the dose of one-fourth of a grain, and is
said by Leroi to be very efficacious in restoring and
establishing the force of young persons exhausted by
sensual indulgence, and of even prolonging the life of
the aged. *Weickard* has recorded several cases of
poisoning by this substance.

648. If Phosphorus be heated in a confined portion
of very rare air, it enters into combination with oxygen,
and produces three distinct oxides, each characterised
by distinct properties. The first is a red solid, less
fusible than phosphorus; the second is a white substance, more volatile than phosphorus; the third a
white and fixed body.

649. Phosphorus combines with earthy and metallic
bases, forming a series of compounds termed Phosphurets.

Phosphorous Acid.

650. If Phosphorus be exposed to the action of the
atmosphere, this acid, together with a portion of phosphoric acid are formed. If our object be to obtain the
former in a state of absolute purity a less direct process

must be followed. We must first sublime phosphorus through corrosive sublimate; then mix the product with water, and heat it until it assumes the consistence of syrup. The liquid obtained is a compound of pure phosphorus acid and water, which becomes solid and crystalline on cooling. It is acid to the taste, reddens vegetable blues, and unites with salifiable bases forming *Phosphites*.

651. Phosphorous acid, according to the latest experiments, appears to be constituted of one atom of phosphorus (12) and one atom of oxygen (8.) Its representative number therefore is 20.

PHOSPHORIC ACID.

652. To prepare this acid we have only to burn phosphorus in oxygen gas. This, however, is rather done with a view to prove its nature, than to obtain it in any quantity. For œconomical purposes, calcined bones are decomposed by sulphuric acid.

653. It consists of one atom of phosphorus (12) and two atoms of oxygen (16), so that its representative number is 28.

654. The salts which phosphoric acid forms with different bases are termed *Phosphates*.

Phosphoric acid exists abundantly in animals and vegetables.

PHOSPHURETTED, AND BI-PHOSPHURETTED HYDROGEN GASES.

655. Phosphorus is capable of combining with hydrogen in two proportions, producing two distinct compounds; the one, termed Phosphuretted Hydrogen, or *Bi-Hydroguret of Phosphorus*, not burning spontaneously when brought into contact with the air, but detonating violently when heated with oxygen to about 300°, the other, *Bi-Phosphuretted Hydrogen*,

taking fire immediately on coming in contact with the atmosphere; a phænomenon which may be witnessed by letting the gas escape into the air as it issues from the beak of the retort; its combustion is attended with a circular dense white smoke which rises in the form of a horizontal ring, enlarging as it ascends, and forming a kind of corona. When mixed suddenly with oxygen gas a violent detonation ensues; the same effect is produced on mixing it with chlorine, or with nitrous oxide. By standing it deposits phosphorus, and is converted into phosphuretted hydrogen.

656. The Phosphuretted Hydrogen is composed of one atom of phosphorus and two atoms of hydrogen $(12 + 1 + 1) = 14$. The Bi-Phosphuretted Hydrogen, of two atoms of phosphorus and one of hydrogen $(12 + 12 + 1) = 25$.

657. By the following process Bi-phosphuretted hydrogen may be easily procured for the purpose of experiment. Into a retort, filled with a solution of potass, introduce a portion of phosphorus, and apply the heat of a lamp; in a short time the gas will be liberated, and may be collected over water. This gas may also be obtained by putting into five parts of water half a part of phosphorus cut into very small pieces, with one of finely granulated zinc, and adding three parts of strong sulphuric acid. This affords an amusing experiment; the gas is disengaged in small bubbles, which cover the whole surface of the fluid, and take fire on reaching the air; these are succeeded by others, and a well of fire is produced. The substance termed Phosphuret of Lime yields this gas also in abundance, when brought into contact with water, which it decomposes.

658. Bi-phosphuretted hydrogen gas has been also supposed to be occasionally evolved during the putre-

factive process, and to have given origin to those luminous appearances, which are known under the name of Will o the Wisp's.

METALS.

659. The Metals constitute the most extensive class of hitherto undecompounded bodies. From their numerous chemical relations, their important applications to the arts, and the many powerful remedies to which they give origin, it is difficult to say whether to the philosophical chemist, to the manufacturer, or to the physician, they present the greatest points of interest. The medical student, however, will not be required to pursue in detail the multifarious combinations of each particular metal; a few only possess medicinal virtue, but these must be examined with precision, and the changes which they undergo, by being united with other bodies, will require the most careful investigation.

660. The characteristic properties of the Metals are a high degree of lustre, opacity, combustibility, and the power of conducting electricity and caloric. A considerable degree of specific gravity was also long considered as an essential character of metals, but Sir H. Davy has discovered bodies lighter even than water, which nevertheless agree in all other essential qualities with metals, and must cousequently be arranged with them. Metals differ also considerably in their mechanical properties, such as hardness, fragility, and tenacity or capability of extension; some, for instance, are capable of being extended almost indefinitely under the hammer, and are therefore called *Malleable*, the same metals are likewise *Ductile*, or admit of being drawn out into wire. Others, on the contrary, are

brittle; but even this quality has its degrees. These differences exhibited by the various metals have been assumed as the basis for their classification, and a distinction was formerly established, which conferred the title of Metals upon those only that are malleable, and that of *Semi-metals* upon those characterized by fragility. Such a distinction, however, was highly improper, and has been therefore abandoned. Nor do metals differ less from each other in fusibility. Mercury is always fluid at the ordinary temperature of the atmosphere, while Platinum can scarcely be melted by the most intense heat of our furnaces.

661. In their chemical relations the most striking differences also subsist between the different metals, which may be said to correspond with their qualities of unalterability by common agents, since the metallic pre-eminence of these bodies, popularly speaking, must be inversely as the number and force of their affinities. This has sanctioned the distinction above stated, and those metals which are capable of resisting the action of heat and air, such as gold, silver, and some others, have been dignified by the title of *Perfect*, or *Noble* metals.

662. The most popular, and perhaps the least objectionable, arrangement of these bodies is founded upon the above general fact; so that those least affected by external agents, or which possess the highest quality of unalterability, are made to occupy the first place, while, as we descend in the scale, these qualities diminish in a progressively decreasing series, until we arrive at those which are incapable of retaining their metallic lustre and appearance, even for a few minutes. In a medical point of view, however, that arrangement will be found most eligible which has a more direct reference to those combinations which are valuable as

pharmaceutical agents, or remedial substances; in treating therefore of metals, it will be seen that I have preferred an order which is better calculated to accomplish such an object.

663. Before we descend into the details involved in the history of particular metals, it will be desirable to speak of those general chemical habitudes which characterize the whole class.

664. Oxygen, which forms one-fifth of the atmosphere, and eight-ninths of the aqueous mass, exerts an incessant and extensive dominion over metals; by which they lose their essential characters of lustre, and tenacity or coherence, and are converted into earth-like masses in which not a vestige of metallic character, except that of specific gravity, can be recognised. These compounds are distinguished by the name of *Metallic oxides.* All metals are susceptible of this change, even the noble metals, although they may resist the joint action of heat and air, become oxidized by the voltaic flame, or by the agency of complicated affinities. That the metals possess different attractive powers in regard to oxygen is evinced during the most ordinary operations to which they are submitted; some, for instance, absorb it when exposed to a temperature approaching a red heat, as Iron, Mercury, &c.; others absorb it when in fusion, as Lead, Tin, Antimony; others again at lower, or even common temperatures, as Arsenic, Manganese, Potassium, &c. In like manner some are capable of at once decomposing water at all temperatures, and of combining with its oxygen, while others effect this change only at a red heat; potassium decomposes this fluid with a rapidity which amounts to actual inflammation, while iron filings, if moistened with water, become very gradually oxidized, but if the red hot metal be brought into contact with it, it

produces a rapid decomposition, and hydrogen gas is
disengaged in torrents. (*Exp. 55.*)

665. In certain cases the metal acquires oxygen by
the decomposition of acids, or other metallic oxides.
In general the rapidity with which the metal is oxidized
by the former of these agents is in an inverse proportion to the affinity of the acid base for oxygen. Those,
which have not been proved to contain oxygen, are
remarkably inert in their action on metals, and the
same inactivity belongs to other acids, in which the
oxygen and base are held combined by a powerful
affinity; concentrated sulphuric acid, for instance,
scarcely attacks any of the metals, at common temperatures, because the oxygen and sulphur, of which it
consists, forcibly attract each other. On the contrary,
the nitric acid, which readily abandons a portion of its
oxygen, acts with considerable energy, as we have witnessed in experiment 76. But such of the acids
as in their concentrated state refuse to transfer any
oxygen to the metal, are found to effect this change by
dilution; this is the case with sulphuric acid; the
student may perhaps enquire whether therefore dilution weakens the affinity subsisting between the sulphur
and oxygen;—by no means,—the metal in this case
derives no portion of its oxygen from the acid, but from
the decomposition of the water, as already explained
(455); as a proof that the acid in this case undergoes
no decomposition, it has been shewn that the same
quantity exists in combination with oxide of iron as
was originally submitted to experiment.

666. That one metal may be oxidized at the expense
of another is shewn by experiments 26, 27, 28, &c.;
and again, if we heat the oxide of mercury with metallic
iron, we shall produce metallic mercury and oxide of
iron: and potassium heated with oxide of manganese,

becomes oxidized, and metallic manganese is obtained.

667. Each metal combines with a certain definite quantity of oxygen, and where the same unites in more than one proportion, in the second, third, and other compounds, it is a multiple of that in the first, consistent with the laws of definite proportions. These different results are distinguished by prefixing the Greek numerals as already explained, *page 252*; although in some cases we give them appellations derived from the colours which they assume; thus we speak of the *black* and *red* oxide of mercury, the *white* and *black* oxide of manganese, &c. These different oxides of the same metal are, moreover, not only characterised by different colours, but by a distinct train of chemical properties, and especially by different habitudes with respect to the acids, as will be shewn when we come to speak of metallic salts.

668. Metallic oxides are decomposed, and surrender their oxygen with very different degrees of facility. Some undergo this change by mere exposure to heat, as those of gold, mercury, &c.; others require the joint action of heat, and some inflammable body, as hydrogen, phosphorus, or carbon. The solutions of perfect metals at common temperatures are decomposed by the above agents, and the metal losing its oxygen is precipitated in a metallic form, producing many beautiful appearances.

669 When an oxide is converted into a metallic form, it is said to be *reduced* or *revived*; and for this purpose charcoal is usually employed in the arts. The oxide is mixed with a portion of inflammable matter, and exposed to an intense heat, when carbonic acid, and metal are the results. In order, however, to obtain the latter in a coherent mass, and not in the small grains which would otherwise be formed, some sub-

stance is generally added, which is capable of being melted, and of allowing the metal to subside through it. Substances of this kind are called *Fluxes.**

670. To shew the power which carbonaceous matter possesses of deoxidizing metals, the student may perform the following simple but interesting experiment.

> *Exp.* 80.—Expose a common red wafer† to the flame of a candle; small ignited globules will be observed to fall from it, and if these be collected on a card they will be found to consist of metallic lead. In this case the flour of the wafer is converted into charcoal, which immediately acts upon the red oxide of lead, with which it is coloured, and reduces it.

671. Metallic oxidation is a process of great importance in a medical, as well as chemical point of view. Metals must undergo this change before they can unite with acids, in order to form salts, or before they can exert any activity on the human body.

672. Many of the metallic oxides have an attraction for water, and form with it either *solutions*, or *hydrates*.

---

* The most useful substance of this kind to the experimental chemist is what is termed " *Black Flux*," which at once effects the reduction of many of the metallic oxides. It consists of charcoal and sub-carbonate of potass, and is best prepared by projecting into a red hot crucible a mixture of one part of *Nitre*, and two of powdered *Tartar*. The mixture remains in fusion at a red heat, and thus suffers the small globules of reduced metal to coalesce into a button.

† In performing this experiment we must select the common red wafers, the superior kinds are coloured with vermillion, which will not answer the purpose. The student may purchase the red wafer which is sold for killing beetles, and which will be found to answer admirably.

673. In some cases the oxides of metals, instead of exhibiting those characters which have been described, are endowed with alkaline, or acid properties, as will be seen when we treat of *Potassium* and *Sodium*, and of *Arsenic*.

674. Metals combine with each other, and form compounds which are termed *Alloys*, except in those cases in which mercury enters as an ingredient, when the results are termed *Amalgams*. It has been a question whether these bodies are true chemical compounds, or merely mechanical mixtures, but the changes which each metal undergoes in its properties leave little doubt of the correctness of the former of these opinions. In some cases, moreover, they are found to unite in definite proportions only; an irresistible proof of their chemical union.

675. Besides oxides, and alloys, metals are also capable of giving rise to an interesting class of compounds by their union with Chlorine, Iodine, Hydrogen, Carbon, Phosphorus, and Sulphur, forming *Chlorides*, *Iodides*, *Hydrogurets*, *Carburets*, *Phosphurets*, and *Sulphurets*.

676. The combinations of metallic bodies with Sulphur, the decompositions which they undergo, and the various new compounds to which they give origin by the action of other bodies, constitute one of the most obscure branches of chemical science, and which is rendered still more difficult of comprehension by the complicated and confused nomenclature which has been employed to express their nature. Vauquelin divides them into three classes, *viz.* 1st, the compounds of metals with sulphur, which alone are with propriety called *Sulphurets*; 2nd, the compounds of sulphur with metallic oxides, termed *Sulphuretted Oxides*; and 3d, those of sulphuretted hydrogen with metallic

2 B

oxides, which have been called Hydro-sulphuretted oxides.

677. The following is a list of the Metals. Those printed in *Italics*, not possessing any interest to the medical student, will scarcely require notice in the present work, and are only introduced into the Synopsis for the sake of its general coherence.

CLASS I. METALS THAT AFFORD ALKALIES, BY COMBINING WITH A CERTAIN PROPORTION OF OXYGEN.

    1. Potassium   3. *Lithium.*
    2. Sodium

CLASS II. METALS THAT PRODUCE EARTHS.

    1. Calcium   5. *Glucinum*   9. *Zirconium*
    2. Barium   6. *Yttrium*   10. *Silicium.*
    3. *Strontium*   7. Aluminum
    4. Magnesium   8. *Thorinum*

CLASS III. METALS THAT PRODUCE ACIDS.

    1. Arsenic   3. *Chromium*   5. *Columbium.*
    2. *Molybdenum* 4. *Tungsten*   6. Antimony.

CLASS IV. METALS THAT AFFORD ONLY OXIDES, WHICH ARE NEITHER CHARACTERISED BY ACID NOR ALKALINE PROPERTIES, AND WHICH REQUIRE THE ACTION OF COMBUSTIBLE MATTER FOR THEIR REDUCTION.

Order 1. *Those that decompose water, but require the assistance of a red heat for that purpose.*

    1. Manganese   3. Iron   5. *Cadmium.*
    2. Zinc   4. Tin

Order 2. Those that do not decompose water at any temperature.

1. *Uranium*	4. *Titanium*	7. *Tellurium*
2. *Cerium*	5. Bismuth	8. Lead.
3. *Cobalt*	6. Copper	

CLASS V. METALS, THE OXIDES OF WHICH ARE REDUCIBLE BY HEAT, WITHOUT THE AID OF COMBUSTIBLE MATTER.

1. Mercury	4. Platinum	7. *Iridium*
2. Silver	5. *Palladium*	8. *Osmium*
3. *Gold*	6. *Rhodium*	9. *Nickel.*

678. By combination with a certain proportion of oxygen, the metals of the first class constitute peculiar bodies which have been long known under the name of the FIXED ALKALIES, and are distinguished by striking and important characters—*viz.* They neutralize acids and give birth to salts. They change the purple colour of many vegetables to a green, the reds to a purple, and the yellows to a brown.† Their taste is acrid and urinous, and they act on animal matter as powerful solvents, or corrosives, combining with it so as to produce neutrality. With oils they unite and give rise to compounds well known by the name of *Soap.* With water they combine in every proportion, and are also extremely soluble in alcohol. As they require a very high temperature for their volatilization, they have been termed, in contradistinction to Ammonia, the *Fixed* Alkalies.

679. With regard to the chemical nature of these bodies Chemists were long in the most complete igno-

---

† See note at page 210.

rance; some imagined that, as oxygen was the acknowledged principle of acidity, its antagonist hydrogen must constitute that of alkalinity, a conjecture which derived some support from the known composition of Ammonia; but no philosopher ever threw out a hint as to the probability of their metallic nature, until the veil was suddenly withdrawn by the brilliant discoveries of Sir H. Davy, who in a memoir,† unequalled for the genius and sagacity which it displayed, announced the fact of their reduction to a metallic form.

POTASSIUM.

680. If a piece of caustic Potass, previously moistened by the breath, be placed between two discs of Platinum, connected with the extremities of a Voltaic apparatus of 200 double plates, four inches square, it will soon enter into fusion; oxygen will separate at the *positive* surface, and small metallic globules will collect at the *negative* end. These form the remarkable metal *Potassium*. By this process, however, although that by which the discovery was effected, the substance can only be procured in very minute quantities; and the one now generally adopted is that discovered by M. M. Gay-Lussac and Thenard in 1808. It consists in melting Potass, and allowing its vapour to come into contact with iron turnings, heated to whiteness in a curved gun-barrel. At this temperature the iron combines with the oxygen of the potass, and the disengaged Potassium collects in the cool part of the tube.

681. Potassium possesses very extraordinary properties. It is lighter than water, its specific gravity being ·865. At common temperatures it is solid, soft, and easily moulded, like wax, by the fingers; but at

---

† Phil. Trans. 1808.

32° it is hard and even brittle. At 105° it fuses, and in a heat little less than that of redness it rises in vapour; if air be present it burns with a brilliant white flame. It is perfectly opaque, and when cut its colour is splendent white like silver, but it instantly tarnishes on exposure to air; in order therefore to preserve it we must enclose it in Naphtha;† so powerful is its affinity for oxygen, and with such energy does it combine with it, that if thrown into water it instantly decomposes it, and burns with a beautiful light of a red mixed with a violet colour, and the water becomes a solution of pure potass. The same phænomenon takes place if it be projected on ice, an experiment which requires some caution, for such is the violence of the action, and the rapidity with which a portion of the liquefied water is converted into steam, that a slight explosion will sometimes occur, and throw the caustic product into the face of the operator. If potassium be placed in chlorine it burns with great brilliancy. It readily acts also on all fluid bodies which contain water or much oxygen or chlorine, whence it becomes an important instrument of analysis. It has been already stated that carbonic acid gas may be thus decomposed. (*Exp.* 71.)

682. Potassium combines with oxygen in different proportions, and forms distinct compounds. The bluish grey crust which forms on its surface by exposure to air was at first considered as an oxide, but it is now acknowledged to consist of metallic Potassium and its *Protoxide.*

683. *Protoxide of Potassium,* or *Potassa,* is formed by the action of water upon Potassium. By

---

† On perfectly colourless and recently distilled Naphtha, Potassium exerts but little action, but it soon oxidates in that which has been exposed to the air, forms an alkali, and produces with the Naphtha a brown soap, which collects around the globules.

observing the quantity of hydrogen evolved during this
process, we at once obtain the proportion of oxygen
which is transferred to the metal; and it appears that
100 parts of Potassium thus unite with 20 of oxygen;
so that the representative number of Potassium may be
stated as 40, for as 20 : 100 : : 8 : 40; and the protoxide will therefore be represented by 48, since it consists of 1 atom of Potassium = 40, and 1 atom of oxygen = 8.   The Potassa, or caustic Potass, however,
as it occurs to the practical Chemist, contains a certain
proportion of water, which the most intense heat is
incapable of separating, it is therefore properly speaking the *hydrated Protoxide of Potassium*, consisting
of 1 atom of water + 1 atom of protoxide, and its representative number will accordingly be 48 + 9 = 57.
Since, however, for chemical or medicinal purposes
this hydrate is never obtained by the direct oxidation
of the metal, its mode of preparation will fall more
properly under the consideration of the salt from which
it is procured.

684.   Hydrated Potass is a white, very acrid and
corrosive substance.   It quickly absorbs water from
the atmosphere, and deliquesces, in which state it forms
the oil of tartar *per deliquium* of the old chemists.   In
solution it constitutes the *Liquor Potassæ*, and, when
cast into sticks for the use of the Surgeon, the *Potassa
Fusa* of the Pharmacopœia.   In this latter state, however, it contains some of the *Peroxide* described below
(687), and consequently evolves oxygen when dissolved
in water.

685.   Potassium unites with. sulphur, and forms a
Sulphuret of Potassium, which has been commonly
called Sulphuret of Potass.   This subject is one of such
considerable obscurity, that it is difficult to offer an
explanation sufficiently simple.   It would appear that,

with some exceptions, the Metals have a stronger affinity than their oxides for sulphur, and, accordingly, several of the oxides, when heated with sulphur, are decomposed, their oxygen being separated in the state of sulphurous acid, and a true metallic sulphuret remaining; such is the case with the oxide of Potassium (*Potassa.*) There would appear to be two † distinct combinations of Sulphur and Potassium, *viz.* a *Sulphuret*, or *Proto-sulphuret*, consisting of one atom of each constituent, $= 16 + 40 = 56$, and a *Bi-Sulphuret* $= 16 + 16 + 40 = 72$. The former of these bodies is obtained by decomposing the sulphate of Potass at a red heat, by means of charcoal; it is, however, difficult to obtain it perfectly pure, for it acts both on glass and platinum. It has a pale cinnabar colour, and a crystalline fracture; attracts moisture from the air, and dissolves into a yellowish fluid. The second sulphuret is produced by fusing carbonate of potass and sulphur together, out of the contact of air, in proportions indicated by the atomic numbers. When Sulphur and Potass are heated together, according to the directions of the London Pharmacopœia, the carbonic acid is expelled from the latter, and three-fourths of the Potass, or oxide of Potassium, are decomposed; its oxygen combines with the sulphur to form sulphuric acid, and this, uniting with the one-fourth of the undecomposed Potass, produces Sulphate of Potass. The Potassium of the decomposed Potass combines also with sulphur, and thus forms *Sulphuret of Potassium*; whence, says Mr. Phillips, the "*Potassæ Sulphuretum*" of the Phar-

---

† Berzelius, by varying the proportions of Sulphur and Subcarbonate of Potass, used in its preparation, obtained what he considers a series of different sulphurets, consisting of 1 atom of Potass with 2, 4. 6, 7, 8. 9, and 10 atoms of Sulphur; but the greater number of these are probably mere mixtures.

macopœia is a compound of Sulphate of Potass, and Sulphuret of Potassium; I suspect, however, from the proportion of sulphur employed, that the product contains a mixture of *Proto-sulphuret* and *Bi-sulphuret of Potassium* in variable quantities.

686. By exposure to air this substance is soon changed, for the sulphur and potassium, both attracting oxygen, give rise to sulphate of potass. It cannot exist in solution, for the moment it comes into contact with water it decomposes it, the oxygen of which forms Potass with the Potassium; and the hydrogen combining with the sulphur, produces sulphuretted hydrogen, part of which escapes, while another part forms, with the excess of sulphur, Bi-sulphuretted hydrogen, which uniting with the base forms an Hydroguretted sulphuret, or which might with more propriety be termed an *Hydro-bi-sulphuret*, for since the compound of sulphuretted hydrogen with a base is called an Hydro-sulphuret, would it not be more perspicuous to designate a similar compound of bi-sulphuretted hydrogen, which differs only in containing double the proportion of sulphur, by the term here proposed? Upon adding an acid to the solution produced by dissolving the sulphuret of Potassium in water, a quantity of sulphur is thrown down, sulphuretted hydrogen is evolved, and a salt of potass remains in solution.

687. *Peroxide of Potassium* is produced by heating the metal in a considerable excess of oxygen. It is also formed during other operations; it exists, for instance, in very small quantities in the *Potassa Fusa* of the Pharmacopœia. It is an orange-coloured substance, fusible at a lower heat than hydrated Potass, and crystallizable on cooling. When thrown into water, oxygen gas is evolved, and it passes to the state of protoxide. It is constituted of 1 atom of Potassium $= 40$,

$+$ 3 atoms of oxygen $= 24$, and its representative number is therefore 64.

688. Potassium unites with chlorine, but as the resulting compound is inseparably connected with the history of the muriate of Potass, it will be expedient to reserve its consideration.

SODIUM.

689. The history of this metal, and its compounds, as well as its mode of preparation, so nearly coincide with what has been already detailed under the head of Potassium, that it will be sufficient to enumerate the circumstances in which these twin metals differ. Sodium is somewhat heavier than Potassium; its specific gravity being ·9. It is also less fusible. It decomposes water with the same effect as Potassium, but not with the same phænomena, for no flame is produced. Its representative number is 24.

Metals of the Second Class.

690. The compounds which result from the oxidation of these metals have long been known by the name of EARTHS, and constitute the great mass of our globe; but since the discoveries which have developed their composition, the grounds of their distinction from other metallic oxides have become more limited. Some of them, as Lime, Baryta, and Strontia, coincide so nearly with alkalies in their taste, and in tinging vegetable colours, forming soluble salts, and dissolving animal matter, that they have been distinguished by the appellation of *Alkaline Earths*; while the other members of this class approach more nearly in their characters to the oxides of other metals, and have been termed *Earths Proper*.

CALCIUM.

691. If quick-lime be electrized *negatively* in contact with mercury, an amalgam is obtained, which affords Calcium by distillation. The metal, however,
has been so imperfectly examined that, at present, we
know but little of its properties. Its oxide (quick-lime)
has been supposed to consist of 100 parts of Calcium,
and 39·4 of oxygen, which will give us 20·3 for the
representative number of the metal; for, as 39·4 : 100
: : 8 : 20·3, and therefore 20 + 8 = 28 will represent
an atom of lime.

692. *Lime*, however, will never be prepared by the
oxidation of its metallic base. The oxide combined
with carbonic acid, constitutes a great mass of the exterior crust of the globe, and by submitting this substance to heat we drive off the carbonic acid, and obtain Lime, or *Quick-lime*, as it is popularly termed
from its corrosive properties, in a state of purity. The
characters of this body are too well known to require
minute description; when pure its specific gravity is
2·3; its colour is grey, but it becomes white on exposure to air, in consequence of the absorption of
water, and a little carbonic acid. It has an acrid,
bitterish taste, and acts as a caustic on animal matter.
It is not volatile, and can be only fused by the intense
heat of voltaic electricity, or of the oxy-hydrogen blow-
pipe. It absorbs water very rapidly, and falls into
powder; during which change a degree of heat is extricated which is supposed by Mr. Dalton to be not less
than 800°, and is sufficient to set fire † to some inflam-

† Several instances stand recorded of fires having been occasioned by the sudden slacking of quick-lime. Theophrastus mentions the case of a ship, which was loaded in part with linen, and
in part with quick-lime, having been burnt by water accidentally

mable bodies. The caloric thus disengaged is that contained in the water, and which is essential to its fluidity; the water, on combining with the lime becomes solid, and it is supposed in a greater degree than when converted into ice. During this operation of *slacking* as it is called, the vapour will be found capable of acting upon test paper; this effect, however, is not to be attributed to the volatility of the lime, but to its being carried up mechanically on the shoulders of the steam. The same fact will explain the smell which attends the operation.

693. *Slacked* Lime is a *dry hydrate,* in which the water would appear to enter in an atomic proportion.

694. When a sufficient quantity of water has been added to reduce lime into a thin liquid, it has been called *Milk* or *Cream of Lime,* and is frequently alluded to in medical and pharmaceutical works.

695. Lime is very sparingly soluble in water; this fluid at 60° Fah. taking up only about 1-752d of its weight; Mr. Dalton, however, has discovered the curious fact that lime is more soluble in cold than in hot water.

696. The aqueous solution of Lime, or *Lime Water* as it is commonly called, has an acrid styptic taste, and exhibits all the alkaline characters. When exposed to the atmosphere, the lime attracts carbonic acid, and falls down as an insoluble carbonate.

697. BARIUM and STRONTIUM were obtained by Sir H. Davy, by distilling amalgams, which were formed with the pure earths, in the manner already stated under

thrown over the latter. A similar accident lately occurred at Edmonton, from a flood making its way among the quick-lime in a Bricklayer's premises, which took fire, and were entirely consumed.

the history of Calcium. Magnesium, the basis of Magnesia, is very imperfectly known; in the attempts made to distil its amalgams, the metal appeared to act upon the glass, even before the whole of the quicksilver was distilled from it. The other metals of this class have never been exhibited in a separate state, but indirect experiment, as well as analogy, authorise us in concluding that they exist as the bases of the several earths from which they have derived their names.

Metals of the Third Class.

698. These metals, of which Arsenic and Antimony alone possess any interest to the medical student, are distinguished from all others in the property which they possess of becoming acidified by a certain degree of oxidation. With other proportions, however, of oxygen some of these metals assume the ordinary characters of *oxides*. This is remarkably exemplified in the history of Antimony, which forms with oxygen two very distinct orders of compounds. In the first, they act the part of a salifiable base, as illustrated by *Tartar Emetic;* in the second, that of an acid, neutralizing the acids and other bases, and giving origin to the salts called *Antimonites,* and *Antimoniates.*

ARSENIC.

699. Arsenic is rarely † found in its metallic state, but is easily procured by art, from its native combinations. If the *white arsenic* of commerce be mixed with " *black flux*" (384 *note*) and put into a crucible, over which another is luted by a mixture of clay and sand, upon the application of a red heat to the under vessel, the upper crucible will become lined with a brilliant metal,

---

† It is sometimes met with in the veins of primitive rocks, accompanying the ores of silver, cobalt, and copper.

not unlike polished steel. This is metallic Arsenic; it is not, however, necessary to operate upon such a quantity of white Arsenic, as the above arrangement will require, in order to obtain experimental proof of the decomposition. By enclosing in a glass tube a very minute quantity of this substance with black flux, and applying the heat of a spirit lamp to the mixture, a sufficient portion of the metal may be obtained for examination.*

700. Metallic Arsenic is extremely brittle, readily fusible, and volatile at 356°. Its specific gravity is 8·31. In close vessels it may be collected unchanged; but when thrown on a red-hot iron, it burns with a blue flame, and a white smoke, yielding at the same time the strong smell of garlic, which it must be remembered belongs alone to the metal.

701. Arsenic unites with oxygen in two definite proportions, and produces two distinct compounds, both of which are possessed of acid properties. By exposure to air, indeed, a bulky blackish powder is obtained, which Berzelius was disposed to regard as a *sub-oxide*, but it is evidently nothing more than a mixture of metallic arsenic and white arsenic.

702. *Arsenious Acid*, or the *white Arsenic* † of

---

* For the delicacy of manipulation required for thus detecting the presence of Arsenic, the reader may consult my Pharmacologia, or my work on Medical Jurisprudence. The object of the present Treatise being to teach the *Elements* of Chemical science, rather than to direct its applications.

† It is brought chiefly from the Cobalt works in Saxony. The ores of that metal contain much Arsenic, which is driven off by long torrefaction. The vapours are collected and condensed into a greyish or blackish powder, which is refined by a second sublimation in close vessels, with a little potass, to detain the impurities. As the heat is considerable, it melts the sublimed flowers into those crystalline masses which are met with in commerce.

commerce, sometimes called the *white oxide,* is the
first product of the metal's oxidation, and may be easily
procured by its combustion. It occurs in shining semi-
vitreous lumps, breaking with a conchoidal fracture;
its taste is acrid and corrosive, leaving an impression
of sweetness; its specific gravity is 3·7; at the temperature 383° it is volatilized, but by a strong heat it
is vitrified into a transparent glass which is capable of
crystallizing in the form of an octohedron. Its vapour
is wholly inodorous, although it frequently will appear
to yield the smell of garlic, on account of the facility
with which it is metallized (*See Arsenicum Album in
the 6th Edition of the Pharmacologia.*) It is soluble
in water, requiring 400 parts of water at 60°, but only
13 at 212°.   Klaproth has, however, shewn, that if
1 00parts of water be boiled on the arsenious acid, and
suffered to cool it will retain three grains in solution,
and deposit the excess in the form of tetrahedral crystals.   It is also soluble in 70 or 80 times its weight of
alcohol, and in oils.   It would appear to fulfil all the
principal functions of an acid; it reddens vegetable
blue colours, and combines with pure alkalies to saturation, giving rise to a class of salts termed *Arsenites.*

703. Arsenious acid would appear to be constituted
of one atom of arsenic and two atoms of oxygen. The
representative number of the metal has been stated as
38, and consequently that of arsenious acid will be$=$
$38+8+8=54.$

704. *Arsenic Acid.* This compound is obtained
by distilling a mixture of 4 parts of muriatic and 24 of
nitric acid off 8 parts of arsenious acid, gradually raising the bottom of the retort to a red heat at the end
of the operation.   It may also be procured by repeated
distillation with nitric acid only. This acid has a
metallic and sour taste; it reddens vegetable blues;

attracts humidity from the atmosphere ; and effervesces strongly with solutions of the alkaline carbonates. When evaporated, it does not crystallize but assumes the consistence of jelly. It is much more soluble than the arsenious acid, two parts of water being sufficient for this purpose. Its union with salifiable bases produce a class of salts called *Arseniates.*

705. It appears that the oxygen in this acid, compared with that in the arseni*ous* is as 3 to 2, its representative number will therefore be (38+8+8+8) =62.

706. Both the above acids are most virulent poisons.

707. Arsenic unites with Chlorine, forming a *Chloride of Arsenic.* This combination may be effected by throwing the metal finely powdered into Chlorine, when the former inflames, and a whitish deliquescent and volatile compound results. It may be also obtained by distilling 6 parts of corrosive sublimate with 1 of powdered arsenic; the chloride in the form of an unctuous fluid passes into the receiver, and was formerly called *Butter of Arsenic.* According to the most probable view of its composition, it consists of one atom of arsenic and two atoms of chlorine ; its representative number will therefore be (38+36+36)=110.

708. There are two *Sulphurets of Arsenic,* both of which are found native, *viz.* a red compound, called *Realgar;* and a bright yellow one, named *Orpiment.* The former of these may be artificially formed by slowly fusing a mixture of metallic arsenic and sulphur, or by heating white arsenic with sulphur. The latter, by dissolving white arsenic in muriatic acid, and precipitating by hydro-sulphuret of ammonia. It is a very curious fact that while these native sulphurets are medicinally inert, the same compounds when artificially produced are extremely virulent.

M. Renault supposed that this remarkable difference of effect was owing to the metal being oxidized
in the latter case, and in its metallic state in the former,
but this conjecture is not supported by experiment.
The atomic constitution of these bodies has not been
satisfactorily established.

709. Arsenic forms alloys with most of the metals,
and they are generally brittle; that however, which it
forms with copper, is white and malleable.

ANTIMONY.

710. The word Antimony is always used in commerce to denote an ore which is the *Sulphuret* of that
metal; sometimes this ore is called *crude* Antimony, to
distinguish it from the pure metal, or *Regulus*, as it
was formerly termed. In order to obtain metallic Antimony, the native Sulphuret must be mixed with two-
thirds its weight of crude tartar (*bi-tartrate of potass*)
and one-third of nitre. This mixture is then to be
projected, by spoonfuls, into a red hot crucible; and
the resulting mass poured into an iron mould greased
with a little fat. The Antimony, on account of its
specific gravity, will be found at the bottom adhering
to the scoriæ, from which it may be separated by a
hammer. Or it may be obtained by fusing two parts
of the sulphuret with one of iron filings in a covered
crucible; to which, when in fusion, half a part of nitre
may be added. In this case the sulphur quits the
antimony and unites with the iron. The metal, however, is still contaminated with foreign matter, it must
therefore be dissolved in nitro-muriatic acid, and the
solution poured into water; when a white precipitate
will take place, which, if dried, and mixed with twice
its weight of crude tartar, and then fused, will surrender the metal in a state of absolute purity.

711. This metal is of a silvery white colour, very brittle, and of a plated or scaly texture; its specific gravity is 6·712. It fuses at 810°, and on cooling crystallizes in the form of pyramids.

712. Antimony combines with oxygen in several definite proportions, but Chemists have differed with each other as to the number and composition of the resulting oxides. According to Proust they may all be reduced to two. Berzelius has described four compounds, but in these he has included the substance obtained by long exposure of the metal to a humid atmosphere, and which he has called a *sub-oxide*, but it does not appear to be a definite compound. We may therefore safely admit the existence of three oxides, the two latter of which, as they combine with salifiable bases and afford a class of salts, must be arranged among acids.

713. *Protoxide of Antimony* may be procured by pouring muriate of antimony into water; washing the precipitate, first with a very weak solution of potass, and afterwards with water; and then drying it; or, by boiling 50 parts of powdered metallic antimony with 200 parts of concentrated sulphuric acid; washing the residuum first with a weak solution of potass, and then with hot water, and drying. It may also be procured by precipitating the well known salt *Tartar Emetic* with pure ammonia, and edulcorating the precipitate with hot water. This oxide is, moreover, the immediate product of the combustion of the metal, and constitutes what were formerly termed the *Argentine flowers of Antimony.* It is fusible at a red heat, and is decomposed by sulphur and charcoal. When acted on by nitric acid it is converted into *peroxide.* It appears to be the only oxide which is capable of acting as a true base with acids; and is that which seems to

possess the greatest degree of medicinal activity; it
accordingly exists in all those antimonial preparations
which are distinguished for their efficacy.*

714. The *Deutoxide* was procured by Berzelius by
the action of nitric acid on the metal, the product of
which was evaporated and ignited; or it may be produced by calcining the protoxide in a platinum crucible: when calcined sufficiently, but not too much,
its colour is snow white. On account of its acid properties, Berzelius bestowed upon it the name of *Sti-
bious* acid, from the Latin appellation of the metal,
Stibium; it is, however, more generally called *Anti-
monious* acid, and the salts to which it gives birth are
accordingly termed *Antimonites. Peroxide of Antimony* is procured by acting for a considerable time
upon the powdered metal, by excess of hot nitric acid,
and exposing the product to a red heat. It has a yellow white colour, is difficultly fusible, and forms salts
with salifiable bases, but not with acids. It has been
termed by Berzelius *Stibic* acid, but *Antimonic* acid
is a preferable name. Its salts are therefore called *Antimoniates.* It constituted the *Diaphoretic Antimony*
of the old Pharmacopœias, but it has little or no
medicinal efficacy.

715. The atomic composition of these bodies is by
no means satisfactorily established. Proust, Berzelius,
and Dr. Thomson have given different views regarding
it; that, however, furnished by the latter philosopher
is more consistent with the general law of chemical
combination, and Dr. Henry thinks we may accordingly assume 44, as the equivalent number of Antimony, and consider the *protoxide* as constituted of
one atom of metal and one of oxygen $(44+8)=52$;

* These will be found enumerated in my Pharmacologia, under
the head "*Antimonii Sulphuretum.*"

the *deutoxide,* of one and one a half (44+12)=56; and the *peroxide,* of one and two (44+16)=60. We have here, however, with regard to the deutoxide, the anomaly of the multiple of the oxygen of the first oxide being 1 ½ instead of an entire number.

716. *A Chloride of Antimony* may be produced by the same processes as those described under the head of *Chloride* of Arsenic. It moreover results from the solution of the protoxide in muriatic acid. This substance was formerly known under the name of *Butter of Antimony,* so called from its consistency. At common temperatures, it is a soft solid, which liquefies by heat, and cystallizes on cooling. On exposure to air it deliquiates, and, when poured into water, a precipitate falls which is a *sub-muriate of the protoxide,* and constitutes a once celebrated preparation, known by the name of *Powder of Algaroth;* it is violently purgative and emetic in doses of three or four grains. From this precipitate the muriatic acid is removeable by a weak solution of potass, and the oxide remains pure (713). Chloride of Antimony probably consists of one atom of each constituent, its representative number will therefore be (44+36)=80.

717. *Sulphuret of Antimony.* It has been already stated that this compound constitutes the crude Antimony of commerce, it may also be artificially prepared by fusing the metal with sulphur. It consists of one atom of each of its constituents; its number will therefore be (44+16)=60.

718. When this Sulphuret is exposed to the joint action of heat and air, the greater part of the sulphur is dissipated; and the antimony, by combining with the oxygen of the air, is converted into protoxide; it is then strongly heated in an earthern crucible, by combining with some of the silica of which it forms a

species of glass, and hence is called *Glass of Anti-mony;* this substance consists of 8 parts of protoxide of antimony, with a variable proportion of silica, some-times amounting to 10 per cent., and one part of sul-phuret. This article is often found in the market mixed with glass of lead, and so great is the resem-blance between these substances, that on inspection, the most experienced eye may be deceived; their specific gravities, however, will afford an easy method of discrimination (50). One part of oxide of antimony and two of sulphuret give an opaque compound, of a red colour inclining to yellow; and called *Crocus Metallorum.* With eight parts of oxide and four of sulphuret, an opaque mass is produced of a dark red colour, called *Liver of Antimony.*

719. Sulphuretted hydrogen is capable of combining with the protoxide of antimony; and of giving origin to compounds which have been long used in medicine; the first of these has been called *Kermes Mineral;* the latter *Golden Sulphur of Antimony,* and is the *Anti--monii Sulphuretum Præcipitatum* of the present Lon-don Pharmacopœia. When the powdered native sul-phuret is boiled in a solution of Potass, various compo-sitions and decompositions arise, and the solution, on cooling, deposits an Hydro-sulphuretted oxide of An-timony, which is *Kermes Mineral.* On the addition of a dilute acid to the cold solution, a precipitate of a brighter colour ensues, and this is the preparation of the Pharmacopœia. In my Lectures, I have usually endeavoured to impress upon the student the nature of the chemical changes which take place by the follow-ing diagram.

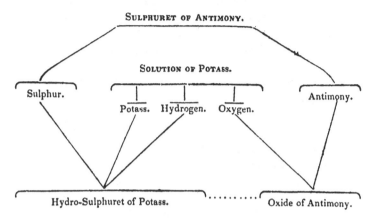

Sulphuret of Antimony and a *Solution of Potass* are the agents employed, which, it will be seen, furnish Sulphur, Potass, Hydrogen, Oxygen, and Antimony. During the boiling, the potass combines with the sulphur of the sulphuret of antimony, and forms Sulphuret of Potass; which, decomposing part of the water, attracts its hydrogen and becomes Hydro-sulphuret of Potass, while its oxygen converts the antimony into an oxide, which latter substance is dissolved by the alkaline hydro-sulphuret. As the solution cools, the affinities by which these bodies are held together undergo a change, and the oxide of antimony, falling down in combination with sulphuretted hydrogen, constitutes *Kermes Mineral.* When sulphuric acid is added a Sulphate of Potass is formed, and sulphuretted hydrogen is disengaged, while the oxide of antimony combines with the remaining sulphuretted hydrogen, and with the excess of sulphur; the difference therefore between the *Kermes* and the *golden sulphur* consists in the latter having a greater proportion of sulphur; if the one be called an *Hydro-sulphuret* of the oxide.

the other may be said to be an *Hydro-bi-sulphuret,*
or an *Hydroguretted sulphuret.* The preparation,
however, of the Pharmacopœia will not be found to be
either one or the other of these compounds, but a mixture of both. Other explanations have been offered,
but I am induced to believe that the one here given
approaches more nearly to the truth.

Metals of the Fourth Class.

720. The compounds which result from the combination of these metals with oxygen, are neither acid,
nor alkaline, but in the strict sense of the term *Oxides,*
and which cannot be reduced to a metallic state, without the aid of the affinity of some combustible body.

MANGANESE.

721. This metal has never been found native; and
such is the force with which it holds a certain proportion of oxygen, that considerable address is required
for its reduction. The black substance, known in commerce by the name of Manganese, is a per-oxide, contaminated, however, with carbonate of lime, the oxides
of iron, copper, and lead, and sometimes with a small
quantity of baryta. If a ball, made with the powder
of this oxide and pitch, be introduced into a crucible,
the empty space of which is filled with powdered charcoal, and submitted to the strongest heat that can be
raised for the space of an hour, we shall obtain metallic manganese. It is of a dusky white colour, very
brittle, and difficult of fusion.

722. Chemists have differed with respect to the
number and composition of the oxides of this metal.
Sir H. Davy admits only two, the olive, and the black;
Mr. Brande enumerates three; M. Thenard four; and
Berzelius, five. There appear, however, to be three

which are well defined. The *Protoxide* consisting of one atom of metal + one of oxygen; the *Deutoxide* of one + one and a half; and the *Peroxide* of one + three. Dr. Thomson, from the analysis of the sulphate, infers the equivalent number of manganese to be 28; that of the protoxide will therefore be $(28 + 8) = 36$; that of the deutoxide $(28 + 8 + 4) = 40$; and that of the peroxide $(28 + 8 + 8 + 8) = 52$. The peroxide is that which alone interests the medical student, and, on account of the facility with which it yields a portion of its oxygen, it proves a valuable chemical and phar-maceutical agent. By intense heat the peroxide passes into the state of deutoxide, but the highest elevation of temperature fails in producing any farther decom-position. This fact depends upon the operation of a law of chemical affinity which has been already con-sidered (269.)

723. Some chemists have supposed that the Perox-ide is susceptible of still farther oxidation, by which it acquires the properties of an acid. This hypothesis is supported by the phænomena exhibited by a substance which is termed *Cameleon Mineral*, and which is ob-tained by igniting together equal parts of black Man-ganese and nitre. This body has the singular property of exhibiting different colours, according to the quan-tity of water added to it.

Exp. 81.—Introduce into a wine glass a small piece of this substance, and pour over it a little more water than may be sufficient to cover it; a *green* solution will result; add more water, and the colour will become *blue*; by farther additions it will successively become *purple*, and *deep purple.*

Exp. 82.—Put equal quantities of this substance into two separate wine glasses, and pour hot water on the one, and cold water on the other. The hot solution will have a beautiful *green* colour, and the cold one that of a *deep purple.*

724. M. M. Chevillot and Edwards conceive that in this compound the Manganese is converted into *Manganesic* acid, which combines with the potass and forms a *Manganesiate.* Forckammer supposes that two acids exist, and that the different colours of the solution depend upon the conversion of the *manganeseous* into the *manganesic* acid.

ZINC.

725. This is a brilliant white metal, the specific gravity of which varies from 6·66 to 7·1. It melts at 680°, and the fused mass, on cooling, crystallizes.

726. There is but one known oxide of Zinc, which is produced either by combustion, or by bringing the vapour of water in contact with it when in the state of ignition.

Exp. 83.—Throw into a crucible, heated to whiteness, a fragment of Zinc; it takes fire, and burns with a beautiful flame of intense brilliance, from which a white oxide, mixed with a little carbonate, sublimes, having a considerable resemblance to carded wool, and hence it was formerly called *Lana Philosophorum.* Dr. Henry takes the equivalent of Zinc to be 33, and hence this oxide will be $(33 + 8) = 41$.

IRON.

727. This metal is more universally diffused than any other with which we are acquainted. Few mineral substances are free from it; it is also found in mineral waters, and in animal and vegetable bodies. It occurs in combination with sulphur, but its oxides and carbonates furnish the immense supplies which are re-

quired for the arts of life. It has never been found native; except in a state of alloy with Nickel, and which is therefore supposed to have a meteoric origin, for a similar alloy is found in *meteoric stones.*

728. The process by which Iron is produced from its ores constitutes the art of *Smelting,* and consists in decomposing them by the action of charcoal at high temperatures.

729. Iron has a blueish-white colour, and admits of a high degree of polish. It is extremely malleable, and although it cannot be beaten out to the same degree of thinness as gold or silver, it is extremely ductile, and so tenacious that a wire only $\frac{23}{1000}$ths of an inch in diameter is capable of supporting a weight of nearly 50 lbs. It requires the most intense heat of our furnaces for its fusion. *Cast Iron* contains oxygen, carbon, sulphur, and silica; and is converted into *wrought* iron by a process called *Pudling.* It consists in stirring the metal, when in a state of fusion in the reverberatory furnace, so as to bring every part in contact with the air and flame.

730. Iron combines with oxygen in, at least, two proportions; constituting the *Protoxide,* or black oxide, and the *Peroxide* or red oxide of iron.

731. The *Protoxide* is produced by burning iron in oxygen gas (*Exp.* 56), or by exposing the metal at an elevated temperature to the action of the air; the small fragments, for instance, which fly from a bar of iron during the operation of forging, consist of this oxide. Iron undergoes the same change, although more slowly, when exposed to the atmosphere, especially if the air be moist, when in common language it is said to rust; so again, when the steam of water is brought into contact with red hot iron, the latter is converted into the black oxide, and a torrent of hy-

drogen gas is evolved (*Exp.* 55). Water also, at the
temperature of the atmosphere appears to oxidize iron
very slowly (*Exp.* 77.); but it has been asserted that
this effect is not produced, unless atmospheric air be
present. In order to obtain this protoxide in a state
of purity, we may precipitate it from a solution of
sulphate of iron by potass, wash the precipitate out
of the contact of air, and then dry it at a low heat.

732. This protoxide constitutes the basis of several
of our medicinal preparations; and which therefore
require to be carefully preserved from the action of the
air, in order to prevent farther oxidation.

733. *Peroxide of Iron.* When the protoxide, or
iron itself, is dissolved in nitric acid, and boiled for
some time, the solution will yield a precipitate by ammonia; which must be washed, dried, and calcined in
a low red heat. This is the *peroxide,* or *red* oxide of
iron, which was formerly called *Saffron of Mars.* It
enters into several medicinal preparations, and appears
to possess activity when combined with bodies that
are capable of promoting its solubility.

734. These two oxides form distinct salts with the
acids, the characters and habitudes of which will be
hereafter considered. There appear also to be two
hydrates, or *hydro-oxides,* corresponding to the two
oxides which are obtained whenever we precipitate
their respective solutions in an acid, by a fixed alkali.
The substance termed *Ochre* is a native hydrate of the
peroxide, mechanically mixed with earthy ingredients.

735. The representative number of iron has been
considered as 28. The *Protoxide* consists of one atom
of metal and one atom of oxygen, its equivalent number will therefore be $(28+8)=36$. The *Peroxide,* of
one atom of metal and one and a half of oxygen
$(28+8+4)=40$; so that the oxygen of the latter is

not a multiplication of that of the former oxide by an entire, but by a fractional number. This, and similar anomalies, is best reconciled by multiplying by 2 the number expressing these proportions, which will make the ratio as 2 to 3, instead of as 1 to 1 $\frac{1}{2}$.

736. *Carburet of Iron.* Iron combines with carbon in various proportions; the most important of these combinations is *Steel,* and it is observed by Dr. Henry that there can scarcely be a more striking example of essential differences in external and physical characters being produced by slight differences of chemical composition, than in the present instance; for steel owes its properties to not more than from $\frac{1}{60}$ to $\frac{1}{140}$th its weight of carbon. This compound combines the fusibility of cast with the malleability of bar iron, and when heated and suddenly cooled it becomes very hard, whence its superiority for the manufacture of cutting instruments. Steel may at once be distinguished from iron by the action of nitric acid; for if a drop of this fluid be placed in contact with the former a black stain is occasioned, in consequence of the precipitation of carbon. *Plumbago,* or black lead, used in manufacturing pencils, and in forming crucibles, is a carburet of iron, in which, however the metal bears a very small proportion to that of the carbon.

737. *Sulphuret of Iron.* Iron unites with sulphur in two definite proportions, the products of which are the *Sulphuret,* or *Proto-sulphuret,* consisting of one atom of metal and one of sulphur $(28 + 16) = 44$, and the *Bi-sulphuret,* consisting of one atom of metal and two of sulphur $(28 + 16 + 16) = 60$. If we obtain by art any sulphuret, in which the proportion of sulphur does not agree with these views, the variety is occasioned by the mechanical admixture of metallic iron. By art, however, the *proto-sulphuret,* or black sul-

phuret, is only to be obtained. Iron and sulphur combine with considerable energy. A mixture of one part
of iron filings, and three parts of sulphur accurately
mixed and melted in a glass tube, will, at the moment
of union, exhibit a brilliant combustion; and a paste
of iron filings, sulphur, and water will burst, after some
time, into flame. If a roll of sulphur be rubbed on a
bar of iron, while of a glow heat, the compound of sulphur and iron falls down in drops, and the bar will be
rapidly corroded. This affords the best method of procuring a sulphuret for the production of sulphuretted
hydrogen (638). The *Proto-sulphuret* is readily
soluble in dilute acids, and gives during solution an
abundance of this gas. The *Bi-sulphuret*, or *yellow
Sulphuret*, is exclusively a natural product, very abundant, and called *Iron Pyrites*. It is nearly insoluble
in diluted sulphuric and muriatic acids, and gives no
sulphuretted hydrogen gas with acids.

738. The principal use of pyrites, is in the formation of *green vitriol* (sulphate of iron) for which purpose the ore is gently roasted and exposed to air and
moisture. Some varieties are spontaneously decomposed, and furnish this salt; the process is termed
Sulphatization. Animal matter would appear to exert
an agency the reverse of this; for if a solution of sulphate of iron be kept in contact with it, it is converted
into a sulphuret; this fact apparently explains the conversion of animal substances into pyrites, a phænomenon with which every fossilist is well acquainted.
Vegetables would also seem to possess the power of
deoxidizing some of the compounds of iron, and of
thus rendering them soluble in their juices. In short
these bodies appear to possess the disposition of passing readily and rapidly through a series of changes
and combinations; a property which has perhaps been

bestowed upon them in order more effectually to ensure the general distribution of an element so essential
to animal and vegetable existence, and so active in
producing many important changes in the mineral
œconomy of the globe.

TIN.

739. This metal has been known from very remote
antiquity. It has never been found native; but in the
form of oxide, it occurs abundantly in Cornwall, from
which the metal is obtained by heating it to redness
with charcoal. There are two oxides of Tin, and two
sulphurets of this metal, but they do not possess any
interest to the medical student.

BISMUTH.

740. This metal is found native; and in combination with oxygen, arsenic, and sulphur. To obtain it
pure, the Bismuth of commerce may be dissolved in
nitric acid, and the solution thus obtained be decomposed by water; the resulting oxide should then be
edulcorated, and reduced to a metallic state, by heating in a covered crucible with black flux.

741. Bismuth is a brittle white metal, with a slight
tint of red; and is composed of broad brilliant plates
adhering to each other. Its specific gravity is 9·822.
It is one of the most fusible metals, melting at 476°;
and it forms more readily than most other metals, distinct crystals by slow cooling. It is capable of being al-
loyed with most of the metals, and of forming with some
of them compounds of remarkable fusibility; 8 parts
of bismuth, 5 of lead, and 3 of tin constitute the *fusible metal* of Sir Isaac Newton, and will melt in water
before it reaches the boiling point. This property renders the alloy useful to the anatomist for taking casts,
and for some other purposes.

742. We are acquainted with only one oxide of Bismuth, which is formed during the combustion of the metal. From its analysis we obtain 71 as the equivalent number of Bismuth, and consequently 79 for that of the oxide.

COPPER.

743. Most of the Copper of commerce is obtained from *Copper pyrites,* or *yellow copper ore,* as it is called; which is a compound of sulphur, iron, and copper, in such proportions as to render it probable that it is composed of two atoms of proto-sulphuret of iron, and one atom of per-sulphuret of copper, with portions of arsenic and earthy matter. The sulphur and arsenic are separated by roasting; and the copper is obtained by repeated fusions, in some of which an addition of charcoal is made. Berzelius states that the copper of commerce is always contaminated with a little charcoal and sulphur, together with lead, antimony, and arsenic; from some experiments, however, which I made very carefully some years ago, I am inclined to believe that charcoal is never present. If, for chemical purposes, we require perfectly pure copper, it may be dissolved in strong muriatic acid; and, after adding water, may be precipitated from the solution by a polished plate of iron. The metal thus obtained should be washed, first with diluted muriatic acid, and then with water, and may be either fused, or kept in a divided form.

744. Copper has a fine red colour, and much brilliancy; it is very malleable* and ductile, and evolves a peculiar smell when warmed or rubbed. It melts at

---

* The Dutch Leaf, sold for gilding, is copper beaten into thin leaves.

a cherry red, or dull white heat. Its specific gravity is 8·8. It does not decompose water, which may even be transmitted, in vapour, through a red hot tube of this metal, without decomposition.

745. It is susceptible of only two degrees of oxidation, producing the *red*, or *protoxide* of copper; and the *black*, or *peroxide*.

746. The *Black*, or *Peroxide*, may be formed by calcining the scales which break off from copper by hammering the metal after exposure to heat; or by igniting the salt termed nitrate of copper. It is a useful agent in the ultimate analysis of vegetable matter, as will be hereafter explained.

747. The *Red*, or *Protoxide*, may be produced by dissolving a mixture of metallic copper, and peroxide of copper, in muriatic acid, and then adding Potass to the solution; a hydrated protoxide of an orange colour falls down, which, if quickly dried out of the contact of air, becomes of a red brown.

748. The number 64 has been deduced from the analysis of the Protoxide of copper, as the representative of the metal; and since the Protoxide consists of an atom of copper and an atom of oxygen, its equivalent will be $(64 + 8) = 72$; and that of the Peroxide, which is composed of one atom of metal and two of oxygen $(64 + 8 + 8) = 80$.

749. The oxides of copper combine with ammonia, and form compounds termed *Ammoniurets of Copper.* Peroxide of copper digested in ammonia, forms a bright blue liquid, from which by careful evaporation, the *Ammoniuret* may be separated in the form of fine blue crystals. Protoxide of copper is also soluble in ammonia, but it yields a colourless solution, which, however, becomes blue by exposure to the air, in consequence of farther oxidation.

Exp. 84.—Into a half ounce stoppered phial, filled
with a solution of ammonia, drop a few pieces of
metallic copper; if the bottle be left unstopped,
we shall soon obtain a beautiful blue liquid; if,
however, the stopper be replaced, this colour, in
a short time will disappear; and reappear on
again admitting the air. In this manner we can
alternately produce a blue, and colourless liquid,
as often as we withdraw and replace the stopper.
The theory of this change is obvious; the Peroxide, when excluded from the air, is converted
into a protoxide by the action of the metallic
copper, which again becomes the peroxide by the
action of the atmosphere.

LEAD.

750. This metal is chiefly obtained from the native
sulphuret. It has a bluish-white colour, and when
recently cut or melted, exhibits considerable lustre,
which, however, soon tarnishes; its specific gravity is
11·35; it melts at 599°; when exposed to a red heat,
with free access of air, it smokes and sublimes, and
affords a grey oxide, which collects on surrounding
cold bodies. It is also slowly oxidized by exposure
to the atmosphere at common temperatures; and more
rapidly, when exposed alternately to the action of air
and water.

751. There are three oxides of lead, viz. 1. The
Yellow Protoxide, obtained by heating the nitrate to
redness in a close vessel. It is tasteless, insoluble in
water, but soluble in potass and in acids, forming with
the latter a class of salts to be hereafter described. In
commerce this oxide is known by the name of *Massi-
cot.* When heated it undergoes a species of semi-vitri-
fication, and assumes the appearance of yellow scales

called *Litharge*, which is, to a considerable degree, volatile at a red heat. 2. The *Deutoxide of Lead*, may be obtained by exposing the protoxide to heat. It has a brilliant red colour, and is known in commerce by the name of *Minium* or *Red Lead*. This article, however, as it occurs in the market, contains much foreign matter, such as sulphate of lead, muriate of lead with excess of base, oxide of copper, silex, and a portion of the protoxide. The *Peroxide* is obtained by exposing the red oxide to the action of nitric acid, by which it is resolved into protoxide which is immediately dissolved, and into peroxide, which appears as an insoluble brown substance.

752. Accurate analyses have been made of these three oxides, from which it would appear that the quantity of oxygen which they relatively contain is in the ratio of 1, ·1½, and 2. The equivalent number of the metal is fixed at 104; and consequently that of the *Protoxide* will be $(104 + 8) = 112$; that the *Deutoxide* $(104 + 8 + 4) = 116$; and that of the *Peroxide* $(104 + 8 + 8) = 120$.

753. Pure water has no action on Lead, provided the air be excluded, but the combined influence of these agents converts the lead into a carbonate; a fact which is at once exemplified by the white line which is so constantly visible at the surface of the water standing in leaden cisterns. There is, however, a great difference in the corrosive power of different waters in relation to lead, depending upon the presence of various saline impurities.*

754. If Sulphuretted Hydrogen, or Hydro-sulphuret of ammonia, be added to water containing lead in

---

* This subject is very fully considered in my work on Medical Jurisprudence, vol. ii. p. 338.

solution, we shall immediately obtain an *Hydro-sul-phuretted oxide of Lead,* which is of a deep brown colour ; whence these compounds furnish very delicate tests of the presence of this metal.

755. Lead in its metallic state appears to be inert, but by combination it furnishes a series of active remedies, and virulent poisons, the history of which will fall under the head of metallic salts.

Metals of the Fifth Class.

MERCURY, or Quicksilver.

756. This is the only metal, with which we are acquainted, that retains a fluid form at the ordinary temperature of the atmosphere. Its colour is a brilliant white resembling that of silver, hence the names, *Quicksilver, Argentum vivum, Hydrargyrum.** Its specific gravity is 13·5. At about 39° or 40°, below zero of Fahrenheit, it becomes solid; at 660° it boils, and is converted into vapour, hence it may be driven over by distillation. At ordinary temperatures, however, it is also volatilized (322); a fact, for the knowledge of which we are indebted to Mr. Faraday, and which serves to explain several phænomena, which were previously unintelligible (*Medical Jurisprudence,* vol. 2. p. 456.)

757. *Oxides of Mercury.* There are two definite compounds arising from the union of Mercury and oxygen. The *Black,* or *Protoxide,* is obtained by long agitation of the metal in contact with the atmosphere, or by washing the substance termed *Calomel* with hot lime-water. By neither of these processes, however, is

---

* *Hydrargyrum,* from its fluidity and colour. *Quick* is an old Saxon word signifying *living,* an epithet applied to this metal on account of its mobility.

the Protoxide obtained in a state of absolute purity. In the former case, the product will be mixed with metallic mercury, in the latter, with a small portion of the peroxide. From the manner in which this oxide was usually prepared, Boerhaave gave it the name of *Ethiops per se.* It constitutes the *Hydrargyri Oxydum Cinereum* of the Pharmacopœia, and enters as an ingredient into several Officinal Preparations. It is an insipid powder, perfectly insoluble in water. The *Red, or Peroxide* is obtained by exposing the fluid metal, for several days, to nearly its boiling temperature, in a flat glass vessel with a very long neck,* by which contrivance the free access of air is admitted, while the length of the tube prevents the escape and waste of mercurial vapour, which condenses and falls back into the body of the vessel. This oxide appears in the form of brownish red or flea-coloured scales and crystals. It was formerly called *Præcipitate per se.* It has an acrid metallic taste, and is poisonous; it dissolves very sparingly in water. When distilled alone in a glass retort, it yields oxygen gas, and returns to a metallic state.

758. According to Fourcroy, Thenard, and Dr. Wollaston, the *Protoxide* consists of 100 metal+4 of oxygen; and the *Peroxide* of 100 metal+8 of oxygen. This will make the atom of Mercury to weigh 200; and the equivalent of the Protoxide will therefore be (200+8) = 208, and that of the Peroxide (200+16) = 216.

759. Mercury combines with Chlorine; but as the history of the resulting chlorides is materially involved in that of muriatic salts, it will be advantageous to consider the former, in connection with the latter.

---

* This apparatus was invented by Boyle, and is known by the name of Boyle's Hell.

760. *Sulphurets of Mercury.* By combination with sulphur, Mercury affords two distinct compounds. The *Proto-sulphuret* which is formed by triturating the metal with sulphur, or by pouring the former into the latter when in a state of fusion. It is a black tasteless substance which was formerly called *Ethiops Mineral,* and is the *Hydrargyri Sulphuretum Nigrum* of the present Pharmacopœia. The *Bi-sulphuret of Mercury,* or *Cinnabar* is obtained by heating the proto-sulphuret red hot in a flask, by which a portion of mercury is evaporated, and a sublimate of a steel grey colour is obtained, which, when reduced to a fine powder, assumes a brilliant red colour, and is called *Vermillion* or *Cinnabar,* and constitutes the *Hydrargyri Sulphuretum Rubrum* of the Pharmacopæia. It is inodorous and insipid; unalterable by exposure to air or moisture. When heated to redness in an open vessel, the sulphur is converted into sulphurous acid, and the mercury escapes in vapour. We perceive therefore that when used in fumigations, the effects produced are to be attributed to the volatilization of metallic mercury.

761. The composition of these Sulphurets has been investigated by Guibourt, from whose experiments it would appear that the *black,* or *proto-sulphuret* consists of one atom of metal and one of mercury, $200+16 =216$; and the red or *bi-sulphuret,* of one, and two, $200+16+16=232$.

SILVER.

762. To obtain this metal in a state of purity, we must dissolve the standard silver of commerce in pure nitric acid, diluted with an equal measure of water; and immerse a plate of clean copper into the solution; a precipitate of metallic silver will take place, which is

to be collected on a filter, boiled with a solution of ammonia, and then washed with water, and fused into a button. Or it may be procured by adding to the above solution, a solution of common salt; the precipitate which occurs is to be collected, washed, and dried, and then fused with its weight of sub-carbonate of potass.

763. Silver has a pure white colour, and a degree of lustre inferior only to that of polished steel; its specific gravity, after being hammered, is 10·51. It is so malleable and ductile, that it may be extended into leaves not exceeding the ten thousandth part of an inch in thickness, and drawn into wire considerably finer than a human hair. It melts at a bright red heat, and by slow cooling of the fused mass it may be made to assume a regular crystalline form; by a still higher temperature it may be volatilized. It is difficulty oxidized by the concurrence of heat and air. The *tarnishing* of silver is owing to its union with sulphur.

764. From some late experiments by Mr. Faraday there appear to be two distinct oxides of this metal; the first of which forms spontaneously as a pellicle on an ammoniacal solution of oxide of silver, exposed to the air; this protoxide, according to Dr. Thomson, consist of 3 atoms of silver and 2 atoms of oxygen. It is not capable of combining with the acids.

765. The *Peroxide of Silver* is obtained by decomposing the nitrate (*lunar caustic*) with a solution of lime, or baryta, and washing the precipitate. It is of a dark-olive colour, tasteless, insoluble in water, and when gently heated is reduced to a metallic state. It is composed of one atom of metal and one of oxygen; and as the representative number of silver has been fixed at 110, its equivalent will be $(110 + 8)$ $= 118$.

766. This oxide readily dissolves in ammonia, and a *fulminating* compound is obtained. If we pour a small quantity of the solution of ammonia on the oxide, a portion is dissolved, and a black powder remains, which is the detonating compound. When gently heated it explodes, and nitrogen and water are instantaneously evolved, and the silver is reduced.

767. Silver combines with Chlorine, producing a *Chloride*, which will be considered under the history of the muriate.

768. Silver in its metallic state furnishes the analytical chemist with a convenient material for the fabrication of crucibles, intended for the fusion of bodies with alkalies; for which purpose Platinum vessels cannot be employed. The utmost degree of heat, however, which they can bear is a moderate redness. They ought therefore to be heated in a sand bath.

GOLD.

769. As Gold is not important to the medical chemist, we shall refer the student who may require any knowledge of its history to other works on Chemistry.

PLATINUM.

770. This is an important metal to the experimental chemist, on account of the high temperature which it requires for its fusion, and which exceeds the greatest heat of our furnaces, and from its perfect unalterability by most agents. It is, however, acted upon by nitre and the alkalies, a property which diminishes its utility. When a Platinum crucible is employed it should be always put in a common one, to defend it from the direct action of the coal, the slack of which affixes it-

self to the sides and bottom with so much obstinacy, that it cannot be detached without risk of injury to the vessel.

771. This metal is discovered in small grains in South America, and which are also found to contain generally gold, iron, lead, and four other metals to which the names of Palladium, Rhodium, Iridium, and Osmium have been given. From these grains the pure metal is extracted, by a process which it is unnecessary to detail in this work.

772. If we precipitate a muriate of Platinum by ammonia, and expose the precipitate to a dull-red heat, we shall obtain the metal in a spongy form, and having the singular property of igniting on contact with hydrogen gas. This fact has been applied for the construction of an instrument for the production of instantaneously light; the same substance also affords to the chemist an instrument of considerable value in eudiometry, as already explained (504).

773. The history of metals has been hastily disposed of, since they possess little or no activity, as remedies, unless in combination. This probably depends upon their insolubility, for when they undergo a change in the stomach by which they become oxidized and then united to an acid, they prove active. This is the case with iron, and some others. The oxides would appear, in many instances at least, to possess a power inversely as their degree of oxidation; this is certainly true with respect to Antimony and Iron, the peroxides of which are comparatively inert, except they be in combination with acids. Arsenic undoubtedly affords an exception, but in this case the oxides are soluble. The Peroxide of Mercury is also more active than the protoxide, but it is more soluble.

SALTS.

774. The word *Salt* was originally confined to *common* salt; it was, however, afterwards so generalized as to include every body which is sapid, easily melted, soluble in water, and not combustible. In process of time this term was restricted to three classes of bodies, viz. *acids, alkalies,* and the *compounds* which acids form with alkalies, earths, and metallic oxides. The first two of these classes were called *simple* salts; the salts belonging to the third class were called *compound* or *neutral,* Chemists have lately still farther restricted the term, by tacitly excluding acids and alkalies from the class of salts altogether. At present then it denotes a compound in definite proportions of acid matter, called by Lavoisier the *salifying principle,* with an alkali, earth, or metallic oxide, termed the *salifiable base.* When the proportions of the constituents are so adjusted, that the resulting substance does not affect the colour of infusion of Litmus, or that of red cabbage, it is then called a *Neutral salt;* when, however, the predominance of acid is evinced by the reddening of these infusions, the salt is said to be acidulous; if, on the contrary, the acid matter appears to be in defect, or short of the quantity necessary for neutralizing the base, the salt is then said to be with excess of base. The composite nomenclature employed to denote the constitution of these bodies has been already explained (424).

775. The statement which has been just offered, respecting the constitution of neutro-saline bodies, although in many cases chemically correct, receives from the experiments of Sir H. Davy and Gay-Lussac some important modifications. Many saline bodies have been shewn by these illustrious philosophers to contain

neither acids nor alkalies, but to be compounds of their bases; common salt, for instance, which was long regarded as a *muriate of soda,* is proved to be a *Chloride of Sodium*; the same difficulty applies to the *Prussiates,* and yet these bodies have all the characters which were formerly regarded as peculiar to neutral salts.

776. The solubility of salts in water, is their most important general habitude. In this menstruum they are generally crystallized; and by its agency they are purified and separated from one another, in the inverse order of their solubility. As a medicinal process also, the solution of the salts deserves peculiar attention, on account of its connection with the efficacy of these bodies.

777. Salts have been divided into three classes, derived from the nature of their bases, viz. *Alkaline, Earthy,* and *Metallic salts.* These might now be all comprehended in the same division, since the Alkalies, with the exception of Ammonia, and the Earths have been shewn to be metallic oxides; in the present work, however, it will be convenient to retain the original classification. The genera have been determined by the acids which they contain, and the species by those of the base. As, however, the base generally imparts a very important medicinal character to the salt, I shall prefer a classification founded upon this principle; instead, therefore, of arranging together all the *muriates, sulphates, nitrates,* &c. &c. I shall bring together those salts which have the same base, as salts of soda, salts of potass, ammoniacal salts, calcareous salts, arsenical salts, &c. &c. It will, however, be first necessary to enumerate the general characters which alkaline and earthy salts derive from their peculiar acids.

778. 1. *Carbonates.* When sulphuric acid is poured upon them, they effervesce violently, emitting carbonic acid. If heated strongly, the carbonic acid is driven off, and the base remains. Some of the species require a very violent heat to be thus decomposed, but the operation is facilitated by mixing them with charcoal, which decomposes the carbonic acid altogether. The *alkaline* carbonates tinge vegetable blues green, and have an alkaline taste; those of the earths are insoluble, but dissolve when an excess of acid is added.

779. 2. *Muriates.* They effervesce with sulphuric acid, and evolve white acrid fumes of muriatic acid. They undergo no change when heated with combustibles. They melt and are volatilized at a high temperature. When mixed with nitric acid, they exhale the odour of chlorine. They are soluble in water, and raise the boiling point of that fluid. With the exception of Muriate of Ammonia, none of these bodies can be said to exist in a dry state; by heat the oxygen of the base, and the hydrogen of the acid, pass off, and true *metallic Chlorides* remain.

780. 3. *Sulphates* are insoluble in alcohol, which latter body precipitates them from water in a crystalline form. When heated to redness with charcoal, they are converted into *Sulphurets.* When barytic water, or a solution of any salt containing that earth, is dropt into an aqueous solution of any sulphate, a copious white precipitate falls, which is insoluble in acetic acid. Many of the sulphates combine with an excess of acid, and form acidulous, or super-salts.

781. 4. *Sulphites* possess a disagreeable taste and smell, analogous to that of burning sulphur. Exposed to moist air, they absorb oxygen, and pass into the state of sulph*ates.* They are decomposed by sulphuric acid, which expels sulphurous acid, and the salts are

converted into sulphates. When perfectly pure they are not affected by the solution of baryta.

782. 5. *Nitrates* are soluble in water, and capable of crystallizing by cooling. Sulphuric acid disengages from them white fumes, which have the odour of nitric acid. When heated along with muriatic acid, chlorine is exhaled. They are decomposed by heat, and yield at first oxygen gas; and when mixed with combustible matter produce, at a red heat, inflammation and detonation.

783. 6. *Chlorates*, formerly termed *Oxy-muriates*, or *Hyper-Oxy-muriates*, are soluble in water, and some of them in alcohol; when raised to a low heat, they give out a great quantity of oxygen gas, and are converted into common muriates or chlorides. When mixed with combustibles, they detonate with much greater violence than the nitrates; this detonation is also produced by friction and percussion; and sometimes takes place spontaneously. By the action of sulphuric, nitric, or muriatic acid, they evolve yellow or green fumes.

784. *Phosphates.* When heated along with combustibles, they are not decomposed, nor is phosphorus obtained. Before the blow-pipe they are converted into a globule of glass, which in some cases is transparent, in others, opaque. They are soluble in nitric acid without effervescence, and are precipitated from that solution by lime water. By sulphuric acid they are partially decomposed; and their acid, which is thus separated, when mixed with charcoal and heated to redness, yields phosphorus (645). After being strongly heated, they often phosphoresce. Like the Sulphates, they readily combine with an excess of acid, and form acidulous salts.

785. , *Prussiates*, or *Hydro-cyanates.* These salts

are all alkaline, even when a great excess of acid is employed in their formation; and they are decomposed by the weakest acids; they have, however, no permanency, unless they be united with some metallic oxide, and exist in the state of triple salts; mere exposure to the air is sufficient to decompose them. They have no useful properties.

The salts formed by the acetic, benzoic, citric, malic, tartaric, and other vegetable and animal acids, will be considered under the history of organic Chemistry.

1. *Salts with a base of Potass.*

786. These salts are soluble in water, and afford no precipitates with pure or carbonated alkalies. They produce a precipitate in muriate of platinum, which is a triple compound of potass, oxide of platinum, and muriatic acid. They are not changed by sulphuretted hydrogen, nor by ferro-prussiate of potass. Added to sulphate of alumina, they enable it to crystallize, so as to form Alum.

787. CARBONATE OF POTASS, or *Sub-carbonate of Potass.* This salt is known in commerce in different states of purity, under the names of *wood-ash*, *potash*, and *pearl-ash*. It is obtained in an impure state by the incineration of vegetables, and hence Potass has been termed the *Vegetable alkali.** This salt is imported from Russia, America, Treves, Dantzic, and

---

† On the other hand, *Soda* has been termed the *mineral* alkali, as being the base of rock salt. These distinctions, however, originally established by Avicenna, are not founded in truth. Potass, so far from being the exclusive product of vegetation, exists as a constituent part of *Granite*, which forms the foundation of our globe. It has also been discovered in various minerals. And although Potass be procured from vegetables, so is *Soda*, and vegetables probably derive the former from the soil in which they vegetate.

Vosges, and as it is a matter of great consequence to the merchant to ascertain the purity of the article, various easy methods of assaying it have been proposed, none, however, are so simple and efficacious as that proposed by Dr. Ure. He takes a tube divided into a hundred parts, in which marks are placed, with the words carbonate of soda, carbonate of potass, &c.; when any of the above substances are to be assayed, sulphuric acid of the specific gravity 1·146, is to be introduced up to the mark, and filled up to the hundredth part with water; this liquid is then to be dropped on a given weight of the sample, until its saturation is effected; when every measure of acid used will denote a grain of the alkali.

788. This substance has been termed *Salt of Tartar*, since by calcining *Tartar* (crystals of *bi-tartrate of potass*) we obtain it in a state of considerable purity; a process which may be advantageously adopted, when we wish to procure the salt for experimental purposes. In this operation the tartaric acid is decomposed and converted into the carbonic. In this manner the tartar will be found to yield about one-third its weight of dry carbonate. Or the tartar may be mixed with about an eighth of purified nitre, and wrapt up in paper in the form of cones, which may be placed on an iron dish, and set on fire; the residuary mass is to be lixiviated and evaporated to dryness. Or purified nitre may be mixed with a fourth of its weight of powdered charcoal, and projected into a red hot crucible, the contents of which are to be poured, when in fusion, into an iron dish. The carbonate, thus obtained, amounts to rather less than one half the nitre which has been employed. Carbonate of potass, when exposed to the air, attracts so much moisture as to deliquesce and pass rapidly into a liquid state, but the

water thus absorbed may be again expelled by a heat of 280°. When submitted in a crucible to a high temperature, it fuses; but none of its carbonic acid is expelled. It dissolves very readily in water, which at the ordinary temperature, takes up more than its own weight; the resulting solution has the specific gravity 1·54, and contains 48·8 per cent. by weight, of carbonate, or eight atoms of water to one of salt; and although the taste of this salt is much milder than the pure alkali, it still turns to green the blue infusion of vegetables, on which account it was for a long time termed a *Sub*-carbonate, but as it consists of one atom of potass and one of carbonic acid, it is now, in compliance with the atomic nomenclature (424) called a *Carbonate*. Its representative number will be (48+22) = 70.

789. The Bi-Carbonate of Potass, or, as it was formerly called, the Carbonate, a term still retained in the Pharmacopœia, is formed by passing a current of carbonic acid into a solution of the preceding salt; after this treatment, if the solution be slowly evaporated, it forms regular crystals which will be found to be the salt in question. Its taste is much milder than the carbonate, although still alkaline, and therefore in the original sense of the term a *sub*-salt. It is unchanged by exposure to the air, and requires for its solution, four times its weight of water at 60°; when boiling, that fluid dissolves five-sixths of its weight, but the salt is partly decomposed during the process, as is manifest from the escape of carbonic acid gas. By calcination at a low heat it is converted into the carbonate. According to Dr. Wollaston, the quantity of carbonic acid in the bi-carbonate is exactly double that in the carbonate, a fact which has been already proved by experiment 38. The atomic composition

therefore of this salt is one atom of potass $= 48$ and two atoms of carbonic $(22 + 22) = 44$, and one atom of water $= 9$, in all, 101, which last is its representative number.

790. The carbonate and bi-carbonate of potass are both decomposed by lime, which deprives them of carbonic acid; hence, by the use of this earth, we may procure caustic potass, or the *hydrated prot-oxide of Potassium*. The best process consists in boiling, for half an hour, in a clean iron vessel, carbonate of potass obtained by calcining tartar (788), with its weight of pure quick-lime, first slaked, in water. The ley is strained through clean linen, and concentrated by evaporation in a silver dish; the dry mass is then put into a bottle, and as much pure alcohol poured upon it as may be necessary to dissolve all that is soluble in that fluid. This alcoholic solution must be decanted, and distilled in an alembic of pure silver fitted with a glass head. When the alkali is in fusion it is to be poured into a silver dish, and when cold, broke into pieces, and preserved in well stopped phials.

791. MURIATE OF POTASS. This salt, which in old Pharmacy was called *febrifuge, and digestive salt of Sylvius,* and *regenerated tartar,* may be obtained by dissolving either hydrate or carbonate of potass to saturation in muriatic acid; as soon, however, as the solution is evaporated to perfect dryness, we are assured by the late researches of Sir H. Davy and other Chemists, that the salt is converted into Chloride of Potassium; the hydrogen of the muriatic acid uniting with the oxygen of the potassa, and forming water, which is volatilized. When, however, this substance is again dissolved in water, it becomes once more muriate of potass. In the state of dry salt it is composed of one atom of potassium, and one atom of chlorine, whence

its representative number will be $(40+36)=76$. But
in solution, when it must exist as a muriate, it may be
regarded as constituted of an atom of muriatic acid $=$
37, and an atom of potass $=48$, its equivalent number
will accordingly be $=85$. This view of the conversion
of one body into another by the mere action of water
is so contrary to all our preconceived and popular
opinions, that it is difficult for the student to give immediate credence to its truth; that a body in our hands
should be chloride of potassium, and in our mouths
muriate of potass, is a startling assertion, but the change
is not more extraordinary than that which takes place
by the action of water on an alkaline sulphuret (686),
the truth of which is at once announced to us by the
escape of sulphuretted hydrogen, of which our senses
can take cognizance. It would be vain to speak of the
taste of chloride of potassium, since it is impossible to
subject it to the test of this sense without converting it
into the muriate; the solution, however, is bitter and
disagreeable. The crystals are cubical in their form,
undergo little change on exposure to the air, and are
soluble in three times their weight of water at 60°, and
in a rather less proportion at 212°.

792. SULPHATE OF POTASS. This is the *sal de
duobus* of the older chemists, and may be formed
directly by saturating sulphuric acid by potass, and
crystallizing the solution. It is a refuse product, also,
of several chemical operations carried on upon a large
scale in the processes of the arts. It crystallizes in six-
sided prisms, terminated by six-sided pyramids with
triangular faces. Its specific gravity is 2·047. Its
taste is bitter; when thrown on a red hot iron it decrepitates, and is volatilized by a strong heat, first running
into fusion. It does not contain any water. Water at
60° takes up only one-sixteenth of its weight; but

when boiling it dissolves one-fifth, or by continuing the application of heat even one-fourth. This salt consists of one atom of acid $+$ one atom of base; its representative number therefore is $(40 + 48) = 88$.

793. Bi-Sulphate of Potass, or Super-sulphate, is formed by adding sulphuric acid to a hot solution of sulphate of potass. It is also formed during the operation of preparing nitric acid (569), and remains in the retort after the distillation. It has an intensely sour taste, and is much more soluble in water than the neutral sulphate, one part being dissolved by two of water at 60°, and in less than an equal weight at 212°. It is insoluble in water. According to Dr. Wollaston it contains just twice as much acid as the sulphate. It is therefore constituted of one atom of base with two atoms of acid $(48 + 40 + 40) = 128$; as it contains however an atom of water, we must add 9, making its representative number $= 137$.

794. Sulphite of Potass is formed by passing sulphurous acid into a solution of potass, and evaporating out of the contact of air. Rhomboidal plates are obtained white, of a sulphureous taste, and very soluble. By exposure to air they pass into sulphate of potass. This salt has the property of preventing the fermentation of syrup.

795. Nitrate of Potass. *Nitre.* * *Saltpetre.* This salt may be procured by saturating potass, or its carbonate, with nitric acid, but it occurs so abundantly in soils in the East Indies, that it may be purchased at a much less expense than that which would attend its preparation. In the state, however, in which it is imported it is extremely impure, containing besides

---

* The *Nitrum* of the ancients was Soda, thus Jeremiah ii. 22. " Though thou wash thee with *nitre,* and take much soap, yet thine iniquity is marked before me."

other substances, a considerable proportion of common
salt; in which state it is called *rough nitre*; for the
purposes of chemistry, it requires to be purified by
solution in water and re-crystallization; and it then
acquires the name of refined nitre, or refined Salt Petre.
In Germany and France it is artificially produced in
what are called *Nitre Beds.*† It crystallizes in six-
sided prisms, usually terminated by dihedral summits.
It is dissolved by seven times its weight of water at 60°,
and in its own weight at 212°; this solubility, however is considerably increased by the presence of common salt. It fuses at a moderate heat, and being cast
in moulds forms what is called *Sal Prunelle.** At a
red heat it is decomposed, and if it be distilled in an
earthen retort, or in a gun barrel, in consequence of
the decomposition of its acid, a large volume of oxygen

---

† It consists in throwing animal substances, such as dung, or
other excrements, with the remains of old mortar, or other loose
calcareous earths, into ditches dug for that purpose, and covered
with sheds, open at the sides, to keep off the rain. Occasional
watering, and turning up from time to time, are necessary to
accelerate the process, and to increase the surfaces to which the
air may apply; but too much moisture is detrimental. After a
succession of many months nitre is found in the mass, and may be
extracted by lixiviation. If the beds contained much vegetable
matter, a considerable portion of the nitric salt will be common
salt-petre, but in other cases, the acid will for the most part be
combined with calcareous earth, and the compound is made to
yield nitre by mixture with sub-carbonate of Potass. The nitrogen of the animal matter and the oxygen of the air appear in this
process to combine and form nitric acid, which then unites to the
potass furnished by the decomposed vegetables, or with the calcareous matter present in the mixture.

* So called from the resemblance which this substance originally bore to a Plum, it having been usual in Germany to colour
it purple.

gas will be evolved, and the alkali remains, which will corrode the earthen vessel in which it is heated. It is rapidly decomposed by carbonaceous matter in a high temperature, the products of which are nitrogen and carbonic acid gases, and a residuum of sub-carbonate of potass which was formerly termed *Clyssus* of nitre. By sulphuric acid it is decomposed, as we have already seen (568,) and nitric acid is evolved. It consists of one atom of nitric acid = 54 and one atom of potass = 48. Its representative number will therefore be 102.

796. CHLORATE OF POTASS; formerly, *Hyper-oxymuriate of Potass.* This salt may be formed by transmitting chlorine through a solution of potass, in Woolfe's bottles (176.) In this process the water is decomposed; its oxygen unites with one portion of the chlorine, forming chloric acid, while its hydrogen unites with another portion, and produces muriatic acid, and hence chlorate and muriate of potass are contemporaneously generated, and must be separated by crystallization, which is easily accomplished since the former salt will be the first to crystallize. Its crystals are in shining rhomboidal plates; its taste is cooling and rather unpleasant; its specific gravity is 2·; 16 parts of water at 60° dissolve one of it, and $2\frac{1}{2}$ of boiling water; in alcohol it is very sparingly soluble. By exposing it to a strong heat it fuses, and then gives off the purest oxygen. The effects of this salt on inflammable bodies are very powerful, and give origin to some of the most striking and brilliant experiments in Chemistry.

Exp. 86.—Mix a little sugar with half its weight of the chlorate, and pour over the mixture a small

quantity of concentrated sulphuric acid, a sudden
and vehement inflammation will ensue.*

Exp. 87.—To one grain of the powdered salt in a
mortar, add half a grain of phosphorus, it will
detonate, with a loud report, on the slightest
triture.

Exp. 88.—Put into an ale glass one part of phosphorus and two of the chlorate, and nearly fill
the vessel with water. Then by means of a glass
tube, or funnel, reaching to the bottom of the
glass, pour in three or four parts of sulphuric
acid; the phosphorus will take fire, and burn under the surface of the water.
See also experiment 2, page 84.

797. These phænomena depend upon the decomposition of the chloric acid; when sulphuric acid is
poured upon mixtures of this salt and combustibles,
intent ignition ensues in consequence of the evolution
of Euchlorine.

798. The above experiments require great caution,
and the salt ought never to be kept in a state of mixture with carbon, sulphur, or other inflammables, as
spontaneous explosion might thus occur.

799　Chlorate of Potass consists of one proportion
of chloric acid = 76 and one of potass = 48; its equivalent therefore is 124.

800. PRUSSIATE, OR HYDRO-CYANATE OF POTASS.
This salt may be produced by the mixture of its constituents, but such is its instability, that the weakest

* Matches coated with this mixture, by means of a little gum,
are sold with a small bottle of sulphuric acid, for the purpose of
producing instantaneous light. We must however take care to
keep the acid in a well stopt bottle, or it will become too dilute
to answer the purpose.

acids, even the carbonic, decompose it; in the form, however, of a triple salt with a metallic base, a more permanent compound is produced, the most useful of which is the following.

801. FERRO-CYANATE, or TRIPLE PRUSSIATE OF POTASS. If a solution of the above salt be digested with protoxide of iron, a portion of the latter body is dissolved, the solution becomes yellow, and if more hydro-cyanic acid be added, we obtain a neutral and crystallizable fluid, which will afford the salt in question. The more usual method, however, of forming it, is·by decomposing the pigment well known by the name of *Prussian blue,* by means of Potass. This body is a combination of Hydro-cyanic acid and peroxide of iron, together with a portion of Alumina.* In order to remove this latter substance, we should heat the Prussian blue of commerce with an equal weight of sulphuric acid which has been diluted with five or six parts of water, and afterwards well wash it with distilled water. It may then be powdered and added in successive portions to hot liquid hydrate of potass, until its colour ceases to be destroyed. The solution is then to be filtered, evaporated, and crystallized. The Prussian blue, by this treatment, is decomposed, it loses its blue colour, and the hydro-cyanic unites with the potass, and a portion of the oxide of iron.

---

* Prussian Blue is manufactured in several parts of Great Britain. Equal parts of sub-carbonate of Potass and various animal matters, such as hoofs and horns, dried blood, &c. are heated red hot in large iron stills; the fused mass is then laded out into iron pans, where it concretes into solid blocks, technically called *metals*; upon these when cold six or eight parts of water are poured. The solution is filtered, and mixed with a liquor containing two parts of alum and one of green vitriol (*sulphate of iron*). A precipitate falls, at first of a dingy green

802. This salt assumes the form of fine large transparent crystals, of a lemon yellow colour, and free from taste and smell; its specific gravity is 1·833. Water at 60° dissolves nearly one-third of its weight, and boiling water nearly an equal weight. It is decomposed at a high temperature.

803. To the Chemist this salt is of great value as a test of the presence of iron, and other metals. It decomposes all metallic solutions excepting those of gold and platinum. The following table, extracted from Dr. Ure's valuable Dictionary of Chemistry, presents a view of the colours of the precipitates thus produced.

Solutions of	*Give a*
Manganese	White precipitate.
Protoxide of Iron	Copious white.
Deutoxide of Iron	Copious clear blue.
Tritoxide of Iron	Copious dark blue.
Tin	White.
Zinc	White.
Antimony	White.
Cobalt	Grass green.
Bismuth	White.
Protoxide of Copper ...	White.
Deutoxide of Copper ...	Crimson brown.
Lead	White.
Deutoxide of Mercury ..	White.
Silver	White, becoming blue.

---

hue, but which, by copious washing with very dilute muriatic acid, acquires a fine blue tint. In this process the animal matter furnishes the Prussic acid, it being a new product arising from the reunion of some of its elements.

Salts with a base of Soda.

804. CARBONATE or SUB-CARBONATE OF SODA is obtained by the combustion of *marine* plants,* the ashes of which afford, by lixiviation, the impure salt termed in commerce *Barilla,* which is imported from Spain and the Levant in hard porous masses, of a speckled brown colour; in which state it is contaminated by common salt, and other impurities, from which it may be separated by solution in a small portion of water, filtering the solution, and evaporating it at a low heat; the common salt may be skimmed off as its crystals form upon the surface.* When required of great purity, it may be prepared by submitting the acetate to a red heat, which is thus converted into carbonate of soda and charcoal, the former of which is separable by water. Like the sub-carbonate of potass, it has strong alkaline characters. When its crystals are heated they fuse, and are converted into a dry white powder from the loss of their water. They are soluble in half their weight of water at 60°, and in rather less than their own weight at 212°. By exposure to the air they *effloresce.* They consist of one atom of Soda=32, and one atom of carbonic acid= 22; the representative number of the salt therefore, in its dry state, will be 54; but in its crystallized form it contains 11 atoms of water = 99, in which case 153 will be the equivalent number.

805. BI-CARBONATE OF SODA, or *Carbonate of Soda.* This is produced by a process similar to that described for the preparation of bi-carbonate of potass. When the solution becomes perfectly neutral, so as not

---

* Chiefly of the genus *Salsola,* which is said to be endowed with the property of decomposing sea salt, and absorbing the Soda, whence it acquired the name of *Saltwort.*

to affect turmeric paper, minute crystals of bi-carbonate of soda are formed, and being less soluble than the sub-carbonate, fall down in that state ; these are perfectly white, have but a slight, and not an alkaline taste, and are partially decomposed even at a very moderate temperature. Independently of their water of crystal-lization, the proportion of which Mr. Phillips thinks has not been clearly ascertained, the bi-carbonate contains double the quantity of carbonic acid con-tained in the carbonate, and therefore consists of 2 atoms of acid $22 \times 2 = 44$, and 1 atom of soda$=32$, its number is therefore 76. Although Mr. Phillips * thinks he has seen real bi-carbonate of soda, in the state of moist crystals, yet he has never met with any that was dry which had not lost one-fourth of its carbonic acid by exposure to heat ; it is then a white gritty powder, less soluble than the carbonate, like which it possesses an alkaline taste, and turns vegetable yellows brown, but both in a less degree. This salt sometimes crystal-lizes, it is decomposed by a red heat, like the true bi-carbonate, and dry carbonate of soda remains. He considers that this salt, the carbonate of the Phárma-copœia, to be a compound of one atom of carbonate and one atom of bi-carbonate,† combined with water, and to consist of 3 atoms of carbonic acid$=(22 \times 3 = 66)$; 2 atoms of soda $= (32 \times 2) = 64$; and 4 atoms of water $(9 \times 4) = 36$. Its representative number will accordingly be 166.

806. Muriate of Soda, or Chloride of Sodium. *Common Salt.* This salt exists abundantly in nature, in immense fossil masses termed *rock salt*, and in solu-

---

* Translation of the Pharmacopœia.

† Salts constituted of an atom of carbonate and an atom of bi-carbonate, are sometimes called *Sesqui-carbonates*, as being equivalent to an atom and a half of acid and one of base.

tion in water of the ocean. In the fluid form it is a true muriate of soda; but in a dry state it passes into a chloride of sodium; the observations which have been already made upon this subject, under the head of muriate of potass, will equally apply to the chemical history of the present saline body. It crystallizes into solid regular cubes, or, by hasty evaporation, in hollow quadrangular pyramids, which when perfectly pure, are but little changed by exposure to the air. The salt, however, as it usually occurs, is contaminated by muriate of magnesia, to which its tendency to deliquescence is attributable. Dr. Henry has shewn that the various forms under which this body appears, such as that of *stoved salt, fishery salt, bay salt*, &c. arise from modifications in the size and compactness in the grain, rather then from any essential difference in chemical composition. For solution, it requires twice and a half its weight of water at 60°; and hot water takes up very little more, hence its solution is made to crystallize by evaporation, and not by refrigeration (198). By heat it fuses and assumes the form of a solid compact mass. When suddenly heated, as by throwing it on red hot coals, it *decrepitates*. It is decomposed by sulphuric acid (543) as well as by nitric acid. This salt, in its dry state, consist of an atom of sodium=24, and an atom of chlorine=36; its equivalent is therefore 60.

807. SULPHATE OF SODA. *Glauber's Salts.* The production of this salt during the preparation of muriatic acid has been already explained. It forms crystals, the primitive form of which is an oblique rhombic prism. Its taste is bitter; on exposure to air it effloresces; in water it is very soluble, three parts at 60°, dissolving one part; at 212° it dissolves its own weight; when exposed to heat it undergoes watery

fusion. It consists of one atom of acid $=40$, one atom of soda $=32$, and ten atoms of water $=90$; its representative number therefore is 162.

808. Nitrate of Soda. This salt may be formed by saturating carbonate of soda with nitric acid, or by distilling common salt with three-fourths its weight of nitric acid. It crystallizes in the form of rhomboidal prisms.

809. Chlorate of Soda. This is prepared by a process analogous to that employed for the production of the similar salt with the base of potass, with which it nearly agrees in character.

810. Phosphate of Soda, may be obtained by saturating the carbonate of soda with phosphoric acid; when obtained in crystals by evaporation, it always contains an excess of base, and yet Mr. Dalton regards it as a *bi-phosphate*, and that in order to neutralize it the acid must be doubled, whence the neutral phosphate must be a *quadri-phosphate*.

Salts with a base of Ammonia, or Ammoniacal Salts.

811. These salts are distinguished by the following characters:—When treated with a caustic fixed alkali or earth, they exhale the peculiar odour of ammonia; they are generally soluble in water, and crytallizable; they are all decomposed at a moderate red heat; and if the acid be fixed, as the phosphoric, the ammonia comes away pure.

812. Muriate of Ammonia. *Sal Ammoniac.* This was originally fabricated in Egypt from the dung of the camel, and having been found in abundance near the temple of Jupiter Ammon, gained the name of *Sal Ammoniac*. It is now prepared on the large scale by various processes, the greater number of which consist

in obtaining an impure ammonia from animal substances by distillation, combining it with sulphuric
acid, and decomposing this sulphate by muriate of
soda; the muriate of ammonia formed from the mutual
action of these compound salts being sublimed in the
form of a solid dense mass, somewhat ductile, and
semi-transparent.* It is readily soluble in water, three
parts and a half of which, at 60°, take up one of the
salt; during its solution much caloric is absorbed; at
212°, it is still more soluble, and the solution, on cooling, shoots into regular crystals. In its dry state it is
composed of one atom of muriatic acid $=37$, and one
of ammonia$=17$; its equivalent number being 54; but
as it generally occurs it contains one atom of water$=9$,
which will make its number $= 63$.

813. CARBONATES OF AMMONIA. There are at
least two definite compounds of Carbonic acid and
Ammonia, *viz.* the *Carbonate,* consisting of one atom
of carbonic acid and one of ammonia; and the *Bi-
carbonate,* consisting of two of the former to one of
the latter. The *Sub-carbonate* of the Pharmacopœia,
if admitted as a third, must be denominated, as Mr.
Phillips proposes, a *Sesqui-carbonate,* being constituted
of one and a half atom of carbonic acid and one of
ammonia, or rather of three and two atoms; but it will
perhaps be more scientific to regard it as a compound
of an atom of the *carbonate* and an atom of the *bi-
carbonate.* The *Carbonate* may be produced by mingling over mercury one volume of carbonic acid, and
two volumes of ammonia. The *Bi-carbonate,* by im-
pregnating a solution of the common carbonate with
carbonic acid gas; it has no smell and but little taste,

---

* For an interesting account of the manufacture of this salt,
see Parkes's Chemical Essays, vol. 2. p. 437.

and is capable of crystallizing in small six-sided prisms. The *Sub-carbonate*, or *Sesqui-carbonate*, is prepared by subliming muriate of ammonia in contact with carbonate of lime, by which a double decomposition is effected, the muriatic acid attaching itself to the lime, and the carbonic acid to the ammonia. This salt, when recently prepared, has a crystalline appearance, and some transparency, and is hard and compact; its smell is pungent, and its taste sharp and penetrating. It affects vegetable colours like an uncombined alkali. It is soluble in about four times its weight of cold water, and by hot water it is decomposed with effervescence. When exposed to the atmosphere, it loses weight very fast, ceases to be transparent, loses its odour and becomes brittle, and is ultimately converted into the *bi-carbonate.*

814. SULPHATE OF AMMONIA exists in soot, and imparts to that substance its peculiar bitterness. It may be formed directly by saturating the sub-carbonate with sulphuric acid. It crystallizes in long flattened prisms with six sides, terminated by six-sided pyramids.

815. PHOSPHATE OF AMMONIA may be obtained by saturating phosphoric acid with ammonia; it also occurs as a common ingredient in the urine of carnivorous animals. When heated it fuses, swells, and is ultimately decomposed, leaving the acid in a glacial form.

Earthy Salts.

816. Many of these compounds retain the insolubility of their bases in water; others, however, are as soluble as the alkaline salts, but they are precipitated from their solutions by the carbonate of potass, although to this law there are some exceptions.

Salts of Lime, or *Calcareous Salts.*

817. They are generally sparingly soluble in water; the solutions of which are decomposed by the alkaline carbonates, potass, soda, baryta, by oxalic acid, and by carbonate of ammonia, but not by pure ammonia. The insoluble salts of lime are decomposed by being boiled with carbonate of potass, and afford carbonate of lime.

818. CARBONATE OF LIME. It occurs abundantly in the mineral kingdom, constituting immense mountains of limestone, chalk, marble, &c. while, in the animal kingdom, it forms the shells of innumerable animals. When subjected to a strong heat, the carbonic acid is driven off, and the base of lime remains. It is also decomposed by acids with effervescence, and the lime remains in combination with the acid employed. Although carbonate of lime is scarcely soluble in water, yet it becomes so through the medium of carbonic acid, as already shewn by *Exp.* 67. By heating the solution of lime in an excess of carbonic acid, we do not, as in the case of alkaline solutions (789), obtain a bi-carbonate, which is a salt still unknown; but the lime is thrown down in the state of simple carbonate, and the excess of carbonic acid escapes. It is constituted of one atom of lime $= 28$, and one atom of carbonic acid $= 22$. Its equivalent number will therefore be 50.

819. MURIATE OF LIME, or *Chloride of Calcium.* When muriatic acid is saturated with carbonate of lime, evaporated to dryness, and afterwards fused, the muriatic acid is decomposed; its hydrogen, uniting with the oxygen of the lime, escapes in the state of water, and the chlorine unites with the calcium, and forms *Chloride of Calcium.* When this body is exposed to the air, it soon deliquesces, as it is soluble in a very small quan-

tity of water, and becomes a true muriatè. Thus are the chloride and muriate mutually convertible into each other by adding or expelling water. The muriate has an extremely bitter taste, and is susceptible of crystallization, in which state it contains five proportionals of water.

820. SULPHATE OF LIME. *Gypsum.* This salt is found abundantly in nature, under the names of *Gypsum, Plaister of Paris, Alabaster,* &c.; it may also be immediately produced by the union of its elements. If sulphuric acid be added to a concentrated solution of the muriate of lime, a precipitate of sulphate immediately falls, so abundant as to impart an almost solid form to the fluid, and if this be gently calcined, we shall procure sulphate of lime. It is sparingly soluble, requiring 500 times its weight of cold water, or 450 of hot water. By a moderate heat it is fusible; after calcination, it absorbs water rapidly, and forms a cement, which renders it highly useful in taking casts. It is decomposed by alkaline carbonates, a double exchange of principles ensuing. Hence, hard waters, which owe this property to sulphate of lime, curdle soap, the alkali of which is detached by the sulphuric acid, and the oil is set at liberty. It consists of an atom of lime $= 28$, and an atom of acid $= 40$, and its equivalent number is 68.

821. CHLORATE OF LIME may be prepared synthetically; it is deliquescent, of a sharp and bitter taste, soluble in alcohol, and gives out oxygen gas when heated.

822. PHOSPHATE OF LIME. This salt constitutes 86 per cent. of bones, and may be obtained from them, by first dissolving them, when calcined and pulverized, in dilute muriatic acid, and precipitating the solution with pure ammonia. The precipitate, when sufficiently

edulcorated, is pure *Phosphate of Lime*, an insipid white powder, insoluble in water, but soluble in diluted nitric, muriatic, and acetic acids, and again precipitable, unaltered, from those acids by caustic ammonia. At a high temperature, it fuses into an opaque white enamel. It consists of 1 atom of acid $= 28$, $+$ 1 of base $= 28$, the compound atom $= 56$. According to various Chemists the phosphoric acid enters into other atomic proportions with lime, producing, a *Bi-phosphate*, *Tri-phosphate*, *Quadri-phosphate*, *Octo-phosphate*, and *Dodeca-phosphate*.

823. FERRO-CYANATE OF LIME. This compound, which is useful as a test of iron, may be formed by a process like that employed for the preparation of the similar salt with a base of potass.

Salts with a base of Baryta, or Barytic Salts.

824. These are generally insoluble in water, and undecomposable by fire; they are all poisonous, and are decomposed by the alkaline carbonates.

825. CARBONATE OF BARYTA. The relations of carbonic acid to this earth are similar to those which it bears to lime, the observations therefore which are offered under the history of the carbonate of that earth apply to the present salt. It is nearly insoluble in water. It is perfectly tasteless. It occurs abundantly in nature, and is useful only as a source of pure baryta and its salts.

826. MURIATE OF BARYTA, or *Chloride of Barium.* This salt in the state of *chloride* may be formed by heating pure Baryta in chlorine; or, in the state of muriate, by dissolving the carbonate in dilute muriatic acid. It is useful to the Chemist as a test for sulphuric salts, since this earth has a more powerful affinity than

any other base for sulphuric acid, with which it forms an insoluble precipitate.

Salts with a base of Magnesia, or Magnesian Salts.

827. These are generally soluble in water, and bitter; and afford precipitates of magnesia, and of carbonate of magnesia, upon the addition of pure soda, and of carbonate of soda; magnesian salts, when added to ammoniacal salts, containing the same acid, quickly deposit crystals of a triple ammoniaco-magnesian salt.

828. Sulphate of Magnesia. This salt, having been originally obtained from the springs of Epsom in Surrey, still retains the popular term of *Epsom Salts.* For a long time it was procured from the *bittern* remaining after the preparation of common salt from sea water; thus obtained, it was usually mixed with so considerable a quantity of muriate of magnesia, that owing to the deliquescent property of this salt, it was damp. This objection, however, is now overcome by a process in which the muriate is removed (153), and the article thus purified appears in the market under the name of *double refined Epsom Salts.* It has lately been prepared from magnesian limestone, by an ingenious process, invented by Dr. Henry. Sulphate of magnesia crystallizes with great readiness, and although the crystals are usually small, they may be obtained of considerable size by slowly cooling the solution. It is an extremely bitter salt, readily soluble in cold, and still more so in hot water, the former dissolving an equal weight, the latter one-third more; it is unalterable by exposure to air, but when heated it loses its water of crystallization. It is constituted of 1 atom of acid $= 40 + 1$ of magnesia $= 20$, and 7 atoms of water $= (9 + 7)$ 63; so that its representative number is 123.

It is one of our most popular purgatives; for which purpose, and for that of furnishing the carbonate, it is principally used.

829. AMMONIACO-MAGNESIAN SULPHATE. This triple salt is obtained by pouring a solution of pure ammonia into that of sulphate of magnesia, when a portion only of the earth is precipitated, the rest remains in solution; and, by evaporation, the triple salt, consisting of sulphuric acid, magnesia, and ammonia, is formed. It crystallizes in octohedrons.

830. SULPHATE OF MAGNESIA AND SODA. This triple salt may be frequently observed in parcels of sulphate of soda; the rhomboidal figure of its crystals will enable us to recognize it.

831. CARBONATE OF MAGNESIA. The sub-carbonate of the Pharmacopœia is procured by mixing together concentrated and hot solutions of carbonate of potass and sulphate of magnesia; by which a double decomposition is effected; the sulphuric acid and potass forming, by their union, sulphate of potass, which remains in solution, while the carbonic acid and magnesia combine to form an insoluble compound which is precipitated. With respect to the exact composition of this body, some doubts may be entertained. It does not appear to be fully saturated with carbonic acid, whence Berzelius is of opinion that it is a compound of three atoms of carbonate of magnesia with one atom of the hydrate of the same earth. By a very moderate heat the carbonic acid is expelled, and pure magnesia remains; hence this earth is commonly called *Calcined Magnesia.*

832. BI-CARBONATE OF MAGNESIA. If a stream of carbonic acid be transmitted through water, in which common magnesia is kept mechanically suspended, we shall obtain after a few days a congeries of crystals,

which have been considered as a true bi-carbonate.
Dr. Henry, however, does not admit the existence of
such a compound, and regards it as a carbonate, consisting of one atom of carbonic acid, one atom of
magnesia, and three atoms of water. The solution of
magnesia in an excess of carbonic acid, is highly useful in certain calculous disorders.

833. PHOSPHATE OF MAGNESIA. This may be
formed by adding the carbonate of magnesia to phosphoric acid; or, by mixing solutions of sulphate of
magnesia and phosphate of soda; from which, after the
repose of a few hours, large transparent crystals of
Phosphate of Magnesia will appear. These crystals
are soluble in 15 parts of cold, but in a smaller proportion of boiling water.

834. AMMONIACO - PHOSPHATE OF MAGNESIA.
This salt derives its interest from its forming a part of
certain urinary calculi, and from being deposited in
crystals on the sides of vessels in which urine has been
long kept. It may be artificially produced by mixing
solutions of phosphate of ammonia and phosphate of
magnesia, or any other soluble salt, with a base of that
earth.

Salts with a base of Alumina, or Aluminous Salts.

835. These are generally soluble in water. Taste
sweetish, and styptic. Ammonia throws down their
earthy base.

836. SULPHATE OF ALUMINA AND POTASS. *Alum.*
This important salt has been the object of innumerable
researches both with regard to its fabrication and composition. The greater part which occurs in commerce
is factitious, being extracted from various minerals,
called *Alum ores.* The most extensive alum manufactory in Great Britain is at Hurlett near Paisley; the

next in magnitude is that at Whitby. The aluminous slate, which abounds in sulphur, is submitted to a smothered combustion, by which 130 tons yield one ton of alum. The calcined mineral is digested in water, a ley is formed, to which muriate of potass is added, and sometimes urine. After due concentration the liquor deposits crystals of alum. The theory of this process is obvious; by the combustion of the sulphur, sulphuric acid is formed which unites to the alumina of the slate and forms sulphate of alumina, but as this salt requires the addition of an alkali to induce it to crystallize, muriate of potass is added for that purpose. Alum crystallizes in octohedrons, and has a sweetish astringent taste, and the specific gravity of 1.71. It dissolves in water, five parts of which, at 60°, take up one of the salt, but hot water dissolves about three-fourths of its weight. This solution reddens vegetable blues, which shews that the acid must be in excess; it has been therefore called a *super-sulphate.* When mixed with a solution of carbonate of potass, an effervescence is produced, and after the excess of acid is saturated, the alumina falls down in the state of a fine white powder. Alum, on exposure to air, effloresces superficially, but the interior remains long unchanged. Its water of crystallization is sufficient at a gentle heat to fuse it. If the heat be increased, it froths up, and loses 45 per cent. of its weight. The spongy residue is called burnt alum, "*alumen exsiccatum*" of the Pharmacopœia; a violent heat separates a great part of its acid. Considerable differences exist in the statements which have been given of the composition of Alum, as well as of the state of combination in which the potass and sulphuric acid exist. Mr. Phillips considers alum as being composed of 1 atom of bi-sulphate of potass $+2$ atoms of sulphate of alumina, $+22$ atoms of water;

2 F 2

whereas Dr. Thomson does not admit that bi-sulphate
of potass exists at all, but that alum is constituted of 1
atom of sulphate of potass +3 atoms of sulphate of
alumina +25 atoms of water. There is also an *ammoniacal alum*, in which ammonia is substituted for
potass; and a *soda alum*, the form and taste of which
resemble the common variety.

Metallic Salts.

837. These are either soluble in water, from which
they are precipitated by hydro-sulphuret of potass; or
insoluble, but fusible with borax into a coloured glass,
or with charcoal into a metallic button.

Salts of Arsenic, or Arsenical Salts.

838. As the metallic oxide in these compounds,
performs the functions of an acid, we have alkaline,
earthy, and metallic salts of Arsenic. We have moreover two classes of such salts; the one termed *Arsenites*, in which the metal is in its lowest state of oxidizement, constituting *Arsenious* acid; the other called
Arseniates, in which it is in its highest state of oxidizement, or in that of *Arsenic acid*. With the fixed
alkalies the Arsenious acid forms thick *Arsenites*,
which are not crystallizable, but are easily soluble.
They are decomposed by heat, and by all the acids ;
the former volatilizing, the latter precipitating, the
arsenious acid. With ammonia it forms a salt capable
of crystallization; it is, however, very easily decomposed by a gentle heat, when nitrogen is evolved, while
the hydrogen, uniting with part of the oxygen of the
acid, forms water. The *Arsenites of Lime, baryta,
strontia*, and *magnesia*, are difficultly soluble. These
salts may be all formed by boiling the acid in the alkaline or earthy solutions. The *metallic Arsenites* are

best formed by mixing solutions of the alkaline arsenites with the metallic salts. Arsenite of potass, for
instance, produces a white precipitate, or Arsenite, with
the white salts of manganese; a dingy green one with
the solutions of iron; a white one in solutions of zinc
and tin; mixed with a solution of sulphate of copper,
a precipitate of a fine apple-green colour falls, called
from its discoverer *Scheele's Green.* In the solutions
of lead, antimony, and bismuth, it forms a white precipitate. With nitrate of silver, a yellow arsenite which
is very soluble in ammonia. All these respective Arsenites, when heated by a blow-pipe on charcoal, exhale
the alliaceous smell peculiar to Arsenic.

839. The ARSENIATES differ essentially from the
Arsenites. They are all, with the exception of the
alkaline arseniates, insoluble in water; but most of
them become soluble by an excess of arsenic acid.
Phosphoric, nitric, and muriatic acids dissolve, and
probably convert into sub-salts, all the arseniates. The
whole of them, when decomposed at a red heat with
charcoal, yield the characteristic smell of garlic.

840. ARSENIATE OF POTASS may be formed by detonating in a crucible a mixture of arsenious acid and
nitre; it crystallizes in four-sided prisms, terminated
by very short four-sided pyramids. It is soluble in
about five times its weight of cold water. The *Arsenite of soda* is only half as soluble as the preceding salt,
and the liquid has alkaline properties, and when dropped
into most earthy and metallic salts occasions precipitates. The metallic arseniates differ in colour from the
arsenites of the same metals; with nitrate of silver, for
instance, an alkaline arseniate forms a brick-coloured
precipitate; with sulphate of copper, a bluish green
one, &c.

Salts of Zinc.

841. They are soluble; colourless; not precipitated
by any metal, nor by infusion of galls; but are precipitated white by alkalies, (in the state of hydrated
oxide) and by ferro-cyanate of potass, and sulphuretted
hydrogen.

842. SULPHATE OF ZINC. *White Vitriol.* This
metal is readily oxidized and dissolved by dilute sulphuric acid, hydrogen gas is given off (454) and a
colourless solution of sulphate of zinc results, which,
by evaporation, affords crystals in the form of four-
sided prisms, terminated by four-sided pyramids. This
salt, however, is prepared in the large way from some
varieties of the native sulphuret; the ore is roasted,
wetted with water, and exposed to the air, by which
means it undergoes the process of sulphatization (738).
The solution thus obtained is evaporated, and run into moulds. The crystals of this salt are colourless,
usually very small, and not readily distinguishable
from those of sulphate of magnesia; the salt has a disagreeable metallic taste; is not altered by exposure to
air; but, if moderately heated, loses its water of crystallization, and when subjected to a high temperature
is entirely decomposed, the acid being expelled, and
the oxide only remaining; it is soluble in $2\frac{1}{2}$ its weight
of water at 60°, and much more so at 212°. It is composed of 1 atom of acid $= 40 + 1$ of oxide $= 42$, with
6 atoms of water $= 54$; its equivalent number is therefore 136.

843. CARBONATE OF ZINC may be formed by adding carbonate of potass to sulphate of zinc. It occurs
also native, forming one of the varieties of the mineral
called *Calamine.* It consists of 1 atom of carbonic acid
$= 22 + 1$ of oxide $= 42$; its number therefore $= 64$.

Salts of Iron, or Chalybeate Salts.

844. They are, for the most part, soluble in water, and the solution is reddish brown, or becomes so by exposure to air; it affords a blue precipitate with ferro-cyanate of potass, or which will become so by exposure to air; and a black precipitate with hydro-sulphuret of ammonia. Infusion of gall nuts produces a black or deep purple precipitate.

845. SULPHATE OF IRON,—*Green Vitriol; Copperas.* This salt is prepared on the large scale by exposing roasted Pyrites to moisture (738). When metallic iron is acted upon by dilute sulphuric acid, the water is decomposed, the hydrogen evolved, and the oxygen united with the metal so as to form a *protoxide*, which immediately combines with the sulphuric acid, and constitutes sulphate of iron, or *proto-sulphate of iron*, which is obtained in the form of rhombic prisms by evaporation of the solution. These crystals have a strong styptic taste; redden vegetable blue colours; and are soluble in about two parts of cold, and in three-fourths their weight of boiling water, but are insoluble in alcohol. When exposed to a moist atmosphere they become encrusted with a brown substance, consisting of a salt composed of peroxide of iron and sulphuric acid; and which, being soluble in alcohol, may be removed by washing in spirit. When the crystals of the proto-sulphate are moderately heated, 100 parts lose 40 of water, and the residue consists of 1 atom of sulphate + 1 atom of water. Distilled at a stronger heat, they are decomposed, and yield sulphuric acid (628). According to Berzelius this salt is composed of 1 atom of sulphuric acid = 40, 1 of protoxide of iron = 36, and 7 atoms of water = 63. Its equivalent number is therefore 139. When a solution

of this proto-sulphate is heated with access of air, part of the protoxide passes to the state of peroxide, and, combining with a portion of acid, falls down in the form of a yellow powder which, according to Berzelius, is a sulphate of the peroxide, with excess of base, or a *sub-per-sulphate*. This effect, indeed, in a slight degree takes place on dissolving the salt in common water, the atmospheric air contained in which peroxidizes the iron, and produces a turbid solution, whence this salt offers the means of detecting free oxygen in aqueous fluids. For this reason, whenever the salt is administered for medical purposes in solution, the water ought to be previously boiled in order to expel the air.

846. Other sulphates, with a base of *per*-oxide of iron, have been formed, and which from their atomic composition have been termed *per-bi-sulphate*, and *per-quadri-sulphate;* but no sulphate of the protoxide with *excess of acid* has been hitherto discovered.

847. By boiling the green, or proto-sulphate, in nitric acid, and evaporating the solution to dryness, taking care that the heat does not rise so high as to expel the acid, we shall obtain a mass, from which water will extract a salt composed of sulphuric acid, and the peroxide of iron; the solution has a yellowish colour, and does not afford crystals; but when evaporated to dryness, forms a deliquescent mass, which is soluble in alcohol, by which it may be separated from the proto-sulphate. This salt has been commonly termed *oxy-sulphate*, but its more legitimate name would be *sulphate of peroxide of iron*, or, for the sake of brevity, *per-sulphate of iron*. It is constituted of 1 ½ atom of sulphuric $= 60 + 1$ of per-oxide of iron $= 40$; its equivalent therefore $= 100$.

848. CARBONATE OF IRON. We are acquainted only with one carbonate of iron, and that consists of 1 atom of the protoxide $= 36, + 1$ atom of carbonic acid $= 22$. It is therefore strictly speaking a *Proto-carbonate*. The combination is best effected by mingling the solutions of proto-sulphate of iron and carbonate of potass, when a double decomposition will take place; the sulphuric acid of the sulphate of iron combines with the soda of the carbonate, while the oxide of the sulphate of iron unites with the carbonic acid separated from the soda. The sulphate of soda remains in solution; but the carbonate of iron being insoluble is precipitated, and may be separated, and dried with a gentle heat. Such, however, is the eagerness with which the protoxide of iron combines with an additional dose of oxygen, that it is impossible to obtain a perfect carbonate by these means; during the act of desiccation, a portion of the iron becomes peroxidized, and carbonic acid escapes, for a solid *per*-carbonate of iron does not exist. The *Ferri Sub-carbonas* of the Pharmacopœia is a mixture of the carbonate, and peroxide of iron.

849. MURIATES OF IRON. There are two distinct muriates of this metal; the one, in which the iron is in the state of protoxide, being a *proto-muriate;* the other, in which the iron exists as a per-oxide, being a *per-muriate;* the former is green, the latter red; both are deliquescent and uncrystallizable. The green muriate is convertible into the red by simple exposure to the atmosphere. If these salts be evaporated to dryness, they are decomposed, the proto-muriate becoming *chloride*, and the per-muriate, *per-chloride* of iron.

Salts of Bismuth.

850. These salts have been very imperfectly examined; they present, however, but few points of interest to the medical student. Their solutions are colourless, are precipitated white by water, by infusion of galls orange, and by hydro-sulphurets, black.

851. NITRATE OF BISMUTH. If pieces of metallic bismuth be added to nitric acid, a solution will be obtained, from which the nitrate will crystallize in small four-sided prisms; when added to water it is decomposed, and a white substance is precipitated, which constitutes the well known pigment called *Magistery of Bismuth*, or *Pearl White*, it consists of hydrated oxide of bismuth with a small proportion of nitric acid; and the super-natant liquid consists of solution of bismuth with great excess of acid. The precipitate is now employed in medicine under the title of *Bismuthi Sub-nitras;* it is soluble to a considerable extent in pure ammonia, but not so largely in pure fixed alkalies. By sulphuretted hydrogen, and the vapours of putrefying animal substances, it is blackened (639.6).

Salts of Copper,—Cupreous Salts.

852. These salts are nearly all soluble in water, and of a blue or green colour. Ammonia, when added in excess to the solutions, produces a compound of a very deep blue. Hydro-sulphuret of ammonia throws down a black precipitate; and when a plate of polished iron is plunged into a liquid salt of copper, the copper is precipitated in a metallic form.

853. SULPHATE OF COPPER. *Blue Vitriol.* If the peroxide of copper be treated with sulphuric acid, we shall obtain from the solution a crop of crystals of a very beautiful blue colour, the composition of which

has been ascertained to be *bi-per-sulphate*, consisting of 1 atom of per-oxide of copper $=80+2$ atoms of sulphuric acid $=80+10$ atoms of water $=90$; making its representative number $=250$; or, exclusive of water of crystallization $=160$. This salt is soluble in four parts of water at 60°, and the solution is decomposed by pure and carbonated alkalies ; the former, however, redissolve the precipitate; if, for example, we add pure ammonia to the solution, a precipitate occurs, but on a farther addition of the alkali, it is redissolved, and a liquid of a beautiful bright blue colour, results. If the precipitate produced in the first instance be examined it will be found to be a *sub-sulphate*, the same salt is also thrown down by the fixed alkalies; it consists, exclusive of water, of 1 atom of acid $+2$ atoms of peroxide. This salt is present in the *Cuprum Ammoniatum* of the Pharmacopœia.

854. Dr. Thomson has also described a *quadri-sulphate*, consisting of 1 atom of base $+4$ of acid. No proto-sulphate has been hitherto discovered.

855. CARBONATE OF COPPER, is produced by the exposure of the metal to a damp air; or, artificially, by the addition of a carbonated alkali to a solution of copper. It consists of 1 atom of peroxide $=80+1$ atom of carbonic acid $=22$; making its equivalent $=102$; but in this state it must be considered *anhydrous;* in its usual form it contains an atom of water, which will make its number $=111$. *Verditer* is the blue precipitate produced by adding carbonate of lime to a solution of nitrate of copper.

856. MURIATES OF COPPER. There is a *protomuriate*, and *per-muriate*, or rather *bi-per-muriate*, of this metal; the former is produced by exposing plates of copper to the action of muriatic acid; it consists of 1 atom of protoxide $+1$ atom of acid; the latter is

obtained by dissolving the peroxide in muriatic acid. By evaporating these respective muriates to dryness they are converted into corresponding chlorides, *viz.* the proto-muriate into proto-chloride, and the per-muriate into per-chloride.

857. Nitrate of Copper. This salt is obtained by the action of dilute nitric acid on the metal (580), the solution yields prismatic crystals of a fine blue colour; if these crystals be gently heated, a portion of the acid is driven off, and a *sub-nitrate* remains.

Salts of Lead.   Saturnine Salts.

858. Some of these are soluble; others insoluble. The former have a sweetish austere taste, and are characterized by the white precipitate produced by ferrocyanate of potass; the brown, by hydro-sulphuret of ammonia; and the yellow, by hydriodate of potass. The latter, or insoluble salts, are dissolved by soda and potass, and by nitric acid, when the metal is rendered manifest by sulphuretted hydrogen. Heated by the blow-pipe upon charcoal, they afford a button of metal.

859. Nitrate of Lead is obtained by dissolving the metal in dilute nitric acid, and evaporating the solution. It crystallizes in tetrahedral or octohedral crystals, which are soluble in about $7\frac{1}{2}$ parts of boiling water without decomposition. They contain no water of crystallization, and consist of 1 atom of yellow protoxide $= 112 + 1$ atom of nitric acid $= 54$; its equivalent number being 166. A Sub-nitrate may be formed by boiling a mixture of equal weights of nitrate and protoxide of lead in water, filtering while hot, and setting it by to crystallize; it forms pearly crystals, and consists of 2 atoms of protoxide $= 112 \times 2 = 224$

+one atom of nitric acid$=54$; so that its equivalent number $=278$.

860. CARBONATE OF LEAD. *White Lead.* This is produced by adding an alkaline carbonate to a solution of the nitrate of lead. It is tasteless, and insoluble in water; but soluble in alkalies. It consists of 1 atom of protoxide $=112+1$ atom of carbonic acid $=22$. Its representative number therefore is 134.

Salts of Mercury. Mercurial Salts.

861. Some of the salts of mercury are soluble; others insoluble. The former furnish white precipitates with ferro-cyanate of potass, and black ones with sulphuretted hydrogen. A plate of copper immersed into their solutions, occasions the separation of metallic mercury. The latter are volatile at a red heat; and, if distilled with charcoal, afford metallic mercury.

862. SULPHATES OF MERCURY. The sulphuric acid does not act upon this metal, unless it be well concentrated and boiling. For this purpose, mercury is poured into a glass retort with about twice its weight of sulphuric acid. As soon as the mixture is heated, an effervescence is produced, and a portion of the acid is decomposed, and resolved into sulphurous acid and oxygen;, the former being dissipated in the gaseous state, the latter combines with the mercury and converts it into peroxide, and this uniting with the undecomposed acid forms a super-sulphate, or, more correctly speaking, a *bi-per-sulphate of mercury.* When the resulting mass is evaporated to dryness, and then acted upon by water, we obtain two distinct salts; the *bi-per-sulphate,*which affords, by evaporation, small, acicular crystals; and the *per-sulphate,* an insoluble substance, which on being well washed immediately assumes a bright lemon colour, and was formerly call-

ed *Turpeth Mineral.* The former of these salts consists of two atoms of sulphuric acid $=80$, and one
atom of peroxide of mercury $=216$; making its equi-
valent$=296$. The latter consists of one atom of each
constituent; its number therefore is $(216+40) = 256.$
There is besides a *Proto-sulphate of mercury,* consisting of one atom of the protoxide $+$ one atom of the
acid. This is formed by boiling mercury in its weight
of sulphuric acid, and washing with cold water the deliquescent mass so obtained, when the above salt will
be separated, which is white, and nearly insoluble,
requiring, for its solution, nearly 500 parts of water.
It crystallizes in prisms. Its representative number is
$(208+40) 248.$

863. MURIATES OF MERCURY. Under these terms
have been comprehended two of the most important of
the class of mercurial salts, commonly known by the
names of *Corrosive Sublimate* and *Calomel.* Before
the views respecting the composition of muriatic acid,
were developed by Sir H. Davy, *Corrosive Sublimate*
was considered as a compound of Muriatic acid, and
peroxide of Mercury; and was accordingly named
Oxy-muriate of Mercury, which is synonymous with
the less objectionable term *Per-muriate. Calomel* was
regarded as a combination of the same acid with the
protoxide of mercury, and was improperly called *Sub-
muriate;* for whether we apply the term *Sub* to denote
atomic composition, or chemical qualities, it is equally
inapplicable to Calomel, which, in every sense of the
word, is a neutral salt. It is true that less muriatic
acid enters into its composition, than into that of
Corrosive Sublimate, because the protoxide of mercury requires less than the peroxide for its saturation.
These salts, however, are now more justly considered
as compounds of Chlorine and Metallic Mercury, dif-

fering from each other only in the proportions of their elements. Nevertheless, in explaining their nature, and the processes by which they are formed, it will be advantageous to state both theories.

864. *Corrosive Sublimate. Oxy-muriate of Mercury, or Bi-chloride of Mercury.* If we bring a stream of chlorine in contact with boiling mercury, the metallic vapour will burn with a beautiful green flame, and corrosive sublimate will sublime. In this case, according to the theory which regarded chlorine as oxy-muriatic acid, its oxygen combined with the metal, and the liberated muriatic acid then united with the peroxide and formed a per-muriate, or oxy-muriate of mercury; but according to the latter views, the two elements at once unite, and form a bi-chloride of mercury. This body, however, is never prepared by direct combination, but by a process founded on the laws of double affinity. Bi-per-sulphate of mercury, formed by the process already described, (862) is rubbed with common salt, in an earthen mortar; the mixture is then introduced into a glass cucurbit, or subliming vessel, and subjected to a gradually raised heat, by which the salt under examination is formed and volatilized. According to the former theory the mixture acted upon consists of muriate of soda and sulphate of mercury, the elements of which are interchanged, giving rise to *Oxy-muriate of Mercury*, and *Sulphate of soda;* the former of which being a volatile compound rises in vapour, and is condensed on the upper part of the vessel. According to the latter theory, common salt consists of Chlorine and Sodium; when therefore this compound is acted upon by the sulphate of mercury, the Sodium unites with the oxygen of the metallic oxide and forms soda, which combines with the sulphuric acid, and produces

Sulphate of soda, while the chlorine combines with
the metallic mercury, thus revived, and constitutes *Bi-
chloride of Mercury.* According to the old theory
this salt is a *Bi-per-muriate,* and consists of

Muriatic acid....... 20·58, or 2 atoms of acid ... = 56*
Peroxide of Mercury. 79·42 — 1 ———— peroxide = 216

——————   ———
100·   Weight of its atom = 272

If considered, according to the modern theory, as a
Bi-chloride, it consists of

Chlorine........ .. 26·48, or 2 atoms of Chlorine = 72
Mercury .... 73·52 — 1 ——— Metal .. = 200

——————
Weight of its atom = 272

865. Corrosive Sublimate occurs in a white semi-
transparent crystalline mass; its specific gravity is 5·2;
its taste is nauseous and acrid, leaving in the mouth a
permanent metallic flavour. Water, at 60° dissolves
rather more than 1-20th, and boiling water 1-3rd of
its weight. According to the experiments of Dr.
Davy, light has no action upon this salt in its solid
state, but partially decomposes its solution, and *calomel,* or proto - chloride of mercury, is precipitated.
From this solution, the fixed alkalies, and lime water,
throw down a precipitate of red oxide; ammonia, a
white one, which is a triple compound of the acid,
metallic oxide, and alkali, and is identical with the
White Precipitate of the Pharmacopœia. Corrosive
sublimate is much more soluble in alcohol, ether, muriatic acid, and muriate of ammonia; with this latter body
it forms a double salt.

————————————————————————————

* The weight of an atom of Chlorine is 36; and if we regard
this body as a compound of muriatic acid and oxygen, the representative number of the muriatic acid will be 36—8=28.

866. PROTO-CHLORIDE OF MERCURY. *Sub-muriate of Mercury. Calomel.* This salt is prepared by triturating the sulphate with metallic mercury, until the globules of the latter disappear; common salt is then added, and the mixture is submitted to heat in earthen vessels, by which the proto-chloride is formed and separated from the mass by sublimation; it is afterwards ground to a fine powder, and then well washed with distilled water. If we consider corrosive sublimate as a bi-permuriate, Calomel must be regarded as a proto-muriate; in which case, the former is converted into the latter by losing half of its acid and oxygen, by the affinity of the metallic mercury for these bodies. But if we concur with modern chemists, and consider corrosive sublimate as a bi-chloride, the changes that take place during its conversion into calomel, are merely that one-half of the chlorine unites with the fresh portion of mercury added, so that the bi-chloride is converted into *proto-chloride.* Regarded as a *Muriate,* Calomel consists of

Muriatic acid .... 11·86 or 1 atom of acid .... = 28
Protoxide of Mercury 88·14 — 1 —— of protoxide = 208
—————   —————
100·..   Weight of its atom = 236

Considered as a *Proto-chloride,* it is composed of

Chlorine............ 15·25 or 1 atom of Chlorine. = 36
Mercury .... 84·75 — 1 —— of protoxide = 200
—————   —————
100·..   Weight of its atom = 236

867. Calomel is inodorous, insipid, and insoluble in water; its specific gravity is 7·175; by the long continued action of light, it assumes a dark colour, owing to partial decomposition. It is decomposed by the alkalies, and lime, but very imperfectly by their carbonates. By nitric acid it is converted into corrosive sublimate. By the action of the alkalies, a prot-

oxide is separated. If we consider it as a *muriate,* the
explanation of the decomposition is simply that the
muriatic acid has a greater affinity for the alkalies than
for the protoxide, in consequence of which the former
unite, and the latter is excluded; but if we regard it
as a proto-chloride, we must in that case suppose that
water is decomposed, and that its oxygen is transferred
to the metal, and its hydrogen to the chlorine.

868. NITRATES OF MERCURY. There are two distinct nitrates of mercury; the one with the protoxide,
the other with the peroxide, for their bases. Diluted
nitric acid, by acting on the metal in the cold, produces the first; strong acid, assisted by heat, the second. The solution of the *Proto-nitrate* yields, by
evaporation, large transparent crystals; it does not become milky when mingled with water, but the pure
fixed alkalies and ammonia give a greyish-black precipitate. The solution of the *Per-nitrate* is decomposed by water, which precipitates a *sub*-nitrate, while
a super salt remains in solution. If the nitrate be exposed to heat, it loses the greater part of its acid, and
is converted into brilliant red scales, commonly called
Red Precipitate. It is, properly speaking, a nitroxide,
for the peroxide always holds combined a portion of
nitric acid, whence this preparation is more active than
the red oxide prepared from the metal by heat.

Salts of Silver.

869. The soluble salts of silver are recognized by
furnishing a white precipitate with muriatic acid, which
blackens by exposure to light, and which is readily
soluble in ammonia; and by affording metallic silver
upon the immersion of a plate of copper. The salts
insoluble in water are soluble in liquid ammonia, and

when heated on charcoal, before the blow-pipe, they afford a globule of silver.

870. NITRATE OF SILVER. Nitric acid, diluted with two parts of water, readily oxidizes silver, and then dissolves it, and forms a nitrate, consisting of one atom of nitric acid $= 54 + 1$ atom of oxide of silver $= 118$, making the weight of an atom of the salt $= 172$. If the acid employed contain the least portion of muriatic, the solution will be turbid, and deposit a white precipitate; and should the silver contain copper, it will exhibit a greenish hue. The solution of this salt is caustic, and stains animal substances a deep black; hence it has been applied to the staining of human hair. By evaporation, transparent colourless crystals may be obtained, the primary form of which is a *right rhombic prism*. The crystals of this salt fuse when heated, and when cast into small cylinders it forms the *lunar caustic* of Pharmacy, the *Argenti Nitras* of the Pharmacopœia. The solution of nitrate of silver is of great value to the Chemist as a test for muriatic acid (252), by this acid the nitrate is immediately decomposed, and a *chloride* of silver is precipitated. This effect is produced by the operation of double affinity; the hydrogen of the acid passing to the oxygen of the oxide, and the chlorine uniting with the silver. Chlorine also decomposes nitrate of silver, singly, as do the solutions of chlorides and muriates in water. The chloride thus produced, although insoluble in water, is very soluble in ammonia; when heated in a silver crucible it does not lose weight, but fuses; and on cooling, concretes into a grey semi-transparent substance, which has been called *Horn-silver*, or *Luna Cornea*. This substance has the property of absorbing a considerable quantity of ammoniacal gas (325, *note*), which is again driven off by the application of heat. The

solution of nitrate of silver is also decomposed by sulphuric acid and its salts, which produce a nearly insoluble *Sulphate of Silver*, requiring for its solution 90 parts of water at 60°. By adding carbonate of potass to the same solution, we obtain an insoluble *Carbonate of Silver*.

Salts of Platinum.

871. These have been but little examined.

872. MURIATE OF PLATINUM. Nitro-muriatic acid is the readiest solvent of this metal; and we obtain a solution which is of use to the Chemist as a test for the presence of potass, with which it produces a yellow precipitate soluble in water.

873. SULPHATE OF PLATINUM is formed by acidifying the sulphuret by means of nitric acid. It is soluble in water, and according to Mr. E. Davy is a very delicate test for gelatine.

PART III.

ORGANIC CHEMISTRY.

874. Although the substances which we have now to consider are the products of the Animal and Vegetable kingdoms, they will nevertheless be found to consist of the elements of common matter, and might therefore, without impropriety, have been included under the history of inorganic bodies. They are, however, distinguished from the compounds hitherto described, by the increased number of their elementary atoms, and the delicately balanced affinities by which they are held united; they possess therefore a greater variation in qualities, with a less permanent or stable constitution; and hence, without a single exception, they are decomposed by high temperature, and are even liable to spontaneous changes from the reaction of their principles on each other, constituting the various phænomena of fermentation and putrefaction, by which new compounds arise that are serviceable to fresh processes for the support of animal and vegetable existence; so that, as it has been well observed, what to-day delights us in the fragrance of the rose may at a future season nourish us as bread, or exhilarate as wine. These observations, however, apply only to organic bodies which

are deprived of life, for the control of this mysterious principle is paramount to all the powers of chemical affinity, and is constantly active in opposing or modifying them ; nor can art supply any thing equivalent to this energy— " the powers of life," says M. Thenard,† " are not in the hands of man, that principle once lost, organization soon ceases, and is only subservient to His will, from whose bounty we received it ;" and if we view organic products in their passive relations, the complicated proportions of their elements, and the unknown mode of their union, will preclude the necessary conformation of our results by synthesis. For such reasons it becomes a matter of convenience to arrange organic products in separate divisions founded upon their natural history.

875. From the peculiar constitution of these bodies, there are two distinct species of analysis to which they may be subjected. The object of the one being to discover their elementary composition, or the *ultimate* principles into which they may be resolved ; that of the other to investigate the secondary compounds, or *proximate* principles, of which they more immediately consist.

876. The *Proximate Principles* may therefore be defined those compounds which exist ready formed in the organized body, and are separable from it, without farther decomposition, by simple processes of art. By the *Ultimate Elements*, we understand the common elements of matter, which, by combining in nicely adjusted proportions, give origin to the proximate principles. The latter may therefore be compared to the letters, the former to the syllables, which by infinitely

---

† See an Essay on Chemical Analysis, chiefly translated from the Fourth volume of the *Traité* de Chimie Élémentaire of L. J. Thenard, by J. G. Children, Esq.

varied mixtures and combinations, produce the words that compose the book of living nature.

877. The ultimate elements are few in number; vegetable substances seldom contain, as essential, more than three,—Oxygen,—Hydrogen,—and Carbon; and sometimes Azote. Animal matters generally contain all these, but with a much greater proportion of the latter principle. It is true that lime, phosphorus, several of the metals, and the two fixed alkalies, either pure, or more commonly in combination with some of the acids, are not unfrequent associates, but seldom in any considerable proportions. " With four elements then," says Mr. Children, (a brief alphabet for so comprehensive a history) " has a bountiful Omnipotence composed the beautiful volume of the living world."

878. Were we to assign to organic products a place in the *chemical* classification of bodies, it is evident that we must arrange the greater number of those of vegetable origin under the head of *Ternary Oxides* ; while those of the animal kingdom would, for the most part, fall under the division of *Quaternary Oxides*.

Ultimate Analysis of Organic Bodies.

879. In proceeding to the ultimate analysis of an organic product, we can rarely expect to obtain its elements in a pure and insulated state; on the contrary, we shall find that they form new arrangements with each other, and appear as compounds, which never constituted any part of the original structure. These results are called *Products*, in contradistinction to such compounds as are obtained by analysis in a state identical with that in which they pre-existed in the organic body, and which are therefore termed *Educts*.

880. For a long period, ultimate analysis was an extremely rude and imperfect process; and its results necessarily unsatisfactory and erroneous. It consisted

in exposing the body to be examined to *destructive distillation*, or, in other words, to the action of heat in close vessels, with a view to collect the products; from such a process, little could be expected, except some vague and general information; but the extent and value of this may be better appreciated, when it is stated that poisonous and esculent plants were found to yield the same results! At length a gleam of light burst upon this obscure subject from a new method of manipulation proposed by Gay-Lussac, and Thenard. It consisted in distilling the subject, in contact with some body that contained oxygen in so loose a state of combination, as to be driven off at the temperature of ignition, and by uniting with the whole of the *carbon*, to convert it into carbonic acid, and with that of the *hydrogen*, to generate water. Numerous analyses have been already conducted upon this principle, all of which conspire to prove that the elements of organic, like those of inorganic matter, are united in definite proportions; and farther, that the law of simple multiples holds strictly with respect to their combinations. This is a very important step in the advancement of these researches, as it enables us to correct, by calculation, the errors of experiment, and to estimate the proportions of elements, which would otherwise require complicated processes for their discovery.

881. The oxidizing substance first employed by Gay-Lussac was *Chlorate of Potass*; but this has been superseded by *Peroxide of Copper*,* which, being found to afford more accurate results, with a less complicated apparatus, and fewer difficulties of manipulation, is now universally preferred, especially in the analysis of *animal* matter. The instruments employed for the process have undergone successive

---

* Best prepared by igniting the pure nitrate of this metal.

improvements. To Dr. Prout we are indebted for an arrangement which enables us to apply heat to the tube charged with the object of analysis, and to receive the gaseous products, without the expense of a large mercurial cistern.† In this apparatus the heat is applied by means of a spirit lamp, which Dr. Ure considers not only as insufficient but too limited in its action, to ensure that uniform ignition of the whole tube, which is always desirable towards the close of an experiment. He therefore prefers a charcoal furnace. The difficulty, however, has been lately removed by Mr. Cooper, who has invented a most ingenious lamp, containing several wicks, disposed at intervals, so that by kindling and extinguishing any number of them, the intensity as well as the locality of the heat, can be regulated as the progress of the experiment may require. It would be inconsistent with the plan and object of this work to enter into minute detail, I shall therefore rest satisfied with presenting the student with an outline of the most improved process, and of the valuable discoveries which have followed its adoption.

882. A tube of crown glass, about 9 or 10 inches long, and 3-10ths of internal diameter, having a syphon-formed end for introduction under mercury, should be employed for the reception of the materials to be acted upon; from three to five grains of the solid matter, (which should be previously well dried by placing it in fine powder under an exhausted receiver (350) along with sulphuric acid) are to be first triturated in a glass mortar, and then mixed and again well rubbed and thoroughly incorporated with about 200 grains of the peroxide of copper. This mixture is then to be transferred into the glass tube, over which 20 or 30 grains

---

† A sketch of this apparatus, together with a detailed account of its mode of application, is to be seen in Dr. Henry's excellent work on Experimental Chemistry. Edit. 9. vol. 2. p. 167.

of the peroxide may be placed; and the remainder of
the tube filled with perfectly dry amianthus. Heat
may now be applied to the tube. The water given out
will be absorbed by the amianthus, and the quantity
will be announced by the increase of its weight. The
gaseous products, having been received over Mercury,
may also be measured with the nicest precision. The
whole of the operation, however, will require the most
scrupulous care, and the utmost skill of manipulation.
An embarrassment may also arise in the estimate of the
quantity of water formed, from the fact of peroxide of
copper rapidly attracting it from the atmosphere; it
should therefore be exposed for some time after its
ignition to the air, and the increase of its weight accurately noted, for which an allowance must be afterwards made. The same difficulty occurs in bringing
the various organic objects of research to a state of
thorough desiccation before they are mixed with the
peroxide. Dr. Ure therefore proposes to expose them
also to the air, after having made them as dry as possible, and to note, as in the case of the peroxide, the
quantity of water which they imbibe, that it may be
allowed for.

883. Such being the mode of manipulation, let us
inquire into the results which are produced. If the
substance submitted to experiment consist of carbon
and some incombustible body; the volume of carbonic
acid developed, will announce the proportion of the
former ingredient; for this purpose we have first to
find the weight of the carbonic acid, which is easily
accomplished by the rule of proportion, since 100
cubical inches will weigh 46·5 grains, at a mean of the
barometer and thermometer. Of this weight six parts
in 22 are pure carbon. Let us next suppose that we
are operating on a compound of carbon and hydrogen;
calculating, by the same method, the quantity of car-

bon which it contains, we have only to deduct this weight from the weight of the body under experiment, and the difference will give us that of the remaining element, hydrogen; at the same time the quantity of water produced should be ascertained, of which one part in nine is hydrogen; if the experiment has been successfully performed, the coincidence of these results will testify the highest degree of accuracy. But we will suppose that the substance, besides hydrogen and carbon, contains oxygen; in this case, we shall find on summing up the results, that the weight of the carbon and hydrogen will fall short of the original weight; if no other product has been formed, beside water and carbonic acid, the deficiency may very safely be placed to the account of oxygen; at the same time the quantity of oxygen lost by the oxide should be examined, which in such a case would fall short of that contained in the carbonic acid, and in the water. Should Azote have existed in the body, it will pass over in a gaseous state, and remain after the carbonic acid has been removed by a solution of potass, and the oxygen by any eudiometric substance.

884. In estimating the quantity of the evolved gases, it must be remembered that they are saturated with water, for this an allowance must be made by means of a formula by Dr. Ure, which is published in the Philosophical Transactions for 1818. In certain cases where the quantity of hydrogen is small, the same distinguished Analyst employs calomel instead of peroxide of copper, when the muriatic acid gas obtained will demonstrate its presence.

885. If the body to be operated upon is a fluid, Dr. Ure incloses it in a small glass bulb capable of holding three grains of water, and having a small pointed orifice; having first expelled the air by heat, the bulb is filled by immersing its orifice in the liquid,

it is then carefully weighed, and placed at the bottom of the tube, and covered with the requisite quantity of peroxide of copper.

886. By a series of experiments upon different vegetable bodies, conducted upon the principles above described, M. M. Gay-Lussac and Thenard have been enabled to establish some general views of great importance, viz.

1. A Vegetable substance is always *acid,* when the oxygen which it contains is to the hydrogen, in a proportion *greater* than is necessary to produce water; or, in other words, where the *Oxygen is in excess.*

2. A Vegetable substance is always *resinous, oily,* or *alcoholic,* where the oxygen is to the hydrogen in a *less* proportion than in water, or where there is an *excess of Hydrogen.*

3. A Vegetable substance is neither acid nor resinous, but saccharine, mucilaginous, &c. where the oxygen and hydrogen enter into its composition in the same proportions as in water, or where there is *no excess of either.*

887. To this general law, however, there would appear to be some exceptions. After the Proximate Principles of Vegetables have been considered, I shall introduce a table, collected from the best authorities, to exhibit the results of their ultimate analyses.

Proximate Principles of Vegetables.

888. These principles exist in vegetables either in a state of mixture, or slight combination, and are capable of being easily separated from each other. Upon one or more of such compounds the medicinal virtues of a plant depend; hence the great importance of the analysis of vegetables belonging to the Materia Medica, so far as relates to their proximate principles; the

knowledge it conveys enabling us to employ them with more discrimination, and to submit them to the proper pharmaceutic treatment.

889. The proximate principles are separable from the various plants in which they exist by processes adapted to the circumstances of each particular case. Sometimes they spontaneously exude from the growing vegetable, or are obtained from it by incisions made in its trunk; this is the case with several gums and resins; in some instances, an important principle is separated merely by heat; in this manner, essential oils, camphor, benzoic acid, &c. may be obtained comparatively pure. In other cases, the separation is effected by different solvents, of which the principles are hot and cold water, alcohol, æther, and a few of the acids. Several of these principles, when thus brought into solution, may be recognised by peculiar chemical tests; thus, Starch is precipitated blue by Iodine; Gallic acid, black, by the salts of Iron; Gum, in a curdy state, by sub-acetate of Lead, &c. By such means the proximate analysis of a vegetable is frequently conducted, for the purpose of ascertaining its composition, or of extracting from it some principle that may be rendered subservient to useful objects.

890. The number of proximate principles which are thus capable of being distinguished and separated from each other, is considerable; and of late years the list has been very greatly extended. They may, at present, be said to amount to upwards of forty; of which the following, from their importance, will require separate consideration.

1. Extractive.
2. Colouring Matter.
3. Gum.
4. Jelly.
5. Resin.
6. Starch.
7. Gluten.
8. Albumen.
9. Sugar.
10. Tannin.
    Caoutchouc.
11. Bitter Principle.
12. Fixed Oils.
13. Volatile Oils.

14. Wax
15. Vegetable Acids.
    Acetic.
    Benzoic.
    Citric.
    Gallic.
    Malic.
    Oxalic.
    Tartaric.
    Prussic, &c. &c.
16. Vegetable Alkalies.
    Morphia.
    Cinchonia.
    Quina, &c. &c.
17. Other Principles.
18. Lignin.

VEGETABLE EXTRACTIVE, OR EXTRACT.

891. The existence of a distinct principle, under this name, has been doubted by M. Thenard, Dr. Bostock, and Dr. Ure; and it is probable that future discoveries may resolve it into other known bodies. But until this has been effected, it will be necessary to consider it as a peculiar educt. The student must, however, be careful not to confound this supposed principle, with those parts of vegetables which are soluble in water, and may be obtained in a solid form by evaporation. These latter preparations, which in the language of Pharmacy are called *Extracts,* must necessarily be mixtures of those various principles which water is capable of separating from the particular plant submitted to the operation. Those Chemists who have maintained the identity of this principle, have exemplified its properties by a substance obtained by evaporating an infusion of Saffron, prepared with boiling distilled water at a temperature below 212°. It is said to be distinguished by the following characters. It is solu-

ble in cold, but more abundantly in hot water, and the solution is of a brown colour; this colour is heightened by alkalies; and when the pure solution of extractive is so diluted as to be deprived of colour, the addition of these latter bodies revive it; so that, under certain circumstances, they may be employed as Tests for detecting its presence. It is insoluble in alcohol and in æther, but it is soluble in alcohol containing a small portion of water. Acids precipitate it from its watery solution. By repeated solutions in water, and evaporations, it acquires a deeper colour, and becomes insoluble and comparatively inert; this change Fourcroy attributed to the absorption and fixation of oxygen. Saussure considered it to depend upon portions of its oxygen and hydrogen combining so as to form water, and consequently leaving an increased proportion of carbon; the former opinion, however, appears to be the most probable. Chlorine would seem to produce at once a similar change upon extractive. When the solution of this principle is slowly evaporated, it affords a semi-transparent mass; but rapid evaporation renders it perfectly opaque; when exposed to the atmosphere, it slowly imbibes water, and is imperfectly deliquescent. There is another character by which *Extractive* may be distinguished; depending upon the affinities which it exerts to alumina, and several metallic oxides; the latter of which precipitate it from its watery solution, their oxides forming with it insoluble compounds. We see therefore the danger of introducing such bodies into medicinal prescriptions, where the activity of the vegetable ingredient depends upon extractive.

892. From the characters above described it might be presumed that sufficient evidence of its existence, as a distinct and independent proximate principle, had been obtained; but in pursuing its history through the series of plants in which it is supposed to exist, we

find that the characters thus assigned to it are by no means uniform; indeed, there would appear to be almost as many varieties of extract as there are species of plants; the extractive of Saffron, for instance, which is considered as the principle in its least equivocal form, possesses properties which are not to be found in other instances; it is capable of assuming many different colours; its natural hue is yellow, and its aqueous solution becomes colourless by the action of the sun's rays. Sulphuric acid dropped into it imparts a deep indigo blue; nitric acid, a green; nitrate of mercury, a red; &c. whence Bouillon la Grange gave it the name of *Polychroite.* " It may, however, be doubted," says Dr. Henry, " whether these changes are not produced in some substance accompanying the extract, rather than in the extractive matter itself;" and it is by no means improbable that the other modifications of character which it presents may depend upon its union with different principles. The insolubility which it acquires by exposure to air is perhaps its most uniform and important peculiarity, and it should guard the pharmaceutic chemist against subjecting medicinal plants containing it as an essential constituent, such for instance as Senna, Peruvian Bark, &c. to those operations by which this change is liable to be produced.

Colouring Matter.

893. It is a question of great doubt whether the colouring matter* of vegetables resides in a particular proximate principle, different from any other; it would seem to present a too great variety in its relation to

---

* An artificial colouring matter may be made by digesting solutions of gall-nuts with chalk; a green fluid is obtained, which becomes red by the action of an acid; and has its green colour restored by means of an alkali.

chemical agents, to render such an opinion probable; from some substances it is extracted by water; from others, by alcohol or oil, according to the nature of the basis or substratum, in which it resides. It is in general very fugitive, particularly when blue and red. The yellow colour is most permanent.

Gum.

894 The first transition of the sap of plants appears to be into gum, and it therefore exists very abundantly in the vegetable kingdom. It is found in all young plants, in greater or less quantity; and is often so plentiful as to be discharged by spontaneous exudation. It abounds also in their roots, stalks, and leaves, and especially in their seeds. It appears from the experiments of Dr. Bostock, that there is a considerable variety in the chemical properties of different gums; but we take Gum Arabic as an example. It is an inodorous, insipid, and glutinous substance, soluble in water, in every proportion, and forming with it a thick viscid solution; in which state it is termed Mucilage. This, however, when evaporated and dried, is again brittle and soluble. It is insoluble in alcohol, æther, or oil, the former of which precipitates it from water, in opaque white flakes. This affords an easy mode of detecting its existence in vegetables; we have only to boil gently the matter to be examined with water; the gum will be thus dissolved, and if much of that principle be present, the solution will be glutinous. It may be allowed to remain until the impurities have subsided; then be evaporated to the consistence of thin syrup; when the addition of three parts of alcohol will separate the whole of the gum in flakes.

895. Mucilage may be kept for a long time without undergoing any change, but it ultimately becomes sour, owing to a partial spontaneous decomposition, and the

re-combination of part of its principles, so as to form acetic acid. It is precipitated, in a thick curdy form, by sub-acetate of lead;* and in the state of a brown semi-transparent jelly by the red sulphate of iron. The same effect is also produced by other salts; those, however, containing mercury and per-oxide of iron, are the most efficient in this respect. The oxides of copper, antimony, and bismuth are, also, acted upon by it; for it prevents water from precipitating them from their acid solutions, in the state of sub-salts. The effects, however, of re-agents on a solution of gum have been found to vary considerably in the different species, and many of them are rather to be attributed to the lime, and other foreign matters present, than to the pure principle. *Silicated Potass* is considered as the most sensible test of gum, as it throws down a white flaky precipitate from very dilute mucilage. Dr. Duncan, however, found that this effect is only produced by solutions of the lighter coloured specimens, which have different properties from those of darker colour. Dr. Bostock is also inclined to attribute the precipitation to the action of the lime present in the gum.

896. Gum is soluble in pure alkalies, and in lime water, from which it is precipitated unchanged by acids. It is also soluble in weak acids, but these agents in a concentrated state decompose it. Sulphuric acid, for instance, converts it into charcoal, acetic acid, and water; nitric acid exerts a peculiar action upon it, and produces results which are sufficiently distinct and characteristic to identify this proximate principle; a disengagement of nitric oxide first takes place, and a solution is obtained which, on cooling, deposits a peculiar acid, to which the name of *Mucic*, or *Saccho-*

---

* The precipitate thus produced was examined by Berzelius, who having found it to consist of gum and oxide of lead, assigned to the former the functions of an acid, and called the compound *gummate of lead.* He found it to consist of gum 100 + oxide of lead 62·105.

lactic acid, has been given ; some *malic* acid is also, at the same time, formed. If the heat be long continued, we shall obtain a quantity of *oxalic* acid, amounting in weight to nearly one half of that of the gum employed. Chlorine, transmitted through a solution of gum, changes it into *citric* acid.

897. When gum is submitted to destructive distillation, an acid substance passes over, which was formerly considered as a distinct body, under the name of *Pyro-mucous* acid. Late experiments, however, have proved it to be no other than the acetic acid, holding in solution a portion of essential oil, and some ammonia ; this last product would seem to indicate the presence of nitrogen in the gum ; but Dr. Henry is inclined to consider it as an accidental ingredient. After the distillation, a residuum of lime, and phosphate of lime, will be found in the retort.

898. Independent of the above earthy constituents Gum may be considered as constituted of carbon, oxygen, and hydrogen, in atomic proportions which have not hitherto been satisfactorily estimated.

899. *Cerasin* is a principle which was long considered as a species of gum ; it exudes from the cherry tree (whence its name), and is known in Pharmacy by the name of *Gum Tragacanth,* being the produce of the Astragalus Tragacantha. It differs from gum in some essential circumstances ; it is, for instance, strictly speaking, insoluble in water ; when plunged into cold water, it imbibes that fluid, swells, and forms a thick mucilage ; in boiling water it would appear to dissolve, but on cooling it again separates in the form of a gelatinous mass. Its habitudes with nitric acid resemble those of gum ; but by destructive distillation it furnishes a greater proportion of ammonia ; whence we are authorised in concluding that it contains more azote.

900. Various Gums occur in commerce, which ap-

pear to be mixtures of these principles in different proportions; that variety, for instance, termed *Dominica Gum*, consists of three parts of cerasin, and one of gum.

VEGETABLE JELLY.

901. This well known principle is obtained from the recently expressed juice of various acid fruits, by gentle evaporation. It is a tremulous soft coagulum, almost colourless after it has been well washed, and of an agreeable sub-acid taste. In cold water it is scarcely soluble, but in hot water it is abundantly dissolved; and when the solution cools, it again assumes a gelatinous form. By long boiling, however, it loses this property of coagulating; hence the necessity, in preparing jelly from certain fruits, of not submitting the expressed juice to protracted ebullition. Its solution in water is precipitated by infusion of galls. It seems probable that jelly is merely gum combined with some vegetable acid; for by exposing it on a sieve, an acid liquor drains off, and a hard transparent gum-like substance remains.

STARCH, FECULA, or FARINA.

902. This important principle exists chiefly in the white and brittle parts of vegetables, particularly in tuberose roots, and the seeds of gramineous plants. It may be mechanically separated from these parts by the processes of pounding or grating, and elutriation with *cold* water; by the former, the fecula will be disengaged from the fibrous matter in which it is involved, and by the latter it will be completely separated, the fibrous part subsiding, and leaving the starch diffused through the water in the form of a fine white powder; or the pounded or grated substance may be placed in a hair sieve, and the starch washed through with *cold* water, leaving the grosser matters behind.

903. In preparing Starch in the large way, the

grain is steeped in water until it becomes soft; it is then put into coarse linen bags, which are pressed in vats of water; a milky juice exudes, and the starch falls to the bottom of the vessel. The vats are then allowed to remain undisturbed for some time, by which the supernatant liquid undergoes a slight fermentation and becomes sour; this is an essential part of the process, since the acid thus formed dissolves some of the impurities in the deposited starch. The sediment is then collected, washed, and dried in a moderate heat, during which it splits into the columnar fragments which we meet with in commerce, and which are generally tinged slightly blue, by a preparation of cobalt.

904. Pure Starch is not soluble in water, unless when heated to 160°; and if the temperature be raised to 180°, the solution coagulates into a thick tenacious transparent jelly. This fact is one of great practical importance, and will suggest to the operator the necessity of regulating the heat in all those processes in which the perfect solution of this principle is required; the neglect of this precaution has frequently proved a source of loss and vexation to the brewer. By evaporation at a low heat, this jelly shrinks, and at length forms a transparent and brittle substance like gum, but is not soluble in cold water. It seems to be a hydrate, or a compound of starch and water. Starch by solution would appear to undergo some change, it cannot again be reduced to its original state, and it is found to have become more nutritive and digestible. The solution of starch in a large quantity of water is precipitated by the Sub-acetate of Lead: but not by any other metallic salt; it is also precipitated by the infusion of Galls, but the precipitate is re-dissolved on heating the mixture to 120°, and deposited again on cooling. Dr. Thomson considers this phænomenon of alternate solution and precipitation, by change of tem-

perature, as characteristic of Starch. The most distinctive property of starch, however, is that of forming with Iodine combinations of various colours; when the proportion of Iodine is small, the colour is violet; when somewhat greater, blue; and when still greater, black. The colour is manifested even at the instant of pouring a solution of Iodine, into a liquid which contains Starch mechanically diffused through it. Hence these bodies afford excellent tests for reciprocally detecting each other.

905. Starch is insoluble in alcohol and in æther, but the pure liquid alkalies act upon it, and convert it into a transparent jelly, which is soluble in alcohol. By acids this compound is decomposed, and the starch is recovered. Sulphuric acid, when concentrated, dissolves starch slowly; sulphurous acid is disengaged, and a considerable quantity of charcoal deposited.

906. By a high temperature, or by roasting, starch undergoes a very important change. At first it becomes yellow, then reddish brown; it afterwards swells, and exhales a penetrating smell. If the process be stopped, a substance is the result, which is known in commerce by the name of *British Gum;* in this state it is no longer sensible to the action of Iodine, and possesses many of the habitudes of gum, although it gives only oxalic acid, without a trace of *mucic* acid, when treated with nitric acid.

907. The Starch obtained from potatoes differs perceptibly from that of wheat; it is more friable; is composed of ovoid grains about twice the size of the other; it requires a lower temperature to reduce it into a jelly with water; it is soluble in more dilute alkaline leys, and is less readily decomposed by spontaneous fermentation. It also contains more hygrometric water. The experiments, however, of Berzelius have shewn that its chemical composition is not different from that

of the Starch procured from the Cerealia, In addition to these varieties of Starch, the following are met with in commerce, viz. INDIAN ARROW ROOT, obtained from the *Maranta Arundinacea*; SAGO, prepared from the pith of the palm tree, *Cycas Circinalis*, the brown colour of which is occasioned by the heat used in drying it; while its granular form is imparted to it before it is completely dry, by forcing it through apertures of the proper size. CASSAVA is prepared from the roots of the *Iatropha Manhiot*, a South-American plant, the juice of which is so virulent, that the Indians employ it for poisoning their arrows, and yet the starch which subsides from it, when well washed and dried is made into a nutricious bread. TAPIOCA is the same substance, under a different form, which it assumes on drying. SALOP is derived principally from the *Orchis Mascula*. There is a great variety of other plants* which, by appropriate treatment, might be made to yield a considerable proportion of nutricious fecula.

908. The different analyses which have been given of this principle are somewhat at variance with each other; according to Gay-Lussac it consists of $43 \cdot 55$ carbon, $+49 \cdot 68$ oxygen, $+6 \cdot 77$ hydrogen.

909. Starch is convertible into sugar; but this phænomenon will more properly fall under our notice when we treat of the nature and composition of this latter principle.

GLUTEN.

910. If a piece of *dough*, made with the flour of wheat, be repeatedly washed in cold water, we shall

---

* Starch has been found in the following plants:— Burdock (*Arctium Lappa*), Deadly Nightshade (*Atropa Belladonna*), Bistort (*Polygonum Bistorta*), White Bryony (*Bryonia alba*), Meadow Saffron (*Colchicum Autumnale*), Dropwort (*Spiræa Filipendula*), Buttercup (*Ranunculus Bulbosus*), Figwort (*Scrophularia nodosa*), Dwarf Elder (*Sambucus Ebulus*), Common Elder (*Sambucus nigra*), Henbane (*Hyoscyamus niger*), Water Dock (*Rumex Aquaticus*), Wake Robin (*Arum Maculatum*), Flower de.luce, or Water flag (*Iris Pseudacorus*), &c. &c.

resolve it into three distinct principles; into a mucila-ginous saccharine matter which is readily dissolved; into Starch, which will be mechanically carried off by the fluid; and into *Gluten,* which remains behind as a tough elastic substance. It has no taste, nor does it melt or lose its tenacity in the mouth; its colour is grey, and when dried becomes brown and brittle. It is very slightly soluble in cold water, requiring more than 1000 parts of that fluid for its solution; when this solution is heated, the gluten separates in the form of yellow flakes; it is also precipitated by acetate and sub-acetate of lead, by muriate of tin, and by several other reagents. It is insoluble in alcohol and æther. When kept moist, it ferments and undergoes a species of putrefaction, exhaling a very offensive odour; at the same time a species of acid is developed. In this state it is partially soluble in alcohol, and the solution may be applied to the purposes of varnish. All acids dissolve gluten, and alkalies precipitate it, but considerably changed, and deprived of elasticity. It undergoes a similar change when dissolved in pure alkalies, and is precipitated by acids. When suddenly heated it shrinks, then melts, blackens, and burns like a piece of horn. If it be submitted to destructive distillation, it yields the same products as animal matter, *viz.* water impregnated with carbonate of ammonia, and a considerable quantity of fœtid oil; a result which announces the presence of a considerable proportion of azote.

911. Gluten exists in the vegetable, from which it is procured in a dry pulverulent form, and owes its tenacity and adhesive qualities to the water which it imbibes during the process of elutriation. It is found in a great number of plants, and appears to be one of the most nutritive of the vegetable substances, and wheat is supposed to owe its superiority to other grain from the circumstance of its containing it in larger

quantities; its presence, moreover, contributes to the porosity or *lightness* of the bread manufactured of it, as will be more fully explained under the head of *Panary fermentation.* Sir H. Davy found that the wheat of different countries, and even that of our own country, sown at different seasons, contained different proportions of gluten. He examined different specimens of North American Wheat, all of which contained rather more gluten than the British. In general that of warm climates abounds most in this principle, and in insoluble parts; and it is therefore of greater specific gravity, harder, and more difficult to grind. The wheat of the south of Europe, in consequence of the larger quantity of gluten it contains, is peculiarly fitted for making *Macaroni,* and other preparations of flour in which a glutinous quality is considered as an excellence.

912. M. Taddei, an Italian chemist, has lately ascertained that the gluten of wheat may be decomposed into two principles, which he has distinguished by the names *Gliadine* (from γλἱα, gluten) and *Zimome* (from ζυμη, ferment.) They are obtained in a separate state by kneading fresh gluten in successive portions of alcohol, as long as that liquid continues to become milky, when diluted with water. The alcoholic solutions being set aside, gradually deposit a whitish matter, consisting of small filaments of gluten, and become perfectly transparent. Being now left to slow evaporation, the *Gliadine* remains behind, of the consistence of honey, and mixed with a little resinous matter, from which it may be freed by digestion in sulphuric ether, in which Gliadine is not sensibly soluble. The portion of the gluten not dissolved by the alcohol is the *Zimome.*

913. *Gliadine* has a faint odour, somewhat similar to that of the honey-comb; when gently heated its smell

is very like that of boiled apples. It becomes adhesive in the mouth and has a sweetish taste. It is moderately soluble in boiling alcohol, which loses its transparency in proportion as it cools, and then retains only a small quantity in solution. It softens, but does not dissolve in cold distilled water. At a boiling heat it is converted into froth, and the liquid remains slightly milky. It is specifically heavier than water. The alcoholic solution is rendered milky by admixture with water, and is precipitated in a flocculent form by the alkaline carbonates. It is scarcely affected by the mineral and vegetable acids. Dry Gliadine dissolves in caustic alkalies and in acids. It swells upon red hot coals, and then contracts in the manner of animal substances. It burns with a lively flame, and leaves a residuum of spongy charcoal, difficult of incineration. In some respects, it would seem to approach the character of resin, but differs from it in being insoluble in sulphuric æther. It is sensibly affected by the infusion of galls. It is capable of itself of undergoing a slow fermentation, and produces it in saccharine substances.

914. *Zimome* appears in the form of small globules, or in that of a shapeless mass, which is hard, tough, destitute of cohesion, and of an ash-white colour. When washed in water it recovers part of its viscocity, and becomes quickly brown, when left exposed to the air. It is specifically heavier than water. Its mode of fermentation is no longer that of gluten; for when it purifies it exhales a fetid urinous odour. It dissolves completely in vinegar, and in the mineral acids at a boiling temperature. With caustic potass, it combines and forms a kind of soap. When put into lime water, or into the solutions of the alkaline carbonates, it becomes harder, and assumes a new appearance without dissolving. When thrown upon red-hot coals, it burns with flame, and exhales an odour similar to that of

burning hair or hoofs. It is found in several parts of vegetables; and produces various kinds of fermentation, according to the nature of the substance with which it comes in contact.

915. To M. Taddei, the discoverer of Zimome, we are also indebted for a simple test for the detection of its presence. When powdered Guaiacum is worked up with gluten, or still better with pure Zimome, a most superb blue colour is produced. This change, however, is not effected, unless the contact of oxygen be allowed. This resinous body therefore furnishes a re-agent, capable of detecting the injurious alteration which flour sometimes undergoes by the spontaneous destruction of its gluten, and also of ascertaining in a general way the proportion of that principle.

Albumen.

916. The existence of this principle in vegetables is a discovery of recent times. It abounds in the juice of the Papaw-tree *(Carica Papaya)*; it is also found in mushrooms, and in the different species of funguses, in the emulsive seeds; in the almond for instance, 30 per cent. have been found, of a substance analogous to coagulated albumen. The juice of the fruit of the Ochra *(Hibiscus esculentus)*, according to Dr. Clarke, contains a liquid albumen in such quantities, that it is employed in Dominica as a substitute for the white of eggs in clarifying the juice of the sugar cane. The chief characteristic of Albumen is its coaguability by the action of heat or acids, when dissolved in water. According to Dr. Bostock, when the solution contains only one grain of albumen to 1000 grains of water, it becomes cloudy on being heated. When burnt it affords similar products to gluten, and probably differs but little from it in composition.

CAOUTCHOUC, or ELASTIC GUM.

917. This principle may be considered, from its
physical properties, as closely allied to gluten. It is
obtained from incisions made in the bark of two different trees, indigenous in Brazil; the *Hævea Caoutchouc,* and *Jatropha Elastica.** The white milky
juice which flows from them immediately concretes
into an elastic substance; and, when it is applied in
successive coats upon clay moulds, it forms the bottles
of *Indian rubber* which are imported into this country.
Its specific gravity is ·9335. It is destitute of smell
and taste, and in its pure state, of colour also. It is
combustible, and burns with a white flame, throwing
off a dense smoke, with a very disagreeable smell. At
a temperature, not much below that required for melting lead, it enters into fusion, and in this state, by a
process not hitherto published, Mr. Hancock of the
Strand, converts it into flat masses, which are convertible into a variety of useful instruments, such as tubes,
bags, &c.; it is also capable, by inflation, of assuming
globules of extreme tenuity. Caoutchouc is insoluble
in water and in alcohol; but it is acted upon by purified æther, and the solution so formed has been applied
to various œconomical uses. It is also soluble in rectified naphtha, and Mr. Mackintosh of Glasgow has
lately formed, by such an agent, a varnish for rendering
cloth water-proof. The oil of Cajeput, however, is the
best solvent of this substance, but the high price of the
article offers a practical obstacle to its adoption. According to a recent analysis of this substance by Dr.
Ure, its essential constituents appear to be carbon and
hydrogen, and to exist in proportions which will au-
thorize us in considering it as a *Sesqui-carburetted*
hydrogen.

---

* It exists also in a great variety of other plants, as *Ficus
Indica, Artocarphus Integrifolia,* and *Urceola Elastica.*

Sugar.

918. This principle, in its purest form, is prepared from the expressed juice of the *Arundo Saccharifera*, or Sugar cane, by a series of processes which may be found described and explained in the various works on general chemistry.

919. There would appear, from late experiments, to be several varieties of saccharine matter; there certainly exist *two* very distinct kinds in the juice of the cane; the one capable of crystallizing in six-sided prisms, the other uncrystallizable, and generally highly charged with colouring matter. These substances are familiarly known under the names of *White sugar-candy*, and *Treacle*. These ingredients are, to a certain extent, separated from each other by the processes of evaporation and filtration, in our Indian colonies; the *raw* sugar, however, which is imported, still contains a considerable admixture of the latter, and the process of sugar-refining, as practised in this country, entirely consists in effecting a more complete separation of the two. In what chemical circumstance these varieties of sugar differ from each other has not yet been well ascertained. The presence of a portion of extractive matter has by some been supposed to prevent the crystallization, while others, sanctioned by the results of ultimate analysis, have attributed this defect to an excess of oxygen.

920. Besides the juice of the cane, sugar may be extracted from a great variety of plants, and from ripe fruits. It appears also that a coarse sugar may be procured from grapes, which, although not so white, crystallizes more easily than that obtained from the cane. Chemists have differed widely with respect to the ultimate constitution of sugar; all agree in considering it as a compound of carbon, oxygen, and hydrogen, but

their experiments are at variance with regard to the proportions. It would, however, appear that the difference between the composition of starch and that of sugar is trifling, and that the former principle is accordingly easily converted into the latter by natural as well as artificial operations. This change occurs in the germination of seeds, as well as in the process of malting. M. Kirchoff of St. Petersburgh also discovered the curious fact of the conversion of starch into sugar by the agency of sulphuric acid; and Sir George Tuthill confirmed the statement by the following experiment. He digested a pound and a half of potato starch, six pints of distilled water, and a quarter of an ounce by weight of sulphuric acid, in an earthen vessel at a boiling heat; the mixture being frequently stirred, and kept at an uniform degree of fluidity by the supply of fresh water. In 24 hours there was an evident sweetness, which increased till the close of the process; at the end of 34 hours an ounce of finely powdered charcoal was added, and the boiling kept up two hours longer. The acid was then carefully saturated by recently burnt lime, and the boiling continued for half an hour, after which the liquor was passed through calico, and the residuum repeatedly washed with warm water. This, when dried, weighed seven-eighths of an ounce, and consisted of charcoal and sulphate of lime. The clear liquor, being evaporated to the consistence of syrup, and set aside, was in eight days converted into a crystalline mass, resembling common brown sugar with a mixture of treacle. The saccharine matter weighed one pound and a quarter, and its qualities were considered as intermediate between those of cane and grape sugar. Professor de la Rive, of Geneva, and M. Theodore de Saussure, by a farther investigation of these results, found that, during this process no gas is evolved; that the conversion proceeds equally well in close vessels; that no part of the sulphuric acid is decomposed, and that the weight of the sugar which is obtained exceeds that of the starch employed. Whence it is fair to conclude that the conversion of the starch into sugar is nothing more than its combination with water in its solid state, or rather with its elements. M. Braconnot has recently still farther extended our views concerning the artificial production of sugar and gum. He found that well dried elm dust, shreds of linen, &c. when treated with sulphuric acid, (sp. gr. 1·827) and afterwards diluted with water; and the acid saturated with lime, yielded, by evaporation, a gummy matter, which was convertible into a crystallizable sugar, by a farther boiling with dilute sulphuric acid. Nothing can more satisfactorily illustrate the facility with which one proximate principle is convertible into another; and strange as the statement may appear to persons not familiarized with chemical speculations, it is nevertheless indisputably true, that a pound weight of rags can be easily converted into more than a pound weight of sugar. To the medical student such researches have a particular interest, for they must hereafter point out the changes which aliment undergoes in its conversion into blood, and serve to explain a number of those changes which are perpetually taking place in the laboratory of the living body.

921. The properties of Sugar are interesting to the medical student, as it frequently enters into the composition of his remedies, and is capable of producing changes upon various bodies, with which he is bound to be acquainted. When perfectly pure, it undergoes no change on exposure to air, except that, in a damp atmosphere, it suffers a slight degree of deliquescence. When exposed to heat it fuses, becomes brown, evolves a little water, and is resolved into new arrangements of its component elements. The products of its destruc-

tive distillation are water, acetic acid, carburetted hydrogen, carbonic acid gas, and charcoal; and a little
oil, with a large proportion of pyro-mucous acid,
amounting to more than half the weight of the sugar.

922. Sugar is soluble in an equal weight of cold,
and almost to an unlimited amount in hot water. The
latter solution is termed *Syrup*, from which, by long
repose in a warm place, transparent crystals *(candied
sugar)* in the form of four or six-sided prisms, are deposited. Alcohol dissolves, when heated, about one-
fourth its weight of sugar.

923. The most interesting habitudes of Sugar are
those which it displays with the alkalies and earths.
The former bodies unite with it, and form compounds
destitute of taste. By adding, however, sulphuric acid
to their aqueous solutions, we form an alkaline sulphate, and if this be precipitated by alcohol, we recover
the sugar unchanged. Sugar also combines with some
of the earths; when dissolved in water at 50°, it is capable of dissolving one half of its weight of lime. The
solution is of a beautiful white wine colour, and has
the smell of fresh slacked lime. The lime is precipitated from it by the carbonic, citric, tartaric, sulphuric,
and oxalic acids; and the solution is decomposed, by
double affinity, by the carbonate, citrate, tartrate, and
oxalate of potass, &c. in which case the sugar combines
with the alkaline base, and the acid forms a salt of lime.

924. Sugar decomposes several of the metallic salts,
when boiled with their solutions; sometimes reducing
the oxides to a metallic state, and, in other cases,
merely reducing the oxide to a lower state of oxidation. With oxide of lead, it combines and forms an
insoluble compound to which Berzelius has given the
name of *Saccharate of Lead,* supposing that in this,
and similar instances, the sugar performs the essential
part of an acid. (446. 6.)

925. Sugar appears convertible into gum by the action of phosphurets, sulphurets, and hydro-sulphurets. The same change is also effected by keeping the caustic alkalies for some time in contact with sugar.

926. When concentrated sulphuric or muriatic acid be poured upon sugar, a decomposition arises, and a quantity of carbon is evolved. Nitric acid converts the sugar into an acid, to be hereafter described, called the *Oxalic acid*, and chlorine changes it into *Malic acid.*

TANNIN.

927. The *Tanning* principle, so called from the great importance of its application to the art of tanning skins, constitutes the principle of astringency in vegetable substances. It may be obtained by digesting bruised gall-nuts, grape seeds, oak bark, or catechu, in a small quantity of cold water. The solution, when evaporated, affords a substance of a brownish-yellow colour, extremely astringent, and soluble both in water and in alcohol, but not in æther. This must be regarded as *Tannin* in a very impure state, being mixed with extractive and other foreign matter; nor is it easy to devise any process by which it may be obtained in a purer form; those which have been proposed for this purpose producing an important change in the tanning principle which they were designed to separate. Sir H. Davy states that the purest form of tannin is that derived from bruised grape seeds. The Extract of Rhatany (*Krameria Triandra*) also affords it in a purer form than Catechu.

928. The following are its distinguishing properties. When evaporated to dryness, it forms a brown friable mass, which is soluble both in hot and cold water, affording a deep brown solution; of a sharp bitter taste; although it is insoluble in pure alcohol, the presence of a very small proportion of water enables

this liquid to act as a solvent. From its aqueous solution, it is precipitated by almost all the acids, with
which it forms insoluble compounds. By nitric acid,
however, as well as by chlorine, it is converted into a
yellowish brown matter, soluble in alcohol. Lime
water, and the alkaline carbonates, form also precipitates with it; the pure fixed alkalies separate it from
its concentrated solution; but ammonia does not occasion that effect. Certain metallic salts, such as acetate
of lead, sulphate of iron, tartarized antimony, &c.
throw down the tannin, whence such substances must
be considered as incompatible with its infusions. The
most striking effect, however, is produced upon its solution by that of isinglass, or any other animal jelly,
with which it instantly forms a dense coagulum insoluble in boiling water, and which in fact is *leather*, for
upon such a combination the art of tanning hides entirely depends. This fact naturally suggests a question
of great practical importance to the Physician; whether in cases of debility, where it is expedient to administer medicines containing tannin, we ought, at the
same time, to administer gelatinous food as nourishment. As mere Chemists we can feel no hesitation in
deciding in the negative; but, in another work,* I have
frequently observed that the laws of *Gastric* chemistry
differ from those which regulate the combinations of
the Laboratory; the matter, however, is doubtful, and
under such circumstances, it will be right, if we err,
to err on the safe side, and to prohibit the use of such
dietetic restoratives, during a course of astringent medicines.

929. Until the important discovery of Mr. Hatchett,
Tannin had been known only as a natural production;
this distinguished Chemist, however, succeeded in forming it artificially by digesting charcoal in dilute nitric

---

* PHARMACOLOGIA. Edit. 6. vol. 1. p. 339.

acid, during several days, by which a reddish brown liquor is obtained, furnishing, by careful evaporation, a brown glassy substance with a resinous fracture,* which differs only in one circumstance from natural tannin; *viz.* in resisting the action of nitric acid, by which all the varieties of this natural principle are decomposed.

930. The ultimate elements of tannin, according to the latest experiments, are 6 atoms of carbon + 4 ditto of oxygen + 3 ditto of hydrogen.

The Bitter Principle.

931. It is extremely doubtful whether the bitter taste of certain vegetables can be fairly attributed to the existence of a peculiar and independent principle. In the present state of our knowledge, however, it will be convenient to adopt the distinction, in order that we may better collect and arrange the several facts connected with the subject. In many cases it will be difficult to distinguish the bitter from the extractive matter, in others, as shewn in my analysis of Elaterium,† and in that of the Hop by Dr. Ives,‡ it is capable of complete separation.

932. The bitter principle may be obtained from the wood of quassia, the root of gentian, the strobiles of the hop, the fruit of colocynth, &c. by infusing them for some time in cold water. For our knowledge respecting this principle we are chiefly indebted to Dr. Thomson, I shall, therefore, here enumerate the cha-

---

* There are other processes for the factitious production of this substance, viz. by distilling nitric acid from common resin; or by the action of sulphuric acid on the same substance, as well as on elemi, assafœtida, camphor, &c. M. Chenevix also observed that coffee berries acquired, by roasting, the property of precipitating gelatin.

† Pharmacologia. vol. 2. Art. *Extractum Elaterii.*

‡ Ibid. Art. *Humuli Strobili.*

racters which he has assigned to it. When water, thus impregnated, is evaporated to dryness by a very gentle heat, it leaves a brownish-yellow substance, which retains a certain degree of transparency. For some time it continues ductile, but at last becomes brittle. Its taste is intensely bitter. When heated, it softens, swells, and blackens; then burns away without flaming much, and leaves a small quantity of ashes. It is very soluble in water, and in alcohol. It does not alter vegetable blue colours. It is not precipitated by the watery solution of lime, nor is it changed by alkalies. Neither the tincture nor infusion of galls, nor gallic acid, produce any effect upon it. Of the metallic salts, nitrate of silver and acetate of lead are those only which throw it down. The precipitate from the latter is very abundant; and it therefore furnishes the best test for discovering its presence, provided no other substances be present, by which, also, it is decomposed.

933. By digesting indigo, silk, and a few other substances in nitric acid, an intensely bitter matter is produced, which has been called by Welther, the *Yellow Bitter Principle*. It is susceptible of a regular crystallized form; burns like gunpowder; and detonates when struck with a hammer. On the whole, says Dr. Henry, it appears better entitled to rank as a distinct principle, than that which is extracted, by infusion, from vegetables.

Fixed Oils.

934. These are generally obtained from certain seeds by pressure. They greatly differ from each other in specific gravity, but which is below that of water. Rape oil has a specific gravity of ·913; almond oil that of ·932, &c. They are liquid, almost tasteless, unctuous to the feel, and give a greasy stain to paper, which is not removed by heat. When kept for any

time they undergo a change which is termed *rancidity*, in which state they are viscid, and appear to contain an uncombined acid. This change, however, would appear to be rather owing to the decomposition of the mucilage which they so generally hold dissolved, than to any material modification of the oil itself. They are commonly coloured, but they may be rendered perfectly colourless by digestion with animal charcoal. They are perfectly insoluble in water, but may be suspended and diffused in that liquid by sugar, gum, and other media, giving rise to a mixture, which is known in Pharmacy by the name of an *Emulsion.* They generally solidify at a temperature not so low as that required for the congelation of water; and some few, as Palm oil, &c. are solid at the ordinary temperature of the atmosphere, and have thence been called *Vegetable Butters.* The fixed oils are, for the most part, sparingly soluble in alcohol and æther, with the exception of castor oil, which has been found by Mr. Brande to dissolve in almost any proportion in alcohol of specific gravity ·820, and in æther of ·7563. When congealed oil is pressed between folds of blotting paper, a wax-like substance remains, to which has been given the name of *Stearin,* while the paper absorbs another portion which has been termed *Elain.*

935. The fixed oils unite with the alkalies, and form a compound called Soap.* This useful body is readily soluble in water, and in alcohol, and if the latter solution be concentrated it becomes gelatinous, and by distilling off the alcohol is made to surrender the soap in a beautiful transparent form. The solution is decomposed by acids, and by neutral salts with earthy

---

* The best soaps are made with olive oil and soda. In this country, animal fat is usually employed for the common soaps, to which resin and some other substances are commonly added. *Soft Soap* is a compound of potass with some of the common oils.

bases. Hence *hard* water curdles soap, the acid of the
salt in solution uniting with the alkali of the soap, and
thus liberating the oil. A solution of soap, therefore,
furnishes the Chemist with a useful re-agent for ascertaining the general quality of any water; should it produce a turbid appearance, we may infer that sulphate
of lime, or some other earthy salt exists, which must
render it unfit for the purposes of infusion, &c.

936. Two different changes are effected by boiling
the fixed oils with metallic oxides; if the quantity of
oxide be small, it combines with the mucilage, and at
the same time yields a portion of its oxygen to the oil,
by which it is rendered *drying*; that is, if it be thinly
spread upon a surface, instead of remaining greasy, it
becomes hard and resinous, and is therefore better
adapted to the use of the painter. As the change
which the oil thus undergoes depends upon the abstraction of its mucilage, and, perhaps, the acquisition
of a small portion of oxygen, it may be made to pass
into the same state by mere exposure to air. Its unc-
tuosity is also completely destroyed by setting it on
fire, when boiling, and then extinguishing the flame
by covering the vessel.

937. If the metallic oxide be added to the oil in a
still greater proportion, the mass, when cold, composes
the well known preparation termed *Plaister*. To
obtain this compound the oxide generally employed is
that of Lead: and since this body would undergo
decomposition, were the temperature employed too
high, a portion of water is added, in order to moderate
the heat, by carrying off the excess of caloric in a
latent form.

938. Fixed oils have been stated, in various chemical works, to boil at 600°; but it may be questioned
whether they ever boil, for the process which has been
mistaken for ebullition is, in fact, decomposition; for

if the vapour thus produced be collected and examined, it will be found to have undergone a very material change, and to be acrid and empyreumatic; in this state it was formerly employed in pharmacy, under the name of " *Philosophers' Oil;*" and, since the usual mode of obtaining it was by plunging a brick in oil and then submitting it to distillation, it was also called " *Oil of Bricks.*" At a somewhat higher temperature, a more complete decomposition takes place, and the oil is resolved into water, and olefiant and carburetted hydrogen gases, with small proportions of acetic acid, carbonic oxide, and carbonic acid.

939. Nitric acid acts with great energy on the fixed oils. If the quantity be small they are merely thick-ened by the addition; but if distilled with a large proportion of acid, the oil is decomposed, nitric oxide disengaged, and oxalic acid remains in the retort. If the nitric acid be in that fuming state in which it is usually termed *Nitrous* acid, a violent combustion is produced by its affusion on the oil: a phænomenon which is always ensured by mixing a small quantity of sulphuric acid. If a stream of chlorine be passed through oil, a waxy substance results.

940. Oil has the property of heating spontaneously, and of ultimately bursting into flame, when allowed to remain, in a state of mixture, with various vegetable substances. It sometimes happens that in boiling flowers and herbs in oil, as occurs in several phar-maceutic operations, these herbs after being taken out, dried, and pressed, inflame spontaneously. Care there-fore should be taken, when such substances are thrown aside, that they are not thrown in a heap near other combustible bodies.

941. The ultimate elements of fixed oil are Carbon and Hydrogen, with a small proportion of oxygen. The existence of this latter element was long doubted,

until Sir H. Davy succeeded in forming soap with oil
and Potassium; now to generate the alkali necessary
for this effect, it is obvious that oxygen must have
been furnished by the decomposition of the oil.

Volatile, or Essential Oils.

942. The property which these bodies possess of
being volatilized by a heat below 212° suggested the
former of these names, while their penetrating and
often fragrant smell gave origin to the latter. They
are usually obtained by distilling the vegetables in
which they exist together with a certain proportion of
water.* After this latter fluid is saturated, the oil sinks
to the bottom, or floats on its surface, according to its
specific gravity, which varies considerably in the different species; thus oil of Sassafras has that of 1·094;
of Cinnamon, 1·035; of Cloves, 1·034, while that of
the oil of Turpentine is only ·792. They do not appear
to be capable, like the fixed oils, of combining with
the alkalies; it is true that, by long trituration, such
compounds have been formed, but in such cases the
process appears to have oxidized, and converted them
into a species of resin.† They are very sparingly soluble in water, but, such is their penetrating quality,
that the waters so impregnated possess valuable properties, and are known in pharmacy by the name of
Distilled Waters. In alcohol they are very soluble,
and the compounds are termed *Essences.* Nitric acid,
when poured upon these oils, inflames them. By long
exposure to air they are thickened, a fact which depends upon the absorption of oxygen. When digested

---

* The essential oil from the rinds of the Orange and Lemon
are obtained by expression.

† This fact is well illustrated by the compound, known by the
name of *Starkey's Soap*, and which is produced by a long and
tedious trituration of alkali with oil of turpentine.

with Sulphur, they unite with it, and form a series of compounds which were formerly distinguished by the term of *Balsams of Sulphur*, and which, when exposed to a strong heat, yield a large quantity of sulphuretted hydrogen gas.

943. The fact of these volatile oils passing into a state of vapour at a temperature below 212°, enables us to distinguish them at once from the fixed oils; for if a single drop be placed on white paper, the stain will be removed by holding it before the fire; whereas the greasy spot produced by the latter bodies is permanent.

944. The well known substance, *Camphor,* appears to be an essential oil combined with some acid, or perhaps a combination of the same elements with a larger proportion of carbon. In many of its habitudes it possesses a striking analogy to these oils, and even some of them deposit it on standing. Oil of Turpentine may, moreover, by the action of muriatic acid be converted into a substance very much resembling camphor in its sensible and chemical properties.

945. Under the division of Volatile Oils, we may very properly introduce a class of substances which have been termed GUM-RESINS, from a supposition that they were compounds of the two proximate principles announced by their name. Strictly speaking, however, they neither contain gum nor resin; the peculiar principle, mistaken for the former, possessing the properties of Extractive rather than those of gum; while that, considered as the latter, is a volatile substance intermediate between volatile oil and resin. They are exudations from certain plants, which become hard by exposure to the air; they are usually nearly opaque, brittle, and have sometimes a fatty appearance. They soften by heat, and swell up and burn with flame. They dissolve partly in water, and partly in

alcohol. They are almost solely used in medicine.
Assafœtida, Ammoniacum, Aloes, Gamboge, Myrrh,
&c. are all varieties of gum-resin. When they contain
any portion of Benzoic acid, they have been termed
" BALSAMS."

WAX.

946. This principle is a product of vegetation,
forming the gloss with which the upper surface of the
leaves of many trees is varnished. It may be obtained
by bruising and boiling the vegetable substance in
water, by which the wax is separated, and will concrete on cooling In this manner, it has been obtained
from the berries of the *Myrica Cerifera,* and the leaves
and stem of the *Ceroxylon.* It is, however, well
known that the principal source of this substance is
the hive of the bee. In the ordinary state in which it
occurs, it has both colour and smell; but these may
be removed by exposure to air and light. When
bleached, it has the following properties. Its specific
gravity is about 96°; it is insoluble in water, and
fusible at a temperature of about 150°; by a greater
heat it is converted into vapour, and burns with a
bright flame. Boiling alcohol dissolves about one-
twentieth its weight of wax, four-fifths of which separate on cooling, and the remainder is immediately
precipitated by water. Caustic alkalies convert it into
a saponaceous compound, soluble in warm water.
Fixed oil unite with wax, and form a compound of
variable consistency, which is the basis of *Cerates.*

VEGETABLE ACIDS.

947 The true vegetable acids are those which exist
ready formed in the juices or organs of plants, and
require for their extraction only some mechanical process. Others are also obtained which appear to be
rather *products* than *educts,* and are either entirely

new arrangements of the elementary matter of the plant, or native acids disguised by combination with other vegetable principles. Some of the acids to be described will also be found to be both educts and products, such are the acetic, malic, and oxalic acids.

948. The following are the principal acids to be noticed: they are all distinguished by a sour taste, except the Gallic and Prussic acids; of which the first has an astringent taste, the latter a flavour like that of bitter almonds. They are all soluble in water, and combining with various bases, give origin to peculiar salts. They are all decomposable at a red heat.

1. Acetic.
2. Benzoic.
3. Citric.
4. Gallic.
5. Malic.
6. Tartaric.
7. Oxalic.
8. Prussic.

Acetic Acid.

949. This acid exists in the sap of several vegetables, generally in combination with potass; the juices of the *Sambucus nigra, Phœnix Dactilifera, Rhus Typhinus,* afford it in abundance. It is also an animal acid, for sweat, urine, and even fresh milk, contain it. It is, moreover, a product of art; strong acids, as the sulphuric and nitric, develope it by their action on vegetables; and various vegetable substances furnish it during their decomposition by heat; and, lastly, it is a product of an operation which we have hereafter to describe under the article, *Fermentation.* Obtained from such various sources, it is easy to suppose that the acid will appear in different degrees of strength; and the earlier Chemists, misled by this circumstance, recognised the existence of two acids of different degrees of oxidizement, calling one the Acetic, the other the Acetous acid. It is now, however, admitted that there exists but one acid of this description, and it is termed

Acetic acid. The processes by which it may be procured will be described hereafter. It cannot, however, be obtained in a perfectly pure state; the strongest acid in a concrete form, termed *Glacial*, containing according to the most accurate experiments 21 per cent. of water, which appears essential to its constitution, and can only be separated by combining the acid with an alkaline, earthy, or metallic base. *Glacial* acetic acid exists in a crystallized state at the temperature of 50°, but, if kept perfectly still, it may be reduced several degrees below its crystallizing point in a fluid state, when the slightest agitation of the vessel will instantly occasion it to solidify, and shoot into beautiful crystals, which again liquefy on the application of heat. This acid in its fluid state has a specific gravity of 1·063. It has an extremely pungent smell, and is so corrosive as to blister the skin. When gently heated in a silver spoon it is volatilized, and its vapour may be set on fire. It is the only vegetable acid, except the *Prussic*, that rises in distillation in combination with water. The most probable constitution of Acetic Acid is the following, viz. Carbon 4 atoms $= 24 +$ Oxygen 3 ditto $= 24 +$ Hydrogen 2 ditto $= 2$, which will give us 50 as its representative number.

950. It has been found that the strength of acetic acid is not accurately represented by its specific gravity. If, for instance, to the acid of sp. gr. 1·063 we gradually add water, we shall find, notwithstanding the water is lighter than the acid, that the density of the mixture will increase until it arrives at 1·079; from which point, the addition of water will occasion a regular diminution of specific gravity. In order therefore to form a correct estimate of the strength of any sample it is plain that we must seek for some other test. This it is evident may be derived from the quantity of alkaline or earthy substances required for its

saturation.* It will greatly facilitate our inquiries upon this occasion to remember, that the representative numbers of acetic acid, and pure white marble coincide; it therefore follows that the weight of marble dissolved by a hundred grains of any sample of acid, will at once represent the percentage of real acid in such a sample.

951. Such operations, however, involve a nicety of manipulation which can scarcely be realised in the ordinary routine of trade. To obviate this difficulty, Messrs. J. and P. Taylor invented an instrument termed the *Acetometer*, and which has been adopted by the Excise, for determining the rate of duty on Vinegar. The acid is first saturated with hydrate of Lime, and the specific gravity of the resulting solution is made the measure of the strength of the acid. The Vinegar, called *Proof* or number 24 by the manufacturer, and which is said to contain 5 per cent.* of real acid, is taken as a standard; and, when neutralized by hydrate of lime, the *Acetometer* stands in it at the mark on the stem which is called *Proof.* To keep the stem at the same mark, when immersed in stronger acids saturated with lime, it is loaded with a series of weights, each of which indicates 5 per cent. of acid above proof, up to 35, which of course contains $5 + 35 = 40$ per cent. of real acid.

952. The habitudes of acetic acid vary essentially with its strength. The quantity, for instance, of camphor and essential oils which it is capable of dissolving will greatly increase with its concentration, while if much diluted, as it exists in Vinegar, it will not exert any action over such substances.

---

* I must refer the reader for farther instructions upon this subject to the Table introduced into the sixth edition of my Pharmacologia, and which was drawn up from a series of experiments performed with great care.

+ It contains exactly 4·73 per cent. of real acid; but see my Pharmacologia, l. c.

953. Acetic Acid forms with the different bases a class of Salts called *Acetates*, and which are distinguished by the following characters. They are mostly very soluble, deliquescent, and difficultly crystallizable; when mixed with sulphuric acid, and distilled in a moderate heat, they are decomposed, and acetic acid is disengaged, which is easily distinguishable by its odour. When submitted to destructive distillation they furnish a modified vinegar, which has been termed *Pyroacetic Acid*, or *Spirit*.

954. ACETATE OF POTASS is generally prepared by saturating acetic acid with the carbonate of potass. It usually occurs in the shops in a foliated form, is colourless, and nearly inodorous; its taste is pungent and saline; it is very deliquescent, extremely soluble in water, and soluble in twice its weight of boiling alcohol. It constitutes the *Terra Foliata Tartari*, and *digestive salt of Sylvius* of old Pharmacy. It consists of one atom of each of its components, viz. $50 + 48 = 98$.

955. ACETATE OF AMMONIA. This is a very deliquescent, and soluble salt, and cannot be easily crystallized; I have, however, frequently crystallized it by submitting it to a slow evaporation, in contact with sulphuric acid. In solution, obtained by saturating distilled vinegar with carbonate of ammonia, it constitutes the *Liquor Ammoniæ Acetatis* of the Pharmacopœia, which moreover contains a quantity of carbonic acid diffused through it, which is proved by its throwing down a carbonate of lead, whenever the acetate of that metal is added to it. It is impossible to afford any definite directions for the proportions of acid and alkali to be employed in its preparation, since both ingredients are liable to vary in strength; the best method is to add the alkaline salt to the acetic acid, and to examine the state of saturation with test paper.

956. Acetate of Lead. This salt has been long known in pharmacy under the names of *Sugar of Lead*, and *Salt of Saturn*. It is formed by dissolving the carbonate of lead in acetic acid, and then crystallizing the solution. These crystals are usually very small; but, if they are suffered to form slowly, they may be obtained of considerable size. Their taste is sweet and astringent; they are almost equally soluble in hot and in cold water, viz. to about one fourth the weight of the fluid; the solution is decomposed by mere exposure to the air, the carbonic acid attracting the lead and forming an insoluble carbonate. It is also partly decomposed by water containing any carbonic acid, whence a solution so formed reddens litmus, in consequence of the disengaged acetic acid; a fact which led chemists to regard the salt as a *Super*-acetate. It is, however, now considered as a neutral salt consisting of 1 atom of acid $= 50 + 1$ of oxide $= 112 + 3$ of water $= 27$, which give 189 as its equivalent number.

957. Sub-Acetate of Lead. If 100 parts of the acetate be boiled in water with 150 of the protoxide of lead, a salt is obtained which crystallizes in plates, and is less sweet and soluble than the acetate. It has been called a *Sub-acetate*, or sometimes a *Sub-tri-acetate*, as consisting of 1 atom of acid $+ 3$ of oxide; but the more probable view of its composition is that it is a *sub-binacetate*, and consists of 1 atom of acid $= 50$, $+ 2$ of oxide $= 224$, making its representative number $= 274$. The solution of this sub-salt constitutes the *Liquor Plumbi Sub-acetatis* of the Pharmacopœia, which, however, is a variable preparation. It is even more rapidly precipitated by carbonic acid than the acetate; and from the strong attraction for vegetable colouring matter, Mr. Brande has applied it with great success in his analysis of vinous liquors.

958. ACETATE OF ZINC may be formed either directly by dissolving the metal, or the white oxide in acetic acid ; or by mingling the solutions of acetate of lead and sulphate of zinc; by which a double decomposition takes place, an insoluble sulphate of lead is precipitated, and the acetate of zinc remains in solution. By evaporation it affords a crystallized and beautiful salt. To produce a solution of this salt for medicinal purposes we should add rather more than two parts of the Acetate of Lead to one of the Sulphate of Zinc, in their dry state.

959. ACETATE OF COPPER. Three distinct compounds exist of acetic acid and peroxide of copper, viz. the *Sub-acetate,* consisting of 1 acid $+$ 2 oxide ; the acetate of $1 + 1$, and the *Binacetate* of $2 + 1$. The essential part of that substance known in commerce by the name of *Verdigris,* and which is produced by a long continued exposure of copper to the fumes of acetic acid, is a true *Acetate of Copper,* contaminated, however, with various impurities; when acted upon by water it is decomposed into a green insoluble powder which is a real *sub-acetate,* and into a *binacetate* which remains in solution. This latter salt is also produced by dissolving *verdegris* in distilled Vinegar and evaporating the solution, when it is obtained in regular crystals. Our views respecting the true nature of *Verdegris* were until lately erroneous and confused, but the experiments of Dr. Ure and Mr. Phillips have established the facts above enumerated.

960. ACETATE OF MERCURY. This salt may be obtained by dissolving the peroxide of mercury in acetic acid ; or, by mingling together solutions of nitrate of mercury and acetate of potsas, when a double decomposition takes place, the results of which are acetate of mercury and nitrate of potass. This is the process

directed by the Colleges of Dublin and Edinburgh. The crystals have a silvery whiteness, and an acrid taste; they are scarcely soluble in water, and quite insoluble in alcohol.

961. The decomposition of the metallic acetates furnishes an easy method of obtaining strong Acetic acid. The following are the different processes by which this result has been obtained. 1st. Two parts of fused acetate of potass are mixed with one of the strongest sulphuric acid; the mixture is then slowly distilled in a glass retort, and the acetic acid passes over into a refrigerated receiver. 2nd. Four parts of acetate of lead are treated in the same manner with one part of sulphuric acid. 3rd. Gently calcined sulphate of iron is mixed with acetate of lead, in the proportion of one of the former to two and a half of the latter, and are distilled from a porcelain retort into a cooled receiver. The best process for the strong acid is that first described; and the cheapest, the second and third. The distillation of the acetate of copper or of lead *per se* has also been employed, but in this case the product is contaminated with the *Pyro-acetic spirit* (953.) The College of Dublin has preferred the first process; that of Edinburgh, the third. The London Pharmacopœia formerly contained a formula which consisted in decomposing the metallic acetate by heat alone. The Berlin College direct the decomposition of the acetate of soda by means of the bi-sulphate of potass.

962. The strong acetic acid now met with in commerce is generally obtained from the acetate of lime, which is formed by saturating the acid, which distils from wood during its conversion into charcoal, with quick-lime. As the operation, from its complete success, and general adoption, forms a subject of interest to the medical student, I shall here introduce some account of the process, with a sketch of the apparatus

2 K

employed, which has been copied from a plate in Mr.
Parkes's popular Essays.

963.   The wood, deprived of its bark, is placed
within large cast-iron cylinders, which are fixed horizontally in brick-work, so as to be easily heated red-
hot by fires placed beneath them; an immense quantity
of inflammable gas is produced; and an acid liquor
distils over, which was formerly considered to be of a
peculiar nature, and distinct from all others; it was
named *Pyroligneous* acid, and in commerce is still
known by that appellation.   It is, however, nothing
more than acetic acid holding in solution a quantity of
tar and essential oil.   The cylinder during the distillation has no communication with the external air,
but at two openings, the one placed at its higher, the
other at its lower part, and of these the former is destined to allow of the escape of the acid and the different
gases, the latter to give vent to the disengaged tar
which at once flows into a barrel placed for its reception.   The following sketch will serve to illustrate this
arrangement more satisfactorily.

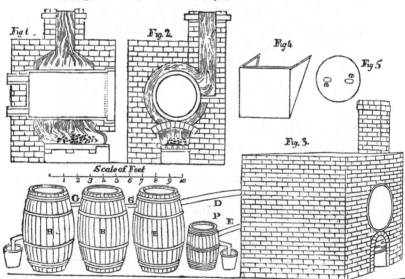

Fig. 1. represents a section of the *Cylinder* in its appropriate furnace; 2, a *Different section*; 3, the *Elevation of the furnace*; 4, the *Outer stopper*; 5, the *Inner shutter*, which is smeared round its edge with clay-lute, and secured to the mouth of the cylinder. D is the *Pipe* for conveying off the acid into the *Casks* H, which are connected by the *Adapters* G; E is the *Tar pipe*, which communicates with the *Tar receiver* P. The escape of the elastic fluids is provided for by bent tubes inserted in the casks, and terminating under the surface of water. The average quantity of impure acid obtained from each cwt. of wood is about four gallons and a half. It is much contaminated with tar; is of a deep brown; and has a specific gravity of 1·025. This crude acid is rectified by a second distillation in a copper still, in the body of which about 20 gallons of viscid tarry matter are left from every 100. It has now become a transparent brown vinegar. By redistillation, saturation with quick-lime, evaporation of the liquid acetate to dryness, and gentle torrefaction, the empyreumatic matter is so completely dissipated, that on decomposing the calcareous salt by sulphuric acid, a pure, perfectly colourless, and grateful vinegar rises in distillation. The strength of which will be proportional to the concentration of the decomposing acid.

BENZOIC ACID.

964. This acid was first described by Blaise de Vigenere in 1608, and has been generally known by the name of *Flowers of Benzoin or Benjamin;* it having been originally obtained by sublimation from a resin of that name. As it continues to be obtained from this substance it has preserved the epithet Benzoic, though known to be a peculiar acid, existing not only in Benzoin, but in various gum-resins, and aromatic

vegetables, as in cinnamon, cloves, &c. It has also been found in the urine of man and herbivorous quadrupeds.

965. There are two principal processes by which this acid may be obtained from *Gum Benzoin*, each of which has been successively adopted by the College; the one, by sublimation, is now considered as the easiest and most œconomical; the other consists in boiling the powdered resin with lime water, and afterwards separating the lime by the addition of muriatic acid, and then subliming the Benzoic acid thus precipitated.

966. When pure, this acid is crystallized in soft, colourless, feathery crystals, which are distinguished by ductility and a certain degree of elasticity; when rubbed in a mortar it passes into a state of paste. It is inodorous, although as generally met with it has a slight, but not disagreeable smell, owing to the imperfect separation of the empyreumatic oily matter. That this odour does not belong to the acid is evident, for we are informed by M. Giese, that on dissolving it in as little alcohol as possible, filtering the solution, and precipitating by water, the acid will be obtained pure, and void of smell, the odorous oil remaining dissolved in the spirit. It is not perceptibly altered by the air, and has been kept for twenty years in an open vessel, without losing any of its weight. None of the combustible bodies have any effect upon it; but it may be refined by mixing it with charcoal powder and subliming; by which it becomes much whiter, and is better crystallized. Its taste is rather acrid, but scarcely sour; it however reddens the colour of litmus. By a moderate heat it is volatilized in white fumes. It requires for solution about 24 times its weight of boiling water, which, as it cools, lets fall $\frac{18}{19}$ths of what it had dissolved. It is much more soluble in alcohol, and, on

exposure of the solution to the air, the alcohol evaporates, and the acid separates in prismatic crystals. With alkaline, earthy, and metallic bases it forms a class of compounds called *Benzoates*.

967. It consists of carbon, oxygen, and hydrogen, but the proportions have not been satisfactorily established.

CITRIC ACID.

968. This acid exists abundantly in the juice of lemons, citrons, limes, and a variety of other fruits. The celebrated Scheele first succeeded in obtaining it in a solid form, and his process has been adopted throughout Europe; there being no other means yet discovered of effecting the object. It has indeed been attempted by the action of alcohol on Lemon juice, but this agent effects the purpose very imperfectly; it will readily separate the mucilage, but it has not the power of removing either the sugar or extractive matter. M. Georgius, a Swedish chemist, proposed to obtain the pure acid, by the application of cold; in this manner the juice may certainly be reduced to an eighth part of its original bulk, by carefully removing the ice as it forms; but after it has been thus condensed, still enough of the extractive matter will remain in it to prevent the crystallization of the acid. After the publication of this method, M. Du Buisson of Paris announced another process for concentrating and preserving lemon juice. This consisted in reducing its quantity by the action of a gradual heat, kept up for a considerable time, which has the effect of thickening the mucilaginous matter, and occasioning its separation in a glutinous mass on the surface of the liquor. When this is removed, the remaining juice may be preserved unaltered for a long period. Both these proposals may be viewed as approximations to-

wards the developement and complete separation of
the acid, as effected by Scheele, whose process we have
now to examine.

969. Scheele's method consists in separating the
real acid by means of carbonate of lime, and then decomposing the citrate of lime by the intermedium of
sulphuric acid. Lemon juice is an aqueous solution
of citric acid mixed with mucilage, extractive matter,
sugar, and a small proportion of Malic acid. When
chalk is added, the citric acid combines with its basis,
and forms citrate of lime, and the carbonic acid is
disengaged. The citrate, being but slightly soluble,
precipitates; the supernatant liquor will contain all
the other elements of the juice, and is to be separated
by means of a syphon. The citrate of lime when heated
in diluted sulphuric acid is decomposed; the sulphate
of lime formed subsides, on account of its insolubility,
while the citric acid separated from the lime remains
in solution, and by evaporation yields crystals.

970. Citric acid is colourless, inodorous, and extremely sour; by exposure to a damp atmosphere the
crystals slightly absorb water. It is very soluble in
water, but the solution becomes mouldy by long keeping, and is ultimately changed into acetic acid. The
crystals, according to Berzelius, consist of one atom
of acid and two atoms of water. When treated with
about three times its weight of nitric acid, the citric
acid is converted partly into the oxalic, of which it
yields half its weight; as the proportion of nitric acid
is increased, that of the oxalic is diminished, till at
length it disappears altogether, and acetic acid appears
to be formed.

971. Citric acid combines with various bases, and
forms a class of salts called *Citrates.* Like the Acetic
acid, it is a compound of oxygen, hydrogen, and carbon; the proportions of which, as deduced from the

analysis of Berzelius, are, of oxygen 1, carbon 4, and hydrogen 2 atoms; the equivalent number therefore of dry citric acid will be 58, and consequently that of its crystals, containing 2 atoms of water, $58 + 18 = 76$.

GALLIC ACID.

972. This acid usually exists in those different vegetable substances which are characterised by astringency, but most abundantly in the excrescences termed *Galls*, or *Nut-galls*. In the experiments of Sir H. Davy, 400 grains of a saturated infusion of galls yielded, by evaporation, 53 of solid matter, composed of nine-tenths tannin, and one-tenth gallic acid. The pure acid may be obtained by the mere exposure of this infusion to the air, which will thus become mouldy, and be covered with a thick glutinous pellicle; under the surface of which, as well as on the sides of the vessels, small yellowish crystals will, after the repose of two or three months, present themselves. These crystals, according to Berzelius, contain both gallic acid and tannin, and, in order to separate the former, they must be dissolved in alcohol, and the solution be cautiously evaporated to dryness. There are, however, several more summary processes for the preparation of this acid, viz. 1. *By Sublimation.* If pounded nut-galls be introduced into a large retort, and heat be slowly and cautiously applied, the gallic acid will rise, and be condensed in the neck of the retort in shining white crystalline plates. But this process requires great care, as, if the heat be carried so far as to disengage the oil, the crystals will be immediately dissolved. 2nd. *By removing the Tannin from the infusion, by muriate of Tin.* For this purpose the muriate must be added, until no more precipitate falls down; the excess of oxide, remaining in the solution, may then

be precipitated by sulphuretted hydrogen gas, and the liquor will yield crystals of gallic acid by evaporation. According to Haussman about three drachms of gallic acid may be thus obtained from an ounce of galls. 3d. *By removing the Tannin and Extractive from the Decoction by means of pure Alumina.* In this case the Alumina is obtained by precipitation from Alum with carbonate of potass·; the precipitate is to be well washed and digested with a decoction of galls for twenty-four hours, taking care to shake the mixture frequently during this interval. The solution, when filtered, will yield the crystallized acid, on gentle evaporation. For this simple process we are indebted to Mr. Fiedler.

973. The deposit which forms in the infusion of galls left to itself, has been found by M. Chevreul to contain, besides gallic acid, another peculiar acid which M. Braconnot has proposed to call *Ellagic* acid, from the word *Galle* reversed. It is probable that this body does not exist ready formed in nut-galls, but is the gallic acid slightly modified. It is insoluble, and carrying down with it the greater part of the gallic acid, forms the yellowish crystalline deposit, of which we have already spoken. Boiling water, however, removes the gallic from the ellagic acid; whence the means of separating them from each other.

974. The pure gallic acid has the following propertiss. Its crystals have an acido-astringent taste; when placed on a red-hot iron, they burn with flame, and emit an aromatic smell, not unlike that of benzoic acid. They are soluble in about 24 parts of cold, or in 3 of boiling water. Alcohol, when cold, dissolves one-fourth, or an equal weight when heated. This acid effervesces with alkaline, but not with earthy, carbonates. By nitric acid it is converted into the oxalic. In its combinations with the salifiable bases, it pre-

sents some remarkable phænomena. It unites with
alkaline solutions without producing any deposit; but
from the watery solutions of lime, baryta, and strontia,
it occasions, at first, a greenish white precipitate, which
changes to a violet hue, and eventually disappears, as
the quantity of acid is increased. Its most distinguishing characteristic, however, is its affinity for metallic
oxides, by which it is enabled to decompose the different metallic salts, and to form a precipitate distinctive of each; but, in order to produce this effect, it
generally requires to be combined with tannin; a state
of union, indeed, in which it almost always exists in
astringent vegetables. The more readily the metallic
oxide parts with its oxygen, the more is it altered by
the gallic acid. To a solution of Gold it imparts a
green hue, and throws down a brown precipitate,
which readily passes to the metallic state, and covers
the solution with a shining gold pellicle. With the
nitric solution of silver, it produces a similar effect.
Mercury it precipitates orange; Lead, white; Bismuth,
yellow; Copper, brown; and Iron of a deep purple or
black. Platinum, Zinc, Tin, Cobalt, and Manganese,
are not precipitated by it. Of these compounds the
Tanno-Gallate of Iron is of the most importance, as
forming the basis of writing ink; and as indicating by
its formation, the presence either of iron or gallic acid,
according as one or the other of these bodies is applied.
In order, however, to produce this characteristic compound, the iron must exist in the state of *per*-oxide,
for the *prot*-oxide does not form a black compound
with these substances. It is true that iron filings will
dissolve in an infusion of galls, with an extrication of
hydrogen gas; but the compound is not black till after
exposure to air, which oxidizes the iron still farther.
Upon the same principle the colour of ink is destroyed
by metallic iron, or by a stream of sulphuretted hydro-

gen. In both these cases the oxide of iron undergoes a partial deoxidation. We are thus also enabled to explain why ink, at first pale, becomes black by exposure to air. As the *Tanno-Gallate* is decomposed by alkalies, the colour of writing ink is discharged by the action of these bodies, the characters becoming brown, and an oxide of iron alone remaining on the paper; but the fresh application of an infusion of galls will take up this oxide, and the characters be again rendered legible. In this manner a spot of ink on linen· is immediately decomposed on the contact of soap, and the oxide of iron being precipitated constitutes an *iron mould*, for the removal of which citric acid is usually recommended; but after it has remained long on the cloth, the iron acquires a degree of oxidation which renders it insoluble in that acid; in which case, it may be washed with a solution of sulphuret of potass, or be wetted with some recent ink, by which the oxide will be sufficiently deoxidized to render it soluble. Mr. R. Phillips, in his elaborate analysis of the Bath water, discovered the very curious and important fact, that carbonate of lime possesses the power of modifying the habitudes of gallic acid with the oxides of iron. He found that it increased the action of tincture of galls upon the *proto*xide of iron; while, on the contrary, it diminished its power in detecting the *per*-oxide.

975. It has been doubted whether gallic acid may not be a modification of the acetic; but Dr. Henry justly observes that there does not appear to be sufficient grounds for denying the distinct nature of a compound, which is distinguished by so many striking peculiarities. It is far more reasonable to consider it as a modification of Tannin, for not only is it generally associated with that principle in nature, but the

latest experiments have shewn, that it differs only from it in chemical composition by containing an atom less of oxygen.

MALIC ACID.

976. This acid was first obtained from Apples, whence its name; it exists, however, in the juice of various fruits mixed with the citric, and occasionally with other acids. It may be obtained most conveniently, and in greatest purity, from the berries of the mountain ash, called *Sorbus,* or *Pyrus Aucuparia;* when first discovered in these berries, it was regarded by Vauquelin as a new acid, and it received the name of Sorbic acid. But it now appears that the sorbic and *pure* malic acids are identical. It also exists in Lemon juice, and, since it forms a *soluble* salt with lime, it may be easily separated during the process of extracting citric acid. Upon the same principle it is to be obtained from various other juices; chalk must be added to saturation, the precipitate washed with boiling water, which takes up the *Malate of lime,* the solution of which may then be decomposed by sulphuric acid.

977. The Malic acid is liquid, and incapable of being crystallized; for when evaporated, it becomes thick and viscid, like syrup. It is very soluble in water, and in alcohol. Nitric acid converts it into the oxalic. It unites with alkalies, earths, and the oxides of metals, forming *Malates.* When heated out of contact of air, it sublimes and assumes new characters, in which state it has been termed *Pyro-malic acid.*

OXALIC ACID.

978. This acid exists in the juice of the Wood Sorrel, *Oxalis Acetosella,* and hence its name. It is, however, an abundant *product* as well as educt of

vegetable matter; and for the various purposes of art it is found more œconomical to form it artificially, than to elicit it from those combinations in which it is produced by nature. Bergman was the first chemist who discovered that a powerful acid might be extracted from sugar by means of the nitric; and Scheele that it was identical with that existing naturally in sorrel. The following is the process usually followed for its preparation.

979. To six ounces of nitric acid in a stoppered retort, to which a large receiver is luted, add, by degrees, one ounce of lump sugar* coarsely powdered. A gentle heat may be applied during the solution, and nitric oxide will be evolved in abundance. When the whole of the sugar is dissolved, distil off a part of the acid, till what remains in the retort has the consistence of syrup, and this will form regular crystals, amounting to 58 parts from 100 of sugar. These crystals must be dissolved in water, re-crystallized, and dried on blotting paper.

980. By this operation, the sugar is converted into oxalic acid by the loss of some carbon and hydrogen, and the acquisition of more oxygen.

981. The crystals as usually found in the shops are too small to exhibit their form with distinctness; but, when of a larger size, Dr. Henry states that they have the figure of a four-sided prism, whose sides are alternately larger, terminated at their extremities by dihedral summits. Their taste is strongly acid, and they act very powerfully on vegetable blue colours; one grain, even, when dissolved in 3600 grains of water, reddens

---

* A variety of other substances afford the oxalic acid, when treated by distillation with the nitric; e. g. Honey; gum arabic; alcohol; animal calculi; acid of cherries; tartaric acid; citric acid; wood; silk; hair; tendons; wool; animal albumen; fecula; gluten; &c. &c.

the colour of litmus paper. When exposed to the air, they effloresce; they dissolve in twice their weight of water at 65°, and in an equal weight of boiling water. They are also soluble in boiling alcohol, which takes up half its weight; and, though sparingly, in æther. By a red heat they are decomposed; at first swelling, and abandoning a large quantity of water; after the operation nothing remains but charcoal.

982. From the resemblance which these crystals bear to those of Epsom salt, many fatal mistakes have arisen, and various measures have been proposed for preventing the recurrence of such accidents;* but as they are hastily dissolved and swallowed by the unwary, it is impossible that any chemical test, however delicate, should answer the purpose. The acid taste is in itself a sufficient mark of distinction; or, without tasting it, if a few drops of water be placed on a slip of the dark blue paper which is commonly wrapped round sugar loaves,† and a small quantity of the suspected crystal be added, if it be oxalic acid, it will change the colour of the paper to a reddish brown. The solution also of a small quantity of this acid in a little water, will effervesce with common whiteing, an effect which is never produced by Epsom salt. From the history of the many cases on record, it appears that this acid produces all the grievous symptoms which characterize the action of a corrosive poison. The instant that the accident is discovered, we should, as quickly as possible, endeavour to form an insoluble oxalate of lime; copious draughts of lime-water, or magnesia and water, should be administered; and vomiting immediately excited.

---

* See our work on MEDICAL JURISPRUDENCE, vol. i. 141, and ii. 315.

† It would be well were Chemists to sell the Oxalic acid in packets covered with such paper.

983. Oxalic acid forms with the salifiable bases a class of Salts termed *Oxalates,* which are distinguished by the following characters. They are decomposed by heat, and form with lime-water a white precipitate, which, after being exposed to a red heat, is soluble in acetic acid. The *earthy* oxalates are very sparingly soluble in water; the *alkaline* oxalates are capable of combining with excess of acid, and thus become less soluble. The number representing the oxalic acid, founded upon the analysis of its salts, is 36; when crystallized, it contains four atoms of water (9×4) which will give us 72 as its equivalent.

984. OXALATE OF AMMONIA. This salt crystallizes in long transparent prisms, of which 45 parts require 1000 of water for their solution. It is of great value as a chemical reagent, for it precipitates lime from almost all its soluble combinations; in using, however, this precipitant, we must take care that the solution to which it is applied contains no excess of acid, for fresh precipitated oxalate of lime is soluble in nitric and muriatic acid. It consists of 1 atom of base $= 17, + 1$ atom of acid $= 36$, which will give us 53 as its equivalent; or, in its crystallized state $(53 + 9)$ 62, for it contains one atom of water.

985. OXALATE OF POTASS forms flat oblique four-sided prisms soluble in three parts of water at 60°. It consists of an atom of each constituent, *viz.* $36 + 48 = 84.$

986. BINOXALATE OF POTASS exists in the juice of the *Oxalis Acetosella,* or of the *Rumex Acetosa,* and is commonly sold under the name of *Salt of Sorrel,* or *Essential Salt of Lemons.* It forms small white parallelopipeds, or rhomboids approaching to cubes; its taste is acidulous mixed with that of bitterness. It contains two atoms of acid to each atom of base; and

accordingly its representative number will be (36 + 36 + 48) 120.

987. QUADROXALATE OF POTASS may be formed by digesting the *binoxalate* in dilute nitric acid, a portion of the alkaline base is taken up, and a salt remains which is a true quadroxalate, being constituted of 4 atoms of acid to one of base, *viz.* $36 \times 4 = 144 + 48 = 192$.

988. The analysis of the above salts furnishes an admirable exemplification of the beautiful law of simple multiples (256.)

989. The *Oxalates of Baryta, Magnesia,* and *Lime,* are very nearly insoluble in water. The latter salt constitutes the principal ingredient of that species of calculus which is termed the *Mulberry* calculus.

TARTARIC ACID.

990. It has been long known that the casks in which certain wines are kept become incrusted with a hard substance, tinged with the colouring matter of the wine, and which has been called *Argal* or *Tartar.* This, when purified by solution and crystallization, constitutes *Cream of Tartar,* and was long considered as a product of fermentation.[*] Boerhaave, Newman, and others, however, shewed that it existed ready formed in the juice of the grape. It has also been found in other fruits, particularly before they become completely ripe. From this salt, Tartaric acid may be procured by a process which is now introduced into the London Pharmacopœia. It consists in boiling cream of tartar in water, and gradually adding chalk to the solution until the effervescence ceases. The mixture is then set aside, to allow the subsidence of

---

[†] An old chemical writer records this opinion by the following quaint metaphor. "Tartar has *Bacchus* for its father ; *Fermentation* for its mother ; and the *Cask* for its matrix.

the precipitate formed, which is afterwards washed, and decomposed by Sulphuric acid, when the Tartaric acid thus liberated will assume a crystalline form. The theory of the process is simply this; Cream of Tartar is a saline compound consisting of potass and Tartaric acid, the latter being in excess so as to form an acidulous salt. When carbonate of lime *(Chalk)* is added to the solution, it is decomposed by the excess of Tartaric acid; carbonic acid is evolved, and a tartrate of lime precipitated. The Cream of Tartar, thus deprived of its excess of acid, is converted into a neutral tartrate, and remains in solution. When the Tartrate of Lime is mixed with sulphuric acid, it is decomposed, owing to the superior affinity of the acid for the lime, and the sulphate of lime precipitating, the Tartaric acid remains in solution. It is, however, evident that, by such a process, that portion only of Tartaric acid is obtained, which exists in the Tartar, in excess above the quantity necessary for the neutralization of its potass. In order to obtain the whole of the acid, we must afterwards add muriate of lime to the supernatant neutral tartrate, by which means it is completely decomposed.

991. Tartaric acid is colourless, inodorous, and very acid and agreeable to the taste, so that it may supply the place of lemon juice. It forms regular crystals of considerable size, the primary form of which is an oblique rhombic prism. They undergo watery fusion at a heat a little exceeding 212°. For solution, the crystals require five parts of water at 60°; but they are much more soluble at 212°. The solution acquires a mouldy pellicle by keeping. The crystals are decomposed by the sulphuric and nitric acids, and if the latter be concentrated, the tartaric will be converted into the oxalic acid; by digestion with water and alcohol it will be changed into the acetic acid. Tartaric

acid, in its *anhydrous* state, consists of carbon 4 atoms + oxygen 5 ditto + hydrogen 2 ditto, which will give us 66 as its representative number. The crystals are composed of 1 atom of acid + 1 atom of water, which will make its equivalent $(66 + 9) = 75$. In referring to citric acid it will be seen that its equivalent is 76. Whence we may consider the saturating power of the crystallized tartaric and citric acids as equal; a fact which should be kept in remembrance by the medical student.

992. Tartaric acid forms with different bases a class of Salts called *Tartrates*, which are characterised by the following properties. They are converted into *Carbonates* by a red heat. Those with earthy bases are scarcely soluble in water; the alkaline tartrates are soluble; but when combined with excess of acid, they become much less so.

993. SUPER-TARTRATE, or BI-TARTRATE OF POTASS. The origin of this salt has been already noticed. It consists of 2 atoms of acid $= (67 \times 2)$ 134, + 1 atom of potass $= 48$, which make its representative number 182; but since it always contains an atom of water, which appears to be essential to the constitution of the salt, we must consider its true equivalent $= (182 + 9)$ 191. It requires for its solution a very large quantity of water, not less than 60 parts at 60° *Fah.* or 14 at 212°. Hence its solution in hot water deposits the salt on cooling, so rapidly and in such quantity as almost to resemble precipitation. It is quite insoluble in alcohol. It becomes very soluble in water by adding to it one-fifth of its weight of borax; or even by the addition of boracic acid. Alum occasions the same effect. Gay-Lussac considers that in many cases this salt acts the part of a simple acid, and even dissolves oxides that are insoluble in the mineral acids, and in the tartaric acid.

994. TARTRATE OF POTASS, from its greater solubility, is very generally known by the name of *soluble
tartar*. It may be obtained by evaporating the solution which remains after adding chalk to the solution
of the bi-tartrate in preparing tartaric acid (990), or it
may at once be prepared, as directed in the Pharmacopœia, by saturating the excess of acid in the bi-
tartrate by carbonate of potass. It has a bitter taste,
is readily soluble in water, and crystallizes; although
it is sometimes found in the shops in the state of
powder. In a moist atmosphere it attracts water, and
is by a red heat decomposed, and converted into sub-
carbonate of potass. The habitudes of this salt with
regard to acids are very interesting, the greater number
of them disturb its solution, and transform it into *bi-
tartrate*. It consists of 1 atom of tartaric acid = 67,
+ 1 ditto of potass = 48, so that its representative
number is 115.

995. TARTRATE OF POTASS AND SODA is formed
by saturating the superfluous acid in the bi-tartrate
with sub-carbonate of *Soda*. It is the *Soda Tartari-
zata* of the Pharmacopœia, and is known in Pharmacy
under the name of *Rochelle Salt*, or *Sel de Seignette*.
It occurs in large and beautiful crystals, the primary
form of which is a right rhombic prism. They are
slightly efflorescent, and dissolve in five parts of water
at 60°, and in a less quantity at 212°. The taste of
this salt is rather bitter. It may either be considered
as a triple salt consisting of 2 atoms of acid (67 × 2)
= 134, + 1 ditto of potass = 48, + 1 ditto of soda
= 32, making its number 214; or we may regard it as
a double salt consisting of 1 atom of tartrate of potass
= 115, + 1 ditto of tartrate of soda = 99. It does
not appear to contain any water of crystallization.

996. The *Tartrates of Ammonia and Soda*, may
be formed by saturating the Tartaric acid with the

respective bases, and if the former be in excess, we shall obtain insoluble bi-tartrates.

997. Of the metallic Tartrates, those of Iron and Antimony are principally interesting to the medical student. The former may be formed, either directly by acting on metallic iron with tartaric acid, or by mingling solutions of *tartrate of potass,* and *proto-sulphate of iron.* The compound forms lamellar crystals, which are sparingly soluble in water. By exposure to air, they pass to the state of *Per-*tartrate. The *Ferrum Tartarizatum* constitutes one of those compounds which Cream of Tartar forms with metallic oxides, and to which, in our present state of knowledge, no name can be more appropriate than that bestowed upon it by the London College, although, according to the new views of Gay-Lussac, it should be called *Cream-tartrate of Iron.* Tartaric acid has no action on antimony, but it dissolves a small portion of its protoxide. The solution scarcely crystallizes; but easily assumes the form of a jelly. With the bi-tartrate of potass, however, oxide of antimony forms a more strongly marked compound, and one of great interest to the medical student.

998. TARTARIZED ANTIMONY. *Cream-Tartrate of Antimony. Tartar Emetic.* Various processes have been described for the preparation of this salt, but that which is the most simple, and perhaps upon the whole the least precarious, is the one directed by the London College. It merely consists in boiling together in distilled water, glass of antimony (718) and cream of tartar, in the state of fine powder, and then evaporating and crystallizing the solution, which according to Mr. Phillips *(Translat: Pharmacop: Lond:)* consists of tartrate of potass, and tartrate of antimony, which combine so as to form a double salt. These views, however, are extremely doubtful. I am rather

inclined to accede to the views of Gay-Lussac as already explained.

999. Tartarized antimony very easily crystallizes, in which state it ought always to be purchased to avoid the adulterations to which it is exposed. The general character of its crystal is stated by Mr. Phillips to be an octohedron with a rhombic base. They are colourless and transparent; their taste nauseous and acrid, and when exposed to the air they slowly effloresce. Boiling water dissolves half the weight of the salt, and cold water not more than a fifteenth part. Sulphuric, nitric, and muriatic acids, when poured into a solution of this salt, precipitate its cream of tartar; and soda, potass, ammonia, or their carbonates, throw down its oxide of antimony. Lime water occasions not only a precipitate of oxide of antimony, like the alkalies, but also an insoluble tartrate of this earth. That produced by the alkaline hydro-sulphurets is wholly formed of *Kermes* (719), while that caused by sulphuretted hydrogen, contains both *Kermes* and Cream of Tartar. The decoctions and infusions of several vegetables, as cinchona, and other bitter and astringent plants, * equally decompose tartar emetic; and the precipitate, in such cases, always consists of the oxide of antimony, combined with the vegetable matter and cream of tartar. It behoves the Physician, therefore, to be aware of such incompatible mixtures. When heated with carbonaceous matter this salt is decomposed, and metallic antimony is obtained. From this phænomenon, and the deep brownish-red precipitate by hydro-sulphurets, this antimonial combination may be readily recognized.

1000. Tartaric acid, when heated to redness, undergoes a peculiar change, and is converted into an acid

---

† See my PHARMACOLOGIA. Art. *Antimonium Tartarizatum.*

termed the *Pyro-tartaric*, which was long confounded
with the acetic.

1001. Such are the principal acids which exist ready
formed in vegetable substances. There are a few
others, obtained only from particular vegetables, as the
Moroxylic, from the white Mulberry; the *Boletic*, from
the *Boletus pseudo-ignarius;* the *Aceric*, from the
Maple; the *Kinic*, from Peruvian Bark; the *Meconic*,
from Opium; the *Igasuric* from the *Strychnus Ignatia*,
&c. many of which are so little interesting to the medical
student that it may be merely sufficient to enumerate
them; although some few will be hereafter described
in connection with the alkalies, with which they are
associated in native combination.

VEGETABLE ALKALIES.

1002. The substances included under this division
are bodies of recent discovery; and there is reason to
believe that the order is no less numerous than that of
vegetable acids, although it is probable that some of
those lately discovered will turn out to be merely disguised modifications of each other. They appear to
exist ready formed in the vegetables from which they
are procured, and to which they would seem to impart
medicinal power; and, although they do not bear a
degree of activity proportional to their concentration,*
still many of them are likely to become important instruments in the hands of the skilful physician.

1003. These bodies have many properties in common, but it is on account of their chemical habitudes
of combination, that they have been received into the
order of alkalies; they possess the power, for instance,
of saturating acids, and of giving origin to a class of

---

† See my PHARMACOLOGIA. Edit. 6. Vol. 1. p. 280; and Vol.
2. Art. Opium, *Note*.

salts, most of which are neutral, although some always contain an excess of acid; and they are generally crystallizable,* and soluble. Since, however, they differ from the bodies long known under the name of alkalies, in having much weaker affinities for acids, and in being decomposable both in their isolated and combined states, at low degrees of heat, their claim to the rank of true alkalies has been doubted, and some chemists have attempted to compromise the question by proposing for them the name of *Alkaloids ;* but an objection founded upon the *degree* and not the *kind* of their action cannot be maintained; they are, without doubt, *salifiable bases,* and, as they generally act on vegetable colours like the known alkalies, there can exist no good reason for excluding them from that class. The termination of *ine,* therefore, which was first given to their names, as Morphine, Cinchonine, Quinine, &c. has been correctly exchanged for that of *a,* in conformity to the nomenclature of the other alkalies, and they are termed Morphi*a,* Cinchoni*a,* Quin*a,* &c. In a pure state they are very sparingly soluble in water, but are in general rendered more so by the presence of resinous matter; and in their peculiar vegetables they are usually combined with some acid. With very few exceptions,† they are all bitter, but the degree of bitterness varies in the different species; that of *Strychnia* is insupportable, while that of *Emeta* is slight. They are inodorous; are not altered by air, or light, but are usually decomposed by a moderate heat; most of them enter into fusion, but at different temperatures, some for instance at below 212° *Fah.*; others not until they are about to be decomposed; *Hyoscyama* will even

---

* They are all crystallizable, with the exception of the *Nitrate of Cinchonia,* and the salts of *Veratria.*

† *Atropia* is insipid; *Pipera* nearly so; the taste of *Veratria* is acrid and burning.

resist a low red heat. They are nearly all highly soluble in alcohol. Æther readily dissolves *Delphia, Veratria, Emeta, Quina,* and *Gentia;* but *Morphia, Cinchonia,* and *Picrotoxa* are very sparingly soluble, and *Strychnia* and *Brucia* are nearly insoluble in that menstruum. All the saline combinations of these bodies with the mineral acids, excepting those of *Picrotoxa,* are exceedingly soluble in water. The *Acetates* also, with a few exceptions, are soluble. All the *Oxalates,* that of *Picrotoxa* excepted, which is the most soluble of its salts, and all *Tartrates* are very sparingly soluble. The action of concentrated nitric acid on these alkaline bodies is very peculiar, converting the greater number of them into artificial *Tannin;* but it appears to peroxidate *Morphia, Strychnia,* and *Brucia,* rendering them less powerful as salifiable bases, and diminishing their medicinal activity. Since the powers of these substances on the animal body are, *cateris paribus,* in the direct ratio of their solubility, the general facts above enumerated should be kept in remembrance by the medical practitioner.

1004. When resolved into their ultimate principles, they are all found to contain oxygen, hydrogen, carbon, and, with one or two exceptions, azot; this latter element was for some time considered as accidental, and as being derived from the ammonia employed in their preparation; but a more careful investigation of the subject has proved it to be essential to their composition. Many attempts have been made to determine the equivalent numbers of these bodies, but upon this point there at present exists great discrepancy, and it will be prudent to wait for more satisfactory analyses.

1005. Having thus afforded a sketch of the *general* characters of these bodies, it will be necessary to descend into the particular history of each individual alkali.

MORPHIA.

1006. Besides Resin, Gum, Extractive matter, Gluten, and some other vegetable principles, Opium contains two peculiar proximate principles, the one of which is an alkaline body, termed *Morphia;* the other, a substance which does not appear to possess the characters of an alkali; it has received the name of *Narcotin.* In these two bodies the specific powers of opium reside, although they are somewhat modified by the state of combination in which they exist. *Morphia* exists in a state of union with a peculiar acid, which has been called the *Meconic acid,* and it moreover appears that this acid exists in such a proportion as to form a super-salt, or a *Super-meconate of Morphia.* Derosne was the first chemist who obtained a crystalline subtance from Opium, which he announced* in 1803, but did not describe its nature or properties. In the following year Seguin discovered another crystalline body in Opium, but he never hinted at its alkaline nature. Sertuerner, at Eimbeck in Hanover, had also, at the same time as Derosne and Seguin, obtained these crystalline bodies, but it was not until the year 1817, that he first unequivocally proclaimed the existence of a new vegetable alkali, and assigned to it the narcotic powers which distinguish the operation of Opium; to this body he gave the name of *Morphia,* and it appears to be the same as the salt of Seguin. The salt of Derosne, now denominated *Narcotine,* is quite a different principle, although it was constantly mistaken for one of the salts of *Morphia,* until M. Robiquet pointed out its distinctive characters.

1007. Various processes have been adopted for obtaining pure *Morphia.* In some, the natural *Meco-*

---

† Annales de Chimie. T. xlv.

nate is decomposed by magnesia, in others, by ammonia.* It is, however, now admitted that the following
improved process of Robiquet † is to be preferred. A
concentrated infusion of opium ‡ is prepared by macerating 300 parts of the pure drug in 100 parts of
common water for five days; to this solution when
filtered, 15 parts of *pure* Magnesia (the *Carbonate* of
that earth must be carefully avoided) are to be added;
then boil the mixture for ten minutes, and separate
the greyish and abundant deposit by a filtre, washing
it with cold water until the water passes off clear;
after which, treat it alternately with hot and cold alcohol (12, 22 Bé) as long as the menstruum takes up
any colouring matter; the residue is then to be treated
with boiling alcohol (22, 32 Bé) for a few minutes,
and the liquor filtered while it still boils; as this solution cools, it will be found to deposit crystals of the
Morphia slightly coloured. By solution in alcohol,
and crystallization being repeated two or three times,
they may at length be obtained colourless. The theory

---

* If we make an acetic solution of opium, and pour ammonia
into it, we shall at once obtain a precipitate which consists of impure Morphia, which, after it has been well washed with water and
alcohol, is to be redissolved by acetic acid, and mixed with ivory
black. This mixture is to be frequently agitated for 24 hours,
and then thrown on the filtre. The liquid passes through quite
colourless, and if ammonia be now dropt into it, pure Morphia
falls in the state of white powder. This process was proposed by
Dr. Thomson, but Dr. Ure observes that this " white powder,"
instead of being *pure* Morphia, contains a considerable quantity
of phosphate of lime derived from the ivory black; its subsequent
solution, however, in alcohol, and crystallization, will render it
pure.

† Codex Medicamentarius, Editus a Facultate Medica Pa-
risiensi. A.D. 1818.

‡ It has been ascertained by Mr. Thomson that good *Turkey*
Opium will yield nearly three times more Morphia than the *East
Indian* variety.

of this process is the following. The Meconiate of
Morphia is decomposed by the Magnesia; and the
Morphia, together with colouring matter and the un-
dissolved Magnesia, falls down; by washing this precipitate with water, and weak alcohol, a small quantity
of Morphia and much colouring matter is taken up.
The concentrated alcohol at a boiling temperature does
not act upon the Magnesia, but dissolves the Morphia,
which it again surrenders, on cooling, in a crystalline
form.

1008. *Morphia,* when pure, is colourless, inodorous, and, when in solution, intensely bitter. Its
crystals have a pearly lustre, and their primary form
is a right rhombic prism. They are insoluble in cold,
and only sparingly soluble in boiling water. They are
soluble in 42 parts of cold, but to a greater extent in
hot alcohol, and in 8 parts of æther; these solutions
are intensely bitter, and turn the syrup of violets green.
Morphia, moreover, exhibits alkaline powers, not only
in combining with acids to form salts, but also, by
decomposing the solutions of metallic salts, precipitating their oxides, owing to its greater affinity for the
acids with which they are combined. It fuses at a
moderate heat, and acquires the aspect of melted sulphur. It is incapable of forming soap with an oxidized
oil. By a strong heat it is decomposed, yielding carbonate of ammonia, oil, and charcoal. It burns readily
in atmospheric air. By Nitric acid it is turned red;
which therefore afford a test of the presence of Morphia. Its medicinal virtues cannot be considered as
falling within the scope of the present work.

1009. *Meconic Acid,* with which Morphia exists in
native combination, may be obtained from opium, by
dissolving, in weak sulphuric acid, the Magnesian
residuum left after the action of the hot alcohol in the
process above described for extracting morphia; to

this solution muriate of baryta is then to be added, which will throw down a precipitate consisting of sulphate and meconate of baryta; digest this with hot and very weak sulphuric acid; and, when the filtered liquor is sufficiently reduced in quantity by evaporation, the *Meconic acid* will shoot, even before cooling, into coloured crystals; these are to be washed with a little water, dried, and sublimed in a flask. This acid is extremely soluble both in alcohol and in water, and the solutions are sour to the taste, and convert vegetable blues to a red colour. It combines with alkalies and forms *Meconates,* several of which crystallize. Its distinguishing character is, that it produces an intensely red colour in solutions of iron oxidized *ad maximum.* When taken into the stomach it does not produce any soporific effect.

1010. *Narcotine* is the salt of Derosne, and is not, as Sertuerner at first imagined, the meconate of morphia, but a distinct and independent principle, in which the exciting powers of opium are supposed to reside, just as its sedative virtues depend upon morphia. The following is the process for its preparation. Evaporate an aqueous solution of opium, until it has acquired the consistence of syrup, and then mix it with sulphuric æther, and shake the mixture; repeat this with fresh portions of æther, as long as it deposits any crystals in distillation. When this solvent produces no further effects, the solution of opium may be evaporated to a proper consistence, when it will furnish the practitioner with an extract possessing all the sedative, without the exciting properties of opium. The theory of this process is obvious; Narcotine being soluble, and Morphia insoluble, in cold æther, it is evident that by the treatment above described, these two principles may be separated from each other. The crystals of this substance have a pearly lustre, are inodorous and insipid. They are soluble in fixed oil, and in acids; slightly soluble in alcohol, but very nearly insoluble in water. They have no action on vegetable colours, and are incapable of neutralizing alkalies.

1011. Besides the Meconic, opium would also appear to contain some acid, which is not volatile, nor has any peculiar effects upon the salts of iron. This subject, however, requires farther investigation.

Salts of Morphia.

1012. *Acetate of Morphia* is formed by saturating the pure base with acetic acid, and allowing the mixture to evaporate slowly to dryness. The difficulty of obtaining it crystallized, on account of its extreme deliquescence, renders it necessary to adopt this mode of preparation. Its crystallization, however, may be facilitated by exposing it to the action of sulphuric acid, as recommended for effecting that of *acetate of ammonia.* (350) It crystallizes in needles, and is extremely soluble, and very active as a medicine.

1013. The *Tartrate* may be obtained by a similar process, and will crystallize in prisms.

1014. The *Sulphate* is prepared by dissolving Morphia in sulphuric acid, previously diluted with water; the solution made hot, and evaporated to a certain point, crystallizes on cooling, in an arborescent form. This salt is soluble in twice its weight of distilled water, and is composed of acid, 22 + morphia 40, + water 38. It has a considerable resemblance to the Sulphate of Quina to be hereafter described, and as this latter salt is now in very general use, the medical practitioner will do well to remember the following simple test by which they may be distinguished from each other, viz. The Sulphate of Morphia, when treated with concentrated nitric acid, becomes red;

whereas no such effect is produced with the Sulphate of Quina.

1015. The *Muriate* crystallizes in plumose crystals, and is much less soluble, requiring 10½ times its weight of distilled water. It consists of acid 35 + morphia 41 + water 24.

1016. The *Nitrate* yields prisms grouped together in stars which are soluble in 1½ times their weight of distilled water. They consist of acid 20, + morphia 36, + water 44.

1017. All the above combinations are decomposed by Ammonia.

ÊMETA.

1018. This alkali constitutes the emetic principle of Ipecacuanha, and was discovered by Pelletier. In that state, however, in which it was first known under the name of *Emetin*, it is impure; but it will be useful in giving the history of this substance to follow the same steps as those which led to its discovery, I shall therefore continue to call the impure body *Emetin*, and reserve the term *Emeta* for the alkali in its perfect state. The following is the process given by M. Majendie for its preparation. Digest powdered Ipecacuanha* in Æther, in order to dissolve its fatty odorous matter. When the powder yields nothing more to this solvent, it must be again exhausted by alcohol. Place the alcoholic tinctures in a water bath, and redissolve the residue in cold water. It thus loses a portion of wax, and a little of the fatty matter which still remained. It is only necessary farther to macerate it on carbonate of magnesia, by which it loses its gallic acid, to redissolve it in alcohol, and to evaporate it to dryness. When thus prepared, Emetine appears

---

† M. Boullay has obtained it from the root of the Violet.

in the form of transparent brownish-red scales. It has
no smell, but a bitter acrid taste. At a heat somewhat
above that of boiling water, it is resolved into carbonic
acid, oil, and vinegar, but it affords no ammonia which
proves that azot is not one of its elements. It is soluble both in water and alcohol, but not in æther. It is
very deliquescent, and incapable of being crystallized.
It is precipitated by proto-nitrate of mercury, and corrosive sublimate, but not by tartarized antimony.
Gallic acid also throws it down from its solution, but
the re-agent that most powerfully affects it, is the sub-
acetate of lead, which completely precipitates it from all
its solutions; if therefore we take a solution of Ipeca-
cuana, purified from its fatty matter, and drop in
sub-acetate of lead, wash the precipitate formed, and
then, diffusing it in water, decompose it by a current
of sulphuretted hydrogen gas, sulphuret of lead will
fall to the bottom, and the Emetin remain in solution,
which by evaporation may be obtained pure.

1019. *Emeta,* says M. Majendie, is to Emetine
what white and crystallized sugar is to moist sugar.
The latter appears to consist of the pure alkaline base
with some acid and colouring matter. By substituting
calcined magnesia, for the carbonate used in the former
process, and taking care to add a sufficient quantity to
take up the free acid which exists in the liquor, and
to unite with that which is combined with the Emeta,
we shall separate the alkali, and precipitate it in combination with the excess of magnesia. This magnesian
precipitate, after being washed by means of a very little
cold water, to separate the colouring matter which is
not combined with the magnesia, must be carefully
dried and digested in alcohol, which dissolves the
Emeta; and this, after it has been separated from the
alcohol by evaporation, must be redissolved in a diluted
acid, and digested with purified ivory black. It must

then be precipitated by a salifiable base. The Emeta, thus purified, becomes less soluble; it is white, pulverulent, and no longer deliquescent; it acts like an alkali upon vegetable colours; it is also dissolved by all the acids, the acidity of which it diminishes, but without entirely destroying it.

CINCHONIA.

1020. It had been long known that an infusion of nut galls will produce a precipitate with the decoctions of Bark, and Dr. Duncan attributed the phenomenon to the presence of a peculiar principle, to which M. Gomez of Lisbon appears to have previously assigned the name of *Cinchonine*, and which in fact was the salifiable base, now termed *Cinchonia*, so disguised with other matter as to have been long mistaken for a resin. M. M. Pelletier and Caventou have the credit of discovering its alkaline nature, and of demonstrating the peculiar state of combination in which it exists in the different species of Bark.

1021. The following is the process by which Cinchonia may be obtained. A pound of *pale* bark, *(Cinchona Lancifolia)* bruised small, is to be boiled for an hour in three pints of a very dilute solution of pure potass. The liquid, after being suffered to cool, is then to be strained through a fine cloth with pressure, and the residuum repeatedly washed and pressed. The Bark thus washed, is to be slightly heated in a sufficient quantity of water, adding muriatic acid gradually until litmus paper is slightly reddened. When the liquid is raised nearly to the boiling point it is to be strained, and the cinchona again pressed. To the strained liquor, while hot add an ounce of sulphate of magnesia, and after this, add a solution of potass, until it ceases to occasion any precipitate. When the liquor is cold, collect the precipitate on a filtre, wash and dry it, and dissolve it in hot alcohol. On evaporation of the alcohol, the cinchonia crystallizes in delicate prisms. Other processes have been recommended, but the one above related appears to be the most eligible. The Cinchonia exists in Bark in combination with an excess of a peculiar acid termed the Kinic acid. It is the object of the process, after removing the various other vegetable principles which bark contains, to decompose this salt, and to obtain the cinchonia in an isolated state.

1022 Cinchonia thus obtained is white, translucent, crystalline, and soluble in 2500 times its weight of boiling water, but a considerable part separates on cooling. Its taste is, owing to its insolubility, scarcely sensible, unless it be dissolved in alcohol or an acid, when it displays the peculiar bitterness of the bark. It is neither fusible, nor volatile at moderate temperatures; it is very soluble in alcohol, less so in æther, and but sparingly in fixed and volatile oils. It restores the colour of litmus, which has been reddened by an acid; unites with all the acids, and gives origin to a class of salts.

1023. The *Kinic acid*, which exists in native combination with the cinchonia may be obtained by macerating the bark in cold water; concentrating the infusion, and setting it aside in an open vessel; when a salt will separate in plates, which is tasteles, soluble in cold water, and insoluble in alcohol; this salt is *Kinate of Lime;* when oxalic acid is added to it, oxalate of lime is precipitated, and the solution by evaporation yields crystals of *Kinic acid* of a brownish colour, and a very acid and rather bitter taste. It is distinguishable from other vegetable acids by forming a soluble salt with lime, and by its not precipitating lead or silver from their respective solutions.

QUINA.

1024. This alkali was discovered by Pelletier and Caventou in *Yellow* Bark *(Cinchona Cordifolia)* from which it may be separated by a process similar to that above described for the preparation of Cinchonia. When dried, this body presents a white porous mass, incapable of crystallization. In water it is as insoluble as cinchonia, but it is much more bitter; it is, however, *very soluble in œther;* it is, moreover, distinguishable from the latter alkali by its habitudes with acids, and from the different properties which characterize its salts. It appears to possess less power of saturating acids than cinchonia, and yet this alkali and its salts are said to be five times more energetic, as medicinal agents, than cinchonia and its compounds.

1025. After having obtained *Cinchonia* from the *Pale,* and *Quina* from the *Yellow* Bark, it became an interesting question to determine to which of those alkalies the *Red* Bark, *(Cinchona Oblongiolia)* which had been considered so eminently febrifuge, was indebted for its virtues. To the astonishment of the experimentalists, they not only obtained from it *Cinchonia,* in threefold quantity, but almost twice as much *Quina* as they had been able to extract from an equal quantity of yellow bark. Ulterior experiments, however, have since shewn, that both these vegetable alkalies exist simultaneously in all the three species of Bark; but that the Cinchonia is, relatively to the *Quina,* in greater quantity in the pale bark; whilst, in the yellow bark, the *Quina* so predominates, that the presence of the Cinchonia might well have escaped notice when small quantities only were submitted to experiment.

1026. Cinchonia has been detected in other vegetables besides the Cinchona; as, for instance, in the root of Cusparia, and in the berries of the Capsicum;

while, strange to say, in the bark of Cascarilla, a substance bearing a much nearer relation in medicinal
efficacy to the Bark, its presence has not hitherto been
recognised. It is stated that experiments have been
lately made by M. M. Robiquet and Petron on the
Bark of the *Carapa,* which has been successfully used
in several parts of America in the cure of agues, and
that they have found in that bark a salifiable basis
analogous to Quina.

Salts of Cinchonia and Quina.

1027. *Sulphate of Cinchonia* is easily crystallizable,
and moderately soluble in water; the primary form of
its crystal appears to be a doubly oblique prism; independently of water of crystallization, it appears to
consist of acid 11·56 + cinchonia 88·44.

1028. *Sulphate of Quina.* This salt is considered
as the most active form of the salifiable principle of
Bark, I shall therefore present the student with the
most approved process for its preparation. Boil for
half an hour two pounds of powdered yellow bark in
sixteen pints of distilled water, acidulated with two
fluid-ounces of sulphuric-acid; strain the decoction
through a linen cloth, and submit the residue to a
second ebullition in a similar quantity of acidulated
water; mix the decoctions, and add by small portions
at a time, powdered lime, constantly stirring it to
facilitate its action on the acid decoction, (half a pound
is about the quantity required.) When the decoction
has become slightly alkaline, it assumes a dark colour,
and deposits a reddish brown flocculent precipitate,
which is to be separated by passing it through a linen
cloth. The precipitate is to be worked with a little
cold distilled water and dried. When dry it is to be
digested in rectified spirit, with a moderate heat for
some hours; the liquid is then to be decanted, and

fresh portions of spirit added till it no longer acquires a bitter taste. Unite the spirituous tinctures, and distill in a water bath until three-fourths of the spirit employed have distilled over. After this operation, there remains in the vessel a brown viscid substance covered by a bitter and alkaline fluid. The two products are to be separated and treated as follows. To the alkaline liquid add a sufficient quantity of sulphuric acid to saturate it; reduce it, by evaporation, to half the quantity, add a small portion of charcoal, and after some minutes ebullition, filter it while hot, and crystals of *Sulphate of Quina* will form. The brown mass is to be boiled in a small quantity of water slightly acidulated with sulphuric acid, which will convert a large quantity of it into Sulphate of Quina. The crystals are to be dried on bibulous paper. Two pounds of Bark will, it is said, yield from five to six drachms of the sulphate; of which eight grains are considered as medicinally equivalent to an ounce of bark.

1029. This salt forms crystals which are remarkable for their satin-like and pearly lustre. It is soluble in cold water, although in a less degree than the sulphate of cinchonia; this property, however, is considerably increased by an excess of acid.

1030. Quina appears to be capable of forming a *super* and *sub* salt with sulphuric salt.

1031. *Nitrate of Cinchonia* is uncrystallizable, and sparingly soluble, and so is the kindred salt with a base of Quina.

1032. *Muriate of Cinchonia* is still more soluble in water than the sulphate; dissolves in alcohol; and crystallizes in delicate prisms. The Muriate of *Quina* has not been examined.

1033. *Acetate of Cinchonia.* This salt crystallizes with difficulty, and simply in plates transparent, and

2 M 2

devoid of lustre; it is much less soluble in water than the sulphate, but an excess of acid dissolves it with tolerable facility.

1034. *Acetate of Quina.* The characteristic of this salt is the great facility with which it crystallizes; it is little soluble in the cold, even with an excess of acid. It thickens in a mass when exposed to cold.

1035. Both Cinchonia and Quina form with the gallic, oxalic, and tartaric acids, insoluble salts; hence it is that the infusion of galls precipitates the decoction of Bark.

STRYCHNIA.

1036. This body was detected by Pelletier and Caventou in the fruit of the *Strychnus nux vomica*, and *Strychnus ignatia*, about the end of the year 1818. M. Majendie has given us the following formula for its preparation. Add a solution of liquid sub-acetate of lead to a solution of the alcoholic extract of the nux vomica in water, until no more precipitate falls down; the foreign matters being thus separated, the *Strychnia* remains in solution with a portion of colouring matter, and sometimes an excess of acetate of lead. Separate the lead by passing a stream of sulphuretted hydrogen through the liquid; filter, and boil with magnesia, which will unite with the acetic acid, and precipitate the Strychnia. Wash the precipitate in cold water; redissolve it in alcohol, to separate the excess of magnesia; and by evaporating the alcohol, the Strychnia will be obtained in a state of purity. Should it, however, be not perfectly white, it must be redissolved in acetic or muriatic acid, and re-precipitated by means of magnesia.

1037. Strychnia obtained by crystallization from an alcoholic solution which has been diluted by means of a small quantity of water, and left to itself, appears under the form of microscopical crystals, forming four-sided prisms, terminated by pyramids with four flattened or depressed faces. Crystallized rapidly, it is white and granular; it is insupportably bitter to the taste, and gives an after-sensation similar to that produced by certain metallic salts; it has no smell; it is not changed by exposure to the air; it is neither fusible nor volatile, for when submitted to the action of heat, it only fuses at the moment of its decomposition and carbonization. It is intensely bitter; it is very slightly soluble in water, requiring 6667 parts of cold, and 2500 of boiling water for its solution, which has an orange colour; water that contains only the 600,000th part of a grain in solution imparts a bitter taste! It is very soluble in alcohol, but insoluble in æther. It is by far the most active poison of all the vegetable alkalies; one-eighth of a grain, blown into the throat of a rabbit, brought on locked jaw in two minutes, and in five minutes proved fatal; one-fourth of a grain produces on a healthy man very sensible effects. It acts as a base to acids, and forms a distinct set of salts, which we have now to consider.

1038. *Sulphate of Strychnia* crystallizes in transparent cubes, soluble in less than ten times its weight of cold water. It is intensely bitter, and the Strychnia is precipitated from it by all the soluble salifiable bases.

1039. *Muriate of Strychnia* crystallizes in very small needles; it is more soluble in water than the sulphate.

1040. *Nitrate of Strychnia* can be obtained only by dissolving Strychnia in nitric acid, diluted with a great deal of water. The saturated solution, when cautiously evaporated, yields crystals of neutral nitrate in pearly needles. It is slightly soluble in water, but insoluble in æther. When concentrated nitric acid is poured upon Strychnia, it immediately strikes an ama‑ranthine colour, followed by a shade similar to that of blood. To this colour succeeds a tint of yellow, which passes afterwards into green.

1041. The Acetic, Oxalic, and Tartaric acids, form with Strychnia neutral salts, which are very soluble in water, and more or less capable of crystallizing. They crystallize best when they contain an excess of acid. The neutral acetate is very soluble, and crystallizes with difficulty.

BRUCIA.

1042. It has been proved by the experiments of M. Pelletier, that the nux vomica, besides Strychnia, contains an alkaline element which has been termed *Brucia,* and which had been previously discovered in the bark of the False Angustura, or *Brucia Antidy‑senterica.* From this latter bark it was extracted by digestion in sulphuric æther, and then in alcohol ; the alcoholic solution was then evaporated, and the dry residuum dissolved in water. This solution was satu‑rated with oxalic acid, and evaporated to dryness. Alcohol, digested on the residue, took up the colour‑ing matter, and left a pure oxalate of brucia. This salt was decomposed both by lime and magnesia, which formed insoluble salts with the oxalic acid, and left the brucia soluble in spirit ; it requires 500 parts of water, at 212°, and 850 at common temperatures, but its solubility is much increased by the colouring matter of the bark.

1043. Brucia crystallizes in oblique prisms, the bases of which are parallelograms ; its taste is exceed‑ingly bitter, acrid, and durable in the mouth ; its poi‑sonous powers are similar to those of Strychnia, but much less energetic ; its intensity, when compared with this latter body, being as 1 to 12.   It combines with the acids and forms both neutral and acidulous salts.

Salts of Brucia.

1044.   *Sulphate of Brucia* crystallizes in long slender needles; it is very soluble in water, and moderately so in alcohol; its taste is extremely bitter.   It is decomposed by the alkalies and alkaline earths.   The bi-sulphate crystallizes more readily than the neutral compound.

1045.   *Muriate of Brucia* forms in four-sided prisms, terminated at each end by an oblique face; it is permanent in the air, and very. soluble in water; it is decomposed by sulphuric acid.

1046.   *Nitrate of Brucia* forms a gum-like mass; but the bi-nitrate crystallizes in acicular four-sided prisms.   An excess of nitric acid decomposes the Brucia into a matter of a fine red colour.

DELPHIA.

1047.   This alkali was discovered in the *Stavesacre,* or *Delphinium Staphysagria.*   M. Majendie has given us the following formula for its preparation.   Boil a portion of the seeds of Delphinium, cleared of their coverings, and reduced to a fine paste, in a little distilled water; pass the decoction through a linen cloth, and filter it.   Add very pure Magnesia, and boil for some minutes; filter again; wash the residue carefully, and digest in highly rectified alcohol.   On evaporating the alcoholic tincture, *Delphia* is obtained in the form of a white powder, which affords some points of crystallization.   Delphia, when thus procured, although crystalline while moist, soon becomes opaque on exposure to the air.   It is inodorous, and has a very bitter, and afterwards acrid taste.   Water dissolves so small a quantity, that it can only be discovered by. the slight bitterness it communicates. It is very soluble in alcohol and æther. It is fusible, and becomes, as it cools, a hard and brittle substance; at a higher temperature it is decomposed and burns without residue. Concentrated nitric acid gives a yellow tint to this alkali, whilst the same acid gives a reddish colour to Brucia and Morphia. It forms with the sulphuric, nitric, muriatic, oxalic, acetic, and other acids, very soluble neutral salts, the taste of which is extremely acrid and bitter. Alkalies precipitate it in the form of a white jelly.

1048. Its effects are much more energetic than those produced by *Stavesacre*. Six grains kill a dog in the space of two or three hours. In the state of an acetate it is still more poisonous.

PICROTOXA.

1049. This alkali constitutes the active ingredient of the *Cocculus Indicus*, (the fruit or berry of the *Menispermum Cocculus*), from a strong infusion of which, ammonia, added in excess, will precipitate a white crystalline powder. This powder, after being well washed in cold water, is partially soluble in alcohol, and by the spontaneous evaporation of this latter liquid thus impregnated, *Picrotoxa* separates in quadrilateral prisms of a silky lustre. It is inodorous, but very bitter. It remains unchanged by the air; is soluble in 25 parts of boiling, and in 50 of cold water; in three parts of alcohol, but much more abundantly in æther. It is extremely virulent, three or four grains of it will kill the largest dog in an hour; it is also very poisonous to fish. Strong sulphuric acid dissolves it, but at the same time chars and destroys it. Nitric acid converts it into oxalic acid. It combines with acids and forms a class of salts, many of which are less active than its base.

ATROPIA.

1050. This was discovered by Brandes in the *Atropa Belladonna*, or Deadly Nightshade. It is to be obtained from a decoction of the leaves of this plant, by first adding to it sulphuric acid, in order to throw down the albumen and similar bodies, by which the solution becomes thinner, and passes with greater facility through the filtre; it is then to be super-saturated with potass; a crystalline precipitate is produced, and if this be well washed in water, and then dissolved in an acid, and re-precipitated by an alkali, furnishes the alkali in question. When perfectly pure it has no taste. Cold water has scarcely any effect upon it, but it dissolves a small quantity when recently precipitated, and boiling water dissolves still more. Cold alcohol dissolves only a small portion, but when boiling it readily acts upon it. Æther and oil of turpentine, even when boiling, have little effect on it. It neutralizes much acid and produces salts, the watery solutions of which give out, by evaporation, vapours which dilate the pupil of the eye, and produce giddiness.

1051. *Sulphate of Atropia* crystallizes in rhomboidal tables and prisms with square bases. It is soluble in four or five parts of cold water.

1052. The muriatic, nitric, acetic, and oxalic acids, dissolve Atropia, and form acicular salts, all soluble in water and alcohol. M. Brandes was obliged to discontinue his experiments on the subject, in consequence of the violent headache, pains in the back, and giddiness, with frequent nausea, which the vapours of this alkali occasioned while he was working on it.

VERATRIA.

1053. This alkaline principle was lately discovered by M. M. Pelletier and Caventou, in the *Veratrum*

Sabadilla, or Cevadilla; in the *Veratrum album*, or White Hellebore, and in the *Colchicum Autumnale*, or Meadow Saffron. By the following process these eminent Analysts obtained the alkali in an isolated form. The seeds of Cevadilla, after being freed from an unctuous and acrid matter by æther, were digested in boiling alcohol. As this infusion cooled, a little wax was deposited; and the liquid being evaporated to an extract, re-dissolved in water, and again concentrated by evaporation, parted with its colouring matter. Acetate of lead was now poured into the solution, and an abundant yellow precipitate fell, leaving the fluid nearly colourless. The excess of lead was thrown down by sulphuretted hydrogen, and the filtered liquor being concentrated by evaporation, was treated with magnesia, and again filtered. The precipitate boiled in alcohol, gave a solution, which, on evaporation, left a pulverulent matter, extremely bitter, and with decidedly alkaline characters. It was at first yellow, but by solution in alcohol, and precipitation by water, was obtained in a fine white powder. The precipitate by the acetate of lead gave on examination, gallic acid; whence it is inferred that the Veratria exists in native combination as a *Gallate*.

1054. Veratria is sparingly soluble in æther, but more abundantly so in alcohol; in cold water it is not more soluble than Morphia or Strychnia; boiling water, however, dissolves 1,000th of its weight, and acquires an acrid taste. It fuses at 122°, and then appears like wax; at a higher temperature it swells, decomposes, and burns. It saturates all the acids, and forms with them uncrystallizable salts, which, on evaporation, take the appearance of gum. The sulphate alone affords rudiments of crystals when its acid is in excess. Nitric acid combines with Veratria; but if added in excess, especially when concentrated, it does not produce superoxidation, as in the case of Morphia and Strychnia; but it very rapidly resolves the vegetable substance into its elements, and gives birth to a yellow detonating matter analogous to the *bitter principle of Welther.* The acetate is the most active of all its salts. Veratria, in a very minute quantity, occasions dreadful sickness and vomiting, and in the dose of a few grains proves fatal. When inhaled into the nostrils, it produces violent and dangerous sneezing, even when the quantity is too small to be weighed.

HYOSCYAMA.

1055. This alkali was extracted by Brandes from the *Hyoscyamus Nigra,* or Henbane. It crystallizes in long prisms, and when neutralized by sulphuric or nitric acid forms characteristic salts.

1056. Such are the principal Vegeto-alkalies which have been discovered as constituting the active ingredients of certain medicinal plants. Others have been extracted from different vegetables, but it will be unnecessary to enter into their history, as farther experiments are required to establish their identity, and to confirm the characters so hastily assigned to them.

OTHER VEGETABLE PRINCIPLES.

1057. In addition to the proximate principles, already enumerated and described, Vegetable analysis has developed others, so indistinct, however, and anomalous in their characters and habitudes, that it may be fairly doubted whether they merit consideration as independent bodies. Such are *Asparagin,* a crystalline substance obtained from the juice of Asparagus; *Ulmin,* a spontaneous exudation from a particular species of Elm; *Inulin,* a species of starch from the *Inula Helenium* or Elecampane; *Fungin,* from the fleshy part of Mushrooms; *Hæmatin,* from the Logwood, or *Hæmatoxylon Campechianum,* of which it forms the colouring matter; *Nicotin,* one of the active principles of Tobacco; *Sarcocoll,* a spontaneous exudation from the *Pænea Sarcocolla,* which greatly resembles gum, but is precipitated by Tannin; *Medullin,* or the pith of the Sun-flower; *Cathartine,* the supposed cathartic principle of Senna; *Piperine,* from the black pepper; *Lupulin,* from the Hop, &c. &c.

LIGNIN.

1058. After we have removed, by the processes already described, all those principles which are soluble in any menstruum, we arrive at the wood, or woody fibre, termed *Lignin,* and which may be regarded as the skeleton of trees, and of the greater number of plants in general; and from whatever vegetable source it is obtained, its chemical properties appear to be essentially the same. It is perfectly destitute of smell, taste, and colour; and its specific gravity is generally inferior to that of water. It is insoluble in water and alcohol at all temperatures. It is acted upon by the pure fixed alkalies, which render it soft and of a brown colour; and if equal weights of potass and saw-dust are heated together, the latter are completely dissolved, and an aqueous solution may be effected, from which a gum-like substance is precipitated by an acid; the same result is obtained by treating *Lignin* with concentrated sulphuric acid, and saturating the product with carbonate of lime. Nitric acid decomposes *Lignin* with the assistance of heat, and oxalic, malic, and acetic acids are formed. When exposed to heat, it yields an acid, formerly termed the *pyroligneous,* but which recent experiments have shewn to be identical with the acetic acid (963.) The ultimate elements of lignin appear to be carbon, 7 atoms, oxygen, 4 ditto, hydrogen, 4 ditto.

1059. Such are the general characters of the most important of the proximate vegetable principles, and it is essentially necessary to the chemical student to become well acquainted with them by study and experiment. It is equally important that he should know the proportions of the ultimate elements of each, since that knowledge will enable him to understand the changes of which they are susceptible, and to explain the theory of their mutual conversion into each other. In order to furnish a more satisfactory view of this subject, I shall here introduce a Synoptical Table of the ultimate composition of the proximate principles.

A TABLE OF ORGANIC ANALYSES,
ACCORDING TO THE LATEST DISCOVERIES OF DR. URE.

Substances.	Carbon.	Oxygen.	Hydrogen.	Azot.	Supposing the Oxygen and Hydrogen to exist as Water.	Excess.
Gum Arabic	35·13	55·79	6·08	3?	54·72	7·15 *Oxygen.*
Starch	38·55	55·32	6·13		55·16	6·3 ditto
Sugar	43·38	50·33	6·29		56·62	0
Resin	75·..	12·50	12·50		15·20	11·20 *Hydrogen.*
Yellow Wax....	80·69	7·94	11·37		8·93	10·39 ditto
Camphor	77·38	11·48	11·14		12·91	9·71 ditto
Oil of Turpentine	82·51	7 87	9·62		8·85	8·64 ditto
Spermaceti Oil..	78·91	10·12	10·97		11·34	9·71 ditto
Castor Oil	74·..	15·71	10·29		17·67	8·33 ditto
Alcohol, sp.g.812	47·85	39·91	12·24		44·9	7·25 ditto
Æther, 0·70	59·60	27·1	13·3		30·5	9·9 ditto
Citric Acid	33·..	62·37	4·63		41·67	25·33 *Oxygen.*
Acetic Acid.....	48·	48·..	4·..		36·..	16·.. ditto
Tartaric Acid ...	31·42	65·82	2·76		24·84	43·74 ditto
Oxalic Acid	19·13	76·20	4·76		42·87	38·09 ditto
Benzoic Acid....	66·74	28·32	4·94		31·86	1·4 ditto

The discordance shewn in the above Table with the analyses of other Chemists evinces the necessity of farther investigation.

OF THE SPONTANEOUS CHANGES OF WHICH CERTAIN VEGETABLE BODIES ARE SUSCEPTIBLE.

1060. It has been already observed, that all the elementary principles of organic nature may be considered as deriving the peculiar delicacy of their chemical equilibrium, and the consequent facility with which it may be subverted and new modelled, to the multitude of atoms grouped together. These atoms, however, have been shewn to be very few in kind, and hence we can easily understand why one proximate principle can be so readily converted into others; the same materials serving for all, and a slight change in their proportions impressing upon the compound a striking change of character. We have thus seen that gum is convertible into no less than four acids by the modified action of nitric acid; that starch is convertible into sugar, and sugar into oxalic acid, &c. Upon the same principle we have now to consider those important changes which spontaneously take place in certain Vegetable bodies, and to which the general term of Fermentation has been applied.

PHÆNOMENA AND PRODUCTS OF FERMENTATION.

1061. Lavoisier was the first philosopher who instituted, on right principles, a series of experiments to investigate the phænomena of Fermentation, and so judiciously were they contrived, and so accurately conducted, that, as Dr. Ure justly observes, his results are comparable to those derived from the more rigid methods of the present day. The generic term *Fermentation* is used to express the spontaneous changes which organic matter undergoes, when combined with sufficient water, and exposed to the ordinary temperature of the atmosphere. It comprises three distinct processes, and the fermentation is said to be *vinous,* *acetous,* or *putrefactive,* according to the result which accompanies it. To these, perhaps, a fourth species may be added, viz. that which accompanies the conversion of flour into bread, and which has been termed the *Panary Fermentation.* We shall consider these different species in succession.

Vinous Fermentation.

1062. Under this name is comprehended every variety of fermentation which terminates in the formation of an intoxicating liquid.

1063. The principal substance employed in *vinous* fermentation is Sugar, and no solution can undergo this process which does not contain it in a sensible quantity ; now, such solutions, though numerous, may be comprehended under two heads, viz. those obtained from the juices of plants, and those obtained from the decoction of seeds. The result of the fermentation of the first is termed *Wine,* that of the second, *Beer.* All these fermented liquors, however, owe their intoxicating quality to one and the same principle, viz. *Alcohol*; and all the varieties of Spirit obtained from them by distillation, and known under the names of Brandy, Rum, Spirits of Wine, Arrack, Gin, and Whisky, differ from each other merely in flavour and strength.

1064. Let us observe the phænomena of Vinous Fermentation, and examine the chemical changes which occur during its progress.

1065. Although no solution is susceptible of vinous fermentation unless it contain sugar, still pure sugar alone is unfermentable, and requires the presence of some *leaven* to enable it to undergo that process. In the native juices of fruits, as well as in imperfectly refined sugar, such a principle exists, whence the addition of a ferment or *leaven*, is not essential for their

fermentation.   Considerable obscurity still exists with
respect to the nature and composition of this fermenting principle, but late experiments render it probable
that it is *Gluten*, or rather *Zimome* (912)

1066. The *Ferment* or *Yeast* is a substance which
separates under the form of flocculi, more or less viscid,
from all the juices and infusions which experience the
vinous fermentation. It is commonly procured from
the beer manufactories, and is hence called the *Barm
of Beer*.

1067. If a mixture of five parts of sugar, twenty of
water, and one of yeast, be introduced into a matrass,
furnished with a doubly bent tube, the end of which
is made to pass under an inverted jar in the hydro-
pneumatic trough, as here represented, the following
phænomena will present themselves.

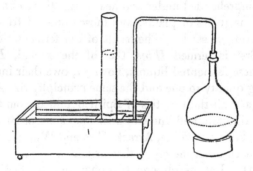

If the materials in the matrass be exposed to a temperature of from 70° to 80°, we shall speedily observe
the syrup to become muddy, and a multitude of air
bubbles to form all around the ferment. These unite,
and attaching themselves to particles of the yeast, rise
along with it to the surface, forming a stratum of froth.
The yeast will then disengage itself from the air, fall
to the bottom of the vessel, to reacquire buoyancy a
second time by attached air bubbles, and thus in suc-

cession; at the same time bubbles of carbonic acid will continue to pass over into the jar of the pneumatic trough. After a certain period, however, this intestinal motion will cease, and the fluid, depositing a sediment, will become clear; it is now no longer sweet, but has acquired a new taste.

1068. Now what is the cause of this fermentation? What are the substances which mutually decompose each other? and what is the nature of the new substance formed?

1069. Lavoisier found that the presence of atmospheric air is not essential to the process; although the liquor when shut up in close vessels does not so readily ferment; this, however, depends upon the accumulation of carbonic acid. No oxygen is absorbed, nor is any water decomposed during the process; so that the sugar alone furnishes the two new products,— the alcohol, and the carbonic acid. Lavoisier, on comparing the composition of sugar with that of alcohol was led to the conclusion that, during the vinous fermentation, part of the carbon, by uniting with the oxygen of the sugar, passes to the state of carbonic acid, and that the remaining carbon, with the hydrogen of the sugar, compose alcohol. If, therefore, it were possible to combine carbonic acid and alcohol, sugar ought to be regenerated. Discoveries subsequent to those of Lavoisier have assigned different proportions to the elements both of sugar and alcohol, but they have left his general theory of the conversion of the one into the other entire; and it will remain as a lasting monument of his sagacity. The part which the ferment plays in this interesting metamorphosis is still involved in much obscurity; it would seem probable that, from its strong affinity for oxygen it abstracts a small proportion of it from the saccharine matter, and the equilibrium between the particles of the sugar being thus broken,

these so re-act on each other, as to be transformed into
alcohol and carbonic acid; at the same time, the yeast
which is already in a state of fermentation, may tend to
equalize the action throughout every part of the liquid.

1070. When we distil any liquor which has undergone the vinous fermentation, we obtain a *spirituous
liquor,* which by rectification affords a diluted alcohol
commonly termed *Spirits of Wine.* It has been doubted
whether the alcohol, obtained by the distillation of
wines, and of other fermented liquors, is an educt or
a product; or, in other words, whether it exists *ready
formed,* or is actually generated during the distillation.
We are indebted to Mr. Brande for setting this question completely at rest, by shewing, in the first place,
that the quantity of alcohol resulting from the distillation of wine, is not altered by a variation of temperature equal to 20 degrees of *Fah.* and in the next place,
by actually separating the alcohol from fermented liquors, without the intervention of heat.   The following is the process by which he effected this object: —
By adding a solution of the acetate of lead to wine, he
precipitated the extractive and colouring matter in
combination with the metallic oxide; he then filtered
the fluid, and thereby obtained a mixture of alcohol,
water, and a portion of the acid of the metallic salt;
the water was separated by hot and dry sub-carbonate
of potass, and the alcohol was left floating at the top,
as a distinct stratum.   The experiment may be conveniently performed in a glass tube, from half an inch
to two inches diameter, which is accurately graduated
into 100 parts.

1071.   By an extensive series of experiments, conducted upon the principle above related, Mr. Brande
was enabled to construct the following Table.

Table of the quantity of Alcohol (*Sp. Gr.* ·825) at 60° *Fah.* in several kinds of Wines, and other liquors.

	Per Cent. by Measure.		Per Cent. by Measure.
Port, average of 6 kinds	23·48	Palm Wine	4·70
Ditto, highest	25·83	Vin de Grave	12·80
Ditto, lowest	21·40	Frontignac	12·79
Madeira, highest	24·42	Coti Roti	12·32
Ditto, lowest	19·34	Rousillon	17·26
Sherry, aver. of 4 kinds	17·92	Cape Madeira	18·11
Ditto, highest	19·83	Cape Muschat	18·25
Ditto, lowest	13·25	Constantia	19·75
Claret, aver. of 3 kinds	14·43	Tent	13·30
Calcavella	18·10	Sheraaz	15·52
Lisbon	18·94	Syracuse	15·28
Malaga	17·26	Nice	14·63
Bucellas	18·49	Tokay	9·88
Red Madeira	18·40	Raisin Wine	25·77
Malmsey Madeira	16·40	Grape Wine	18·11
Marsala	25·87	Currant Wine	20·55
Ditto	17·26	Gooseberry Wine	11·84
Red Champagne	11·30	Elder Wine, Cider, and	
White ditto	12·80	Perry	9·87
Burgundy	11·55	Stout	6·80
Ditto	11·95	Ale	8·88
White Hermitage	17·43	Brandy	53·39
Red ditto	12·32	Rum	53·68
Hock	14·37	Hollands	51·60
Ditto	8·88		

1072. When the intoxicating powers of wine and spirit are compared, it must be admitted that they bear no proportion to the quantity of spirit which they contain; it has for instance been shewn that Port, Madeira, and Sherry, contain from one-fourth to one-sixth their bulk of alcohol, so that a person who takes a bottle of either of them will thus take nearly half a pint of alcohol, or almost a pint of pure brandy! and moreover, it is known that wines which have been found by experiment to contain the same absolute proportion of alcohol, possess very varying powers of intoxicating. We cannot therefore be surprised at the scepticism which has existed upon the subject of the alcohol's pre-existence in vinous liquors. The fact, however, is easily reconciled by the Physiologist, who is well aware of the extraordinary powers of chemical combination, or even of mixture, in modifying the activity of substances upon the living system. I have already commented so largely upon this fact in the first volume of my Pharmacologia, that in this place it is only necessary to state, that in wine, the alcohol would appear to be so intimately combined with extractive matter, that it is incapable of exerting its full effects upon the stomach before it becomes altered in its properties by the powers of digestion. This remark, however, applies only to pure wine; such as are *brandied* contain the alcohol in an uncombined form, and are consequently as powerful as the same quantity of spirit in a similar state of dilution.

ALCOHOL.

1073. To obtain pure alcohol, the *Spirit of Wine* of the shops must be slowly and carefully distilled; but by this process we can scarcely obtain it of a lower specific gravity than ·825. If our object be to procure it in a higher state of concentration, we must distil the *Rectified Spirit* of commerce with one-fourth of its weight of dry and warm sub-carbonate of potass, at a temperature of 200°; and draw over about three-fourths of the original quantity. The dry chloride of calcium (*Muriate of Lime*) may also be employed for the same purpose, but it has been supposed that, in this case, a small portion of æther is generated. It has been also found that spirit of wine of sp. gr. ·867, when inclosed in a bladder, and exposed for some time to the air, is converted into alcohol of sp. gr. ·817, the water only escaping through the coats of the bladder. This fact is well deserving the attention of the anatomist, who has long supposed that the diminution of liquid in his wet preparations entirely depended upon the loss of alcohol.

1074. Alcohol thus obtained is a limpid, colourless liquid, of an agreeable smell, and a strong flavour. Its specific gravity varies with its purity, but it is always lighter than water. The strongest spirit that has hitherto been procured is of sp. gr. 0·796, at the temperature of 60° *Fah.* and it is probably alcohol free from water; the " *Alcohol*" of the Pharmacopœia (sp. gr. ·815) is supposed to contain 7 per cent; " *Rectified Spirits of Wine,*" 15 per cent.; and *Proof Spirit*, 56 per cent. of water. Pure alcohol has never been rendered solid by exposure to any degree of cold, either natural or artificial, but rectified spirit has been lately solidified by cold produced by the rapid evaporation of sulphurous acid (625). The quantity of alcohol and water, in mixtures of different specific gravities, may be at once learnt by referring to Mr. Gilpin's copious tables, published in the Philosophical Transactions (1794), and which, either entire or in the form of an abridgement, have been introduced into the various works on general Chemistry.

1075. Alcohol, sp. gr. ·820, boils at 176°, or in vacuo at 56° ; when of sp. gr. ·800 it boils at 173·5. If water be added, its boiling point is raised. Alcohol sp. gr. ·900 has been found by Mr. Dalton to boil at 182° ; that of ·850 at 170°. In this case, as in others where two volatile liquids are mixed, the boiling point is varied, but not proportionably; for Dr. Henry observes that the mixture boils nearest the boiling point of that which is most volatile ; thus, a mixture of equal parts of alcohol of such a strength as to boil at 170°, and water, ought by the rule of proportion to boil at 194°, whereas in fact it boils at 182°.

1076. Alcohol is extremely volatile, and produces considerable cold by its evaporation (351), the degree of which will be proportional to the strength of the spirit employed; if, for instance, a certain degree of cold be produced by the evaporation of water, and another degree by that of alcohol, a spirit of half strength will give a degree of cold just half way between the two. Alcohol is highly inflammable, and burns away with a blue flame, without leaving any residue; and although the light emitted by its combustion be feeble, the heat evolved is so considerable as to render it subservient to various chemical processes. (367.)

1077. If the products of the combustion of alcohol be carefully collected, we shall obtain carbonic acid, and a quantity of water exceeding the original weight of the alcohol burned. According to Saussure, jun. 100 parts of alcohol will thus afford 136 parts of water; this excess is to be explained by the addition of the oxygen of the atmosphere during its combustion.

1078. Alcohol unites chemically with water in every proportion, and the union is accompanied by an evolution of heat. Equal measures of alcohol and water, each at 50° Fah., give by sudden admixture an elevation of nearly 20 degrees of temperature; and equal measures of proof spirit and water occasion an increase of 9½°. There is also a " *Penetration of dimensions*" (31) upon this occasion; that is to say, the bulk of the resulting liquid is less than that of the two before admixture, and consequently, the specific gravity of the mixture must exceed the mean of those of its ingredients; thus a pint of alcohol and a pint of water, when the mixture has cooled to the temperature of the atmosphere, falls considerably short of two pints.

1079. Alcohol is a very powerful solvent of many substances; and it is on this account that it constitutes a most valuable agent in pharmacy. It dissolves soap, alkaline sulphurets, vegetable extractive, tannin, sugar, volatile oils, camphor, and resins; it dissolves fixed oils but sparingly, except Castor oil, which it dissolves abundantly. It also combines with phosphorus, sulphur, and the pure alkalies, but not with their carbonates. It dissolves the benzoic, camphoric, gallic, oxalic, and tartaric acids. Of the class of salts, alcohol dissolves many copiously, some sparingly, and others not at all. The solubility of each particular salt in this menstruum has been mentioned under its history. It may be here stated generally, that all deliquescent salts are soluble, while the sulphates are insoluble. The solubility, however, of salts, like that of many of the vegetable principles, will depend upon the strength of the alcohol employed; for a mixture of spirit and water appears to possess greater energy as a solvent in many cases, than is possessed by either of them in a separate state. When alcohol, containing certain saline bodies in solution, is set on fire, its flame is often tinged of different colours according to the body; thus, nitrate of strontian tinges it purple; boracic acid and cupreous salts, green; muriate of lime, red; and nitre and oxy-muriate of mercury, a yellow colour.

1080. According to Saussure, jun. alcohol is composed of nearly

Carbon....	52·17 or 2 atoms	= 12
Oxygen ...	34·79 .. 1	= 8
Hydrogen..	13·04 .. 3	= 3
100.00	Weight of its atom	= 23

ÆTHER.

1081. When alcohol is distilled with the more powerful acids, it undergoes an important change, and is converted into a much more highly volatile and inflammable liquid, and which, unlike alcohol, is only miscible in a small proportion with water. This fluid has received the generic name of *Æther*; and the peculiar varieties are distinguished by adding the name of the acid, by the intervention of which they are prepared; we have accordingly, *Sulphuric, Nitric, Muriatic,* and *Acetic* Æther, &c.

SULPHURIC ÆTHER.

1082. To prepare this liquid, let a quantity of alcohol be introduced into a glass retort, and then an equal weight of sulphuric acid be gradually added in divided portions, taking care to agitate the mixture after each addition, and to allow a sufficient interval, to prevent its exceeding the temperature of 120°. The retort is then to be cautiously placed in a sand bath, previously heated to 200°, so that the liquor may boil as quickly as possible, and the æther may pass over into a tubulated receiver, to which another receiver is adapted and kept cold by ice or water. Let the liquor distil, until a heavier fluid begins to pass over, and to appear under the æther at the bottom of the receiver. To the liquor which remains in the retort, again pour twelve ounces of rectified spirit, that æther may distil in the same manner. If, when the æther ceases to be formed, the receiver be removed, and the heat still continued, a black froth boils up in the retort, sulphurous acid is produced abundantly, and the heavier liquor, above mentioned, may be obtained in considerable quantity; this is termed *Oil of Wine.* Sulphuric æther, as thus prepared, is far from pure, it contains alcohol, water, and sulphurous acid; to remove these, a process of Rectification is instituted, which consists in redistilling it in contact with fused potass.

1083. Let us endeavour to explain the changes which the alcohol and sulphuric acid undergo in this operation. It appears that the process of ætherification depends upon abstracting from alcohol one-half of its oxygen, and one-sixth of its hydrogen, which, it will be observed, are in the proportions required to form water, thus

	Alcohol.	*Æther.*	*Olefiant Gas.*
Carbon........	4 atoms	4 atoms	4 atoms
Oxygen........	2	1	0
Hydrogen.....	6	5	4 ....

1084. When a much larger proportion of sulphuric acid is added to the alcohol, than that directed for the preparation of æther, we obtain different results; little or no æther is formed, but *bi-carburetted hydrogen gas*, or *Olefiant gas*, constitutes the principal product (486). In this case double the quantity of oxygen and hydrogen is taken from the alcohol, which we shall see by a reference to the atomic proportions above stated will leave 4 atoms of carbon + 4 atoms of hydrogen, which will exactly constitute 4 atoms of bi-carburetted hydrogen (485).

1085. The black matter which bubbles up in the retort is produced by a farther decomposition of the alcohol, in which its hydrogen, combining with the oxygen of the acid, leaves a residuum of carbon, while a portion of the sulphuric is at the same time converted into sulphurous acid. The carbon in this case exists in a peculiar state of modification, (471) and appears to hold some hydrogen in combination.

1086. If any additional argument were necessary to confirm the above theory of ætherification, it has been lately afforded by a most satisfactory experiment, in which alcohol was converted into æther by the transmission of a current of *Fluoboric* acid* gas through that fluid. Now it is well known that the only principle which this gas could extract from the alcohol is water, or its elements; forming with its hydrogen *fluoric*, and with its oxygen, *boracic* acid.

1087. Sulphuric æther is an extremely light, volatile, and inflammable fluid; its odour is grateful, and its taste pungent. Its specific gravity varies; for medicinal purposes it should never exceed ·750, in which state it contains nearly 25 per cent. of alcohol; it may be procured at ·700, or according to Lovitz at ·632°. Unlike alcohol, it does not combine in any considerable proportion with water, this latter fluid holding dissolved not more than about one-tenth its weight of æther. This fact offers an easy mode of purifying æther from alcohol; for by repeated agitation with water it is deprived of most of the spirit which it may contain, and is brought to a high degree of purity. It is extremely volatile, and during evaporation it occasions a great degree of cold. (351.) By pouring a small stream of æther from a capillary tube, on a thermometer bulb filled with water, the water may be frozen, even in a warm summer atmosphere. Under the usual atmospheric pressure æther boils at 96°, and in *vacuo* at a temperature much below zero. (317.) At — 46° it assumes a solid form. The existence of æther, as a gaseous body, at 96°, may be satisfactorily shewn by the following simple experiment.

Exp. 89. Fill a jar with hot water, and invert it in a shallow vessel containing the same; then, by means of a small glass tube closed at one end,

---

* This gas consists of the elementary bases of the fluoric and boracic acids, called *Fluorine* and *Boron*. The acidifying principle of the former appears to be hydrogen, that of the latter, oxygen.

introduce a little æther. The æther will rise to the top of the jar, and, in its ascent, will be changed into gas, filling the whole jar with a transparent, invisible, elastic fluid. On permitting the water to cool, the æthereal gas is condensed, and the inverted jar again becomes filled with water. The same effect will be produced by placing an inverted jar of cold water with some æther floating on its surface, under the receiver of an air pump.

1088. Æther is highly inflammable, so that its vapour readily takes fire on the approach of flame; the student should therefore be cautious whenever he operates upon it by candle light. Its slow combustion may be easily shewn by passing a few drops into a receiver furnished with a stop cock and pipe, and inverted in water of the temperature of 100°. The receiver will be filled with the gas of æther (*Exp.* 89,) which may be expelled through the pipe and set on fire. It burns with a beautiful deep blue flame. When previously mixed with oxygen gas it detonates on the approach of flame; it explodes also spontaneously with chlorine.

Nitric Æther.

1089. Various processes have been described for the preparation of this compound. It will be only necessary to state the following. To two pints of alcohol contained in a glass retort, add, by degrees, half a pound of nitric acid; and, after each addition, cool the materials by setting the retort in a vessel of cold water. Distil the mixture by a very cautiously regulated heat, till about a pint and a half have come over. In this state the æther is far from being pure, and must be redistilled with the addition of pure Potass, preserving only the first half, or three-fourths that come over.

1090. The great difficulty and danger of the operation depends upon the violent action which alcohol
and nitric acid exert on each other; to obviate which,
Dr. Black contrived a method by which the mixture
was effected with extreme slowness. On two ounces
of strong nitrous acid, contained in a phial having a
conical ground glass-stopper, and a weak spring fitted
to keep the stopper in its place, he poured slowly and
gradually about an equal quantity of water, which, by
being made to trickle down the sides of the phial,
floated on the surface of the acid, without mixing with
it; he then added, in the same cautious manner, three
ounces of alcohol, which, in its turn, floated on the
surface of the water. By such means the three fluids
were kept separate, on account of their different specific
gravities (*Exp.* 1.) and a stratum of water was interposed between the acid and spirit. The phial was
now deposited in a cool place, to allow the acid gradually to ascend, and the spirit to descend, through
the water; this last acting as a boundary to restrain
their action on each other. When this commences,
bubbles of gas rise through the fluids, and the acid
assumes a blue colour, which it again loses in the
course of a few days, and a yellow nitric æther appears
swimming on the surface. By this ingenious method
nitric æther may be formed, without the danger of
explosion. It is, however, evident that such a process,
however ingenious, cannot be conveniently adopted by
the manufacturer.

1091 The " *Spirit of Nitric Æther*," directed in
the Pharmacopœia, is a mixture of nitric æther and
alcohol, in proportions which have not been determined.

1092. The exact nature and composition of Nitric
æther have not been satisfactorily ascertained; no
doubt, however, exists of its being formed of a portion

of each of the elements of the acid and spirit, and that consequently, oxygen, hydrogen, carbon, and azote, all enter into its composition.

1093. Nitric æther, in its ordinary state, is a liquid of a yellowish-white colour. It has an æthereal odour, which is so powerful that its inhalation into the nostrils produces a species of giddiness. It does not redden litmus; its taste is acrid and burning; it boils at 70° *Fah.*: when poured into the hand it immediately boils and produces considerable cold. It is sufficient to grasp in our hands a phial containing it, to see bubbles immediately escape. It takes fire very readily, and burns quite away with a white flame. When agitated with 25 or 30 times its weight of water, it is divided into three portions. One, the smallest, is dissolved; another is converted into vapour; and a third is decomposed. The solution becomes suddenly acid; it assumes a strong smell of apples; and if, after saturating the acid which it contains with potass, it be subjected to distillation, we withdraw the alcohol, and obtain a residue formed of nitrate of potass. We see here that there is a separation of one part of the two bodies which constitute the æther. Left to itself in a well stopped bottle, the æther suffers a spontaneous change, for it becomes perceptibly acid. By distillation, acid is instantly developed, which shews that heat favours its decomposition.

MURIATIC ÆTHER.

1094. To prepare this æther, we may saturate alcohol with muriatic acid gas; or mix together equal bulks of alcohol and concentrated muriatic acid, and heat the mixture in a glass retort connected with a Woulfe'e apparatus; the first bottle should contain water at about 80°; the second should be surrounded with ice. Another process consists in adding, to a

mixture of 8 parts of manganese and 24 of muriate of
soda, in a retort, 12 parts of sulphuric acid, previously
mixed, with the necessary caution, with 8 of alcohol,
and proceeding to distillation. The æther thus obtained requires to be rectified by a second distillation
from potass.

1095. The nature of this body is involved in some
obscurity. Boullay considers it as a compound of
muriatic acid and alcohol, but according to M. M.
Colin and Robiquet, it is a compound of bi-carburetted
hydrogen and muriatic acid. This is rendered probable by the fact that when one volume of muriatic
æther is passed through a porcelain tube, at a dull red
heat, it is resolved into a mixture of one volume of
each of these gases, the sum of whose densities very
nearly corresponds with the specific gravity of the
æthereal vapour.

1096. Liquid muriatic æther vaporizes at 51°;
poured on the palm of the hand it instantly boils and
produces much cold. Its odour is very strongly æthereal; its taste is perceptibly saccharine; its specific
gravity is 0·874. It dissolves in an equal volume of
water, but it does not redden vegetable blues; nor
precipitate nitrate of silver or proto-nitrate of mercury.
It is extremely inflammable, and the odour of muriatic
acid is developed by its combustion.

ACETIC ÆTHER.

1097. This is probably the only true and distinct
æther which can be formed by the action of a vegetable
acid on alcohol, for although the *Benzoic, Malic,
Oxalic, Citric,* and *Tartaric* acids dissolve in alcohol,
yet they separate from it again by distillation, without
any peculiar product being formed, however frequently
we act upon the same quantity of acid and alcohol.
But if, instead of putting the vegetable acids alone in

contact with alcohol, we add to the mixture one of the concentrated mineral acids, we can then produce, with several of them, compounds analogous to the preceding æthers. It has been supposed that the mineral acid in this case acts by condensing the alcohol, and elevating the temperature to such a degree as to determine the requisite chemical action, but the truth of this explanation is extremely doubtful. These observations, however, do not apply to acetic acid, for that fluid can exhibit by itself the phænomena of ætherization.

1098. Acetic æther may be formed by repeatedly distilling concentrated acetic acid with alcohol, and returning the distilled liquor to the charge in the retort. The æther thus produced, may be freed from a redundance of acid, by distillation with a small quantity of potass. Or we may obtain an excellent acetic æther, very œconomically, by taking 3 parts of acetate of potass, 3 of concentrated alcohol, and 2 of sulphuric acid; introducing the mixture into a tubulated retort, and distilling to perfect dryness; then mixing the product with a fifth part of its weight of sulphuric acid, and, by a careful distillation, drawing off as much æther as there was alcohol employed.

1099. Acetic æther is a colourless liquid, having an agreeable odour of sulphuric æther and acetic acid. It does not redden litmus; its taste is peculiar; it is heavier than other æthers, its specific gravity being ·866. It is volatile, boils at 228°, and burns with a yellowish-white flame. During combustion, acetic acid is developed, though none can be discovered in the æther before. It is not changed by keeping. Water at 62° dissolves $7\frac{1}{4}$ parts of its weight.

ACETOUS FERMENTATION.

1100.   When any of the vinous liquors are exposed
to the free access of the atmosphere ; or the fermentative process be suffered to proceed in open vessels,
more especially if the temperature be raised to 90
degrees, a second and very different fermentation is
established ; and, as it terminates in the production
of a sour liquid, called Vinegar, the process has been
termed The *Acetous Fermentation.* If the phænomena
of this operation be carefully examined, we shall find
that a quantity of the oxygen of the air is consumed,
and that little or no gas is evolved, so that, unlike
vinous fermentation, the access of the atmosphere is
indispensable.   The most obvious change which the
vinous spirit undergoes during its conversion into vinegar is the acquisition of oxygen.   It was long supposed,
on the authority of Boerhaave, that the *acetous* is universally preceded by the *vinous* fermentation ; but this
is an error ; Cabbages sour in water, making *sour
crout ;* Starch, in the vats of the starch-maker, and
dough itself, form vinegar, without any previous production of wine ; so again in certain solutions of saccharine matter, this acid is evidently generated without
any previous alcoholic state of the fluid.

1101.   Neither pure alcohol, nor alcohol distilled
with water, is susceptible of this fermentation ; the
presence of some vegetable matter is essential to the process.   The vinegar of this country is usually obtained
from malt liquor, while wine is employed as its source
in those countries where the grape is abundantly cultivated.   The great art of making good vinegar consists
in the skilful regulation of the fermentative temperature.   The qualities of vinegar are well known.   It
consists of a mixture of acetic acid, a little alcohol,

mucilage, colouring matter, and sometimes tartaric and
malic acids; its specific gravity is liable to much variation; when exposed to the air it becomes mouldy and
putrid, in consequence of the mucilage which it contains; this change, however, is much retarded by the
addition of a small quantity of sulphuric acid, whence
the maker is allowed by law to mix one-thousandth of
its weight of that acid with it. The strongest malt
vinegar, which is termed *Proof* Vinegar, is estimated
as containing five per cent. of real acetic acid.* The
habitudes of Acetic acid have been already considered.
(507.)

The Panary Fermentation.

1102. Farinaceous vegetables are converted into
meal by trituration, or grinding in a mill; and when
the husk or bran has been separated by sifting, or *bolt-
ing*, the powder is called *flour*. This is a compound
of a small quantity of mucilaginous saccharine matter,
soluble in cold water; much starch, which is scarcely
soluble in cold water, but combines with that fluid by
heat; and an adhesive grey substance, or gluten.
When flour is kneaded together with water, it forms a
tough paste, containing these principles very little altered, and not easily digested by the stomach. The
application of heat, however, renders the compound
more easy to masticate, as well as to digest, whence we
find in the earliest history a reference to some process
instituted for the purpose of producing this change,
although the discovery of the manufacture of bread,
simple as it may appear to us, was probably the work
of ages.

1103. The only substance well adapted for making
bread, that is to say *loaf* bread, is wheat flour; this

---

* For a farther account of this acid see my Pharmacologia,
vol. 2.

circumstance is owing to the quantity of gluten which
it contains, and which operates in a manner to be presently explained. The first stage of the process consists in mixing the flour with water, in order to form a
paste; the average proportion of which is two parts of
the latter to three of the former; but this will necessarily vary with the age and quality of the flour; in
general, the older and the better the flour the greater
will be the quantity of water required. This flour
paste may be regarded as merely a viscid and elastic
tissue of gluten, the interstices of which are filled with
starch, albumen, and sugar. If then it be allowed to
remain for some time, its ingredients gradually react
upon each other, the gluten probably performing an
important part by its action on the sweet principle, a
fermentation is established, and alcohol, carbonic acid,
and lastly acetic acid, are evolved. If the paste be now
baked, it forms a loaf full of eyes like our bread, but
of a taste so sour and unpleasant that it cannot be
eaten. If a portion of this old paste, or *Leaven* as it
is called, be mixed with new made paste, the fermentation commences more immediately, a quantity of carbonic acid is given off, but the gluten resists its disengagement, expands like a membrane, and forms a multitude of little cavities, which give lightness and sponginess to the mass. We easily perceive therefore why
flour, deficient in the tenacity which gluten imparts to
it, is incapable of making raised bread, notwithstanding
the greatest activity may be given to the fermentative
process by artificial additions. The same view will
explain why the addition of spirit, ammonia, &c. as
practised by some bakers, will tend to lighten the
bread; they furnish volatile matter by the action of
heat, the bubbles of which, being imprisoned in every
part, increase the number of cavities. Where, however, *leaven* is employed, the bread will be apt to be

sour, in consequence of the great difficulty of so adjusting its proportion, that it shall not, by its excess, impart an unpleasant flavour, nor, by its deficiency, render the bread too compact and heavy. It is for such reasons that we employ in this country *Barm*, a ferment which collects on the surface of fermenting beer. It appears that we are indebted to the ancient Gauls for this practice. In Paris it was introduced about the end of the 17th century; the faculty of medicine, however, declared it prejudicial to health, and it was long before the bakers could convince the public that bread baked with *barm* was superior to that with *leaven*. A great question arose amongst chemists as to the nature of this *barm* that could produce such effects, and elaborate analyses were made, and theories deduced from their results; but all these ingenious speculations fell to the ground, when it was found that *barm* dried, and made into balls would answer every purpose; the bakers imported it from Flanders and Picardy in such a state, and yet when again moistened, it fermented bread equally well; the presence therefore of carbonic acid, water, acetic acid, and alcohol, could not be essential, for these ingredients were separated by the process of its preparation; at length it was discovered that gluten, mixed with a vegetable acid, produced all the desired effects. Such is the nature of *leaven*, and such is the compound to which barm is indebted for its value as a panary ferment.

1104. After the dough has sufficiently fermented, and is properly raised, it is put into the oven previously heated, and allowed to remain till it be baked. The mean heat of an oven, as ascertained by Mr. Tillet, is 448°. When the bread is removed from the oven, it will be found to have lost about one-fifth of its weight, owing to the evaporation of water, but this proportion will be varied by the occurrence of numerous circumstances which it is not easy to appreciate. Newly baked bread has a peculiar odour as well as taste, which are lost by keeping; this shews that some peculiar substance must have been formed during the operation, the nature of which is not understood. Bread differs very completely from the flour of which it is made, for none of the ingredients of the latter can now be discovered in it; it is much more miscible with water than dough ; and on this circumstance its good qualities most probably, in a great measure, depend; it is not easy to explain the chemical changes which have taken place; it appears certain that a quantity of water, or its elements, is consolidated and combined with the flour; the gluten too would seem to form a union with the starch and water, and to give rise to a compound upon which the nutritive qualities of bread depend.

1105. Much has been said and written upon the subject of the adulteration of bread; but I am inclined to believe that the evils arising from such a practice have been greatly exaggerated. It is certain, that the inferior kinds of flour will not make bread of sufficient whiteness to please the eye of the fastidious citizen, without the addition of a small proportion of alum. It has been said that the smallest quantity of alum that can be employed for this purpose is from three to four ounces to 240 pounds of flour. It cannot be denied that the habitual and daily introduction of a portion of alum into the human stomach, however small, must be prejudicial to the exercise of its functions, and particularly to dyspeptic persons of a costive habit, and to children. Dr. Ure has given us the following test for the detection of alum. Crumble down the suspected bread when somewhat stale into distilled water, squeeze the pasty mass through a piece of cloth, and then pass the liquid through a paper filtre; by which means a limpid infusion will be obtained. If new bread, or hot water be employed in this experiment it is difficult to obtain the liquid clear. A dilute solution of muriate of baryta dropped into the filtered infusion will indicate by a white cloud, more or less heavy, the presence and quantity of alum. Genuine bread will not give any precipitate by this treatment. The earthy adulterations which have been sometimes introduced into bread must be regarded as a much more serious evil; some years ago the flour in Cornwall was very generally adulterated with the white felspar which is used in the porcelain manufactory, and much mischief arose from its use. It is therefore very necessary for a medical practitioner to be prepared with such knowledge as may detect the fraud. For this purpose the suspected bread should be incinerated at a red heat in a shallow earthen vessel, and the residuary ashes treated with nitrate of ammonia; the earths themselves will then remain characterized by their whiteness and insolubility.

1106. It was found by Mr. E. Davy of Cork that bad flour may be made into tolerable bread by adding to each pound from 20 to 40 grains of the common carbonate of magnesia. The operation of this substance in rendering the bread lighter and better has not been satisfactorily explained; but from my own experience of its effects, I apprehend that it neutralizes an acid which is produced during the fermentation of inferior flour, and becoming itself decomposed by the same action, gives out carbonic acid, and thus contributes to the sponginess of the loaf.

PUTREFACTIVE FERMENTATION.

1107. The spontaneous decomposition of organic matter, when attended with a fœtid smell is termed *Putrefaction.* The process proceeds with most rapidity in the open air; but the contact of air is not absolutely necessary; water, however, is in all cases essential, and it probably undergoes decomposition. Its abstraction, therefore, by drying, or fixation by cold, will counteract the process of putrefaction. It is constantly attended with a fœtid odour, in consequence of the emission of certain gaseous matters, which differ in their nature according to the putrefying substance. Some vegetable substances as gluten, and cruciform plants, emit ammonia; but carbonic acid gas, and carburetted hydrogen gas, impregnated with various vegetable principles are more generally emitted in abundance. The pernicious effluvia, says Sir H. Davy, disengaged in this process seem to point out the necessity of burying putrefying substances in the soil, where they are fitted to become the food of vegetables. The fermentation and putrefaction of organised substances in the free atmosphere are noxious processes; beneath the surface of the ground they are salutary operations. In this case the food of plants is prepared where it can be used; and that which would offend the senses and injure the health if exposed, is converted by gradual processes into forms of beauty and of usefulness; the fœtid gas is rendered a constituent of the aroma of the flower, and what might be poison, becomes nourishment to man and animals.

1108. The student having been made acquainted with the nature and chemical habitudes of the several proximate principles of vegetable substances, will easily understand the processes to be employed for the analysis of any body containing mixtures of these principles. In the present state of our knowledge, it is not possible to give any general rule, that may be applicable to every particular case; but if he has gained by study and experiment a competent knowledge of the proximate principles, he will not find it difficult to vary and modify his processes according to the circumstances of each investigation. He should never omit to record, and distinguish by a letter, every step of the process; by which he will not only prevent confusion, but will be thus enabled, in stating the general results, to indicate the processes by which they were individually obtained, by prefixing to each principle a letter corresponding with that of the process which separated it, as will be hereafter illustrated.

1109. In the examination of a vegetable substance, a given quantity, say 200 grains, is usually first submitted to the action of cold water, in order that every principle soluble in that menstruum may be dissolved, the residue is then to be dried, and weighed; this insoluble part may then be treated successively with hot water, alcohol, æther, acids, (chiefly the acetic or diluted nitric) and a solution of potass or soda, and the substances separated by each of these solvents must be submitted to careful examination. Their respective natures will be ascertained by comparing the results on the application of appropriate tests, with the characters of the various substances detailed in the preceding pages.* Under the article " *Extractum Elateri*" in the 6th edition of my Pharmacologia, the student will find a specimen of this method of analysis. I shall here subjoin, as being one of the most instructive examples with which I am acquainted, the examination of a substance termed *Shell-lac,* by Mr. Hatchett.

* Children's Translation of Thenard's Essay on Chemical Analysis.

A. 500 grains of shell-lac were first treated with successive portions of distilled water, till it ceased to be coloured; the whole of the aqueous solution was then evaporated, and left a deep-red substance, which possessed the general properties of vegetable extractive, and weighed 2·50 grains.

B. The 497·50 grains which remained were then digested with different portions of cold alcohol, until this ceased to produce any effect; the resin which was then separated amounted to 403·50 grains.

C. As the shell-lac had not been reduced to powder, but only into small fragments, these were become white and elastic, and when dry were brittle, and of a pale brown colour; the whole then weighed 94 grains.

D. These 94 grains were digested in diluted muriatic acid; and the acid, after being saturated with a solution of carbonate of potass, afforded a flocculent precipitate (resembling that obtained from solutions of vegetable gluten) which, when dry, weighed 5 grains.

E. Alcohol acted but feebly on the residuum; it was therefore put into a matrass, with three ounces of acetic acid, and was suffered to digest without heat during six days, the vessel being at times gently shaken; the acid thus assumed a pale brown colour, and was very turbid. The whole was then added to half a pint of alcohol, and was digested in a sand bath; by which a brownish tincture was formed, and at the same time a quantity of a whitish flocculent substance was deposited, which being collected, well washed with alcohol on a filter, and dried, weighed 20 grains. This substance was white, light, and flaky, and when rubbed by the nail, it became glossy like wax; it also easily melted, was absorbed by heated paper, and when placed on a coal or hot iron, emitted a smoke, the vapour of which very much resembled that of wax, or rather spermaceti.

F. The solution formed by acetic acid and alcohol being filtered, was poured into distilled water, which immediately became milky, and being heated, the greater part of the resin which had been dissolved assumed a curdy form, and was partly separated by a filtre, and partly by distilling off the liquor; this portion of resin amounted to 51 grains.

G. The filtered liquor, from which this resin had been separated, was saturated with a solution of carbonate of potass, and being heated, a second precipitate of gluten was obtained, which, when well dried, weighed 9 grains.

The 500 grains of shell-lac therefore yielded, by the foregoing analysis, the following proximate principles.

A.	Extractive	2·50
B. F. }	Resin	454·50
D. G. }	Vegetable Gluten	14
E.	Wax	20
		491 ..

1110. It will be seen that neither gum, nor tannin was present: had either of these principles been present it would have been found in solution A, and might have been recognised by its appropriate test. The shell-lac being in small fragments, and not in the state of powder, considerably facilitated the decantation of the solution in alcohol from the residuum; and although in this last, a portion of the resin was protected from the action of the alcohol by being enveloped in the gluten and wax, yet, by the assistance of acetic acid, the remainder of the resin, as well as the whole of the gluten, were dissolved, the wax was obtained in a pure state, and a separation of the resin from the gluten was afterwards easily effected, by the method already described. As therefore Acetic acid is capable of dissolving resin, gluten, and many other of the vegetable principles, it may be regarded as a very useful solvent, in the analysis of vegetable substances.

The student might now proceed to the consideration of Animal Chemistry, but as the present volume has already exceeded the limits assigned to it, and as the extent of this branch of science would require several hundred pages for its investigation, I have determined to defer it until a more favourable opportunity will enable me to treat of the characters and composition of animal products, and to shew the changes which the fluids of the body undergo during the influence of disease.

FINIS.

APPENDIX.

TABLE OF WEIGHTS AND MEASURES.

MEASURES OF LENGTH.

	Inches. Decimals.
1 Line, the 12th part of an Inch.	
3 Barley Corns....................................	1·000
A Palm, or Hand's breadth (Scripture Measure)	3·648
A Hand (Horse Measure)	4·000
A Span (Scripture Measure)......................	10·944
A Foot..............	12·000
A Cubit (Scripture for common purposes)	18·000
A Cubit (ditto for sacred purposes)................	21·888
A Flemish Ell....................................	27·000

A Yard.......................................	3 ft. 0
An English Ell..................................	3 : 9
A Fathom or Toise	6 : 0
A Pole..	16 : 6
A Chain	22 yards.
A Furlong..........................	220
A Mile................................	1760

NEW FRENCH MEASURES.

Eng. Inch. & Decim.

		Eng. Inch. & Decim.		Mil.	Fur.	Yds.	Ft.	Inch.
Millimetre.....	=	0·039						
Centimetre....	=	0·393						
Decimetre.....	=	3·937						
METRE	=	39·371	=	0 :	0 :	1 :	0 :	3·37
Decametre	=	313·710	=	0 :	0 :	10 :	2 :	9·7
Hecatometre...	=	3937·100	=	0 :	0 :	109 :	1 :	1
Kilometre	=	39371·000	=	0 :	4 :	213 :	1 :	10·2
Myriometre ...	=	393710·000	=	6 :	1 :	156 :	0 :	6

MEASURES OF CAPACITY.

Eng. Cubic Inch. & Decim.

		Eng. Cubic Inch. & Decim.		Tons.	Hogs.	WineGall.	Pints.
Millilitre......	=	0·061					
Centilitre	=	0·610					
Decilitre	=	6·102					
LITRE	=	61·028	=	0 :	0 :	0 :	2·113
Decalitre	=	610·280	=	0 :	0. :	2 :	5·135
Hecatolitre....	=	6102·800	=	0 :	0 :	26·42	
Kilolitre	=	61028·000	=	1 :	0 :	12·19	
Myriolitre.....	=	610280·000	=	10 :	1 :	58·9	

The English Ale Gallon contains 282 cubic inches.
The Wine Gallon..............231 ditto.
63 gallons Wine Measure, 54 gallons Beer Measure, and 48 gallons Ale Measure, each make a hogshead; 49 Ale pints contain 1727¼ cubic inches, and may therefore be considered (in round numbers) as a cube foot, which contains 1728 cubic inches; a cubic foot of pure water weighs 1000 ounces.

WEIGHTS.

Eng. Grs. & Decim.

		Eng. Grs. & Decim.		Avoirdupoise.		
				lbs.	oz.	dram.
Milligramme ..	=	0·015				
Centigramme ..	=	0·154				
Decigramme ...	=	1·544		Avoirdupoise.		
GRAMME......	=	15·444		lbs.	oz.	dram.
Decagramme...	=	154·440	=	0 :	0 :	5·65
Hecatogramme.	=	1544·102	=	0 :	3 :	8·5
Kilogramme....	=	15444·023	=	2 :	3 :	5
Myriogramme..	=	154440·234	=	22 :	1 :	2

The pound and the ounce in Troy weight and Apothecaries' weight are the same only differently divided and subdivided, and the proportions of a pound Troy to a pound Avoirdupois is as 14 to 17, the former pound containing 5760 grains and the latter 7000. The Troy and Avoirdupoise ounce are therefore not alike, but 14 ounces 11 pennyweights and 16 grains Troy are equal to a pound Avoirdupoise. The drachm of Apothecaries weight must not be confounded with the drachm of Avoirdupoise, the latter being much the smallest.*

---

* For farther particulars respecting Weights and Measures, see Dr. Kelly's Metrology, 1 vol. 8vo. London.

Appendix.

TABLE

*for reducing the Degrees of Béaum's Hydrometer to
the Common Standard.*

═══

Béaums Hydrometer for Liquids lighter than Water.

Temperature 55° Fahrenheit, or 10° Reaumur.

Deg.	Sp. Gr,	Deg.	Sp. Gr.	Deg.	Sp. Gr.
10	1·000	21	·922	32	·856
11	·990	22	·915	33	.852
12	·985	23	·909	34	·847
13	·977	24	·903	35	·842
14	·970	25	·897	36	·837
15	·963	26 ...	·892	37	·832
16	·955	27	·886	38	·827
17	·949	28	·880	39	·822
18	·942	29	·874	40	·817
19	·935	30	·867		
20	·928	31	·861		

Béaum's Hydrometer for Liquids heavier than Water.

Temperature 55° Fahrenheit, or 10° Reaumur.

Deg.	Sp. Gr.	Deg.	Sp. Gr.	Deg.	Sp. Gr.
0	1·000	27	1·230	54	1·594
3	1·020	30	1·261	57	1.659
6	1·040	33	1·295	60	1·717
9	1·064	36	1·333	63	1·779
12	1·089	39	1.373	66	1·848
15	1·114	42	1·414	69	1·920
18	1·140	45	1·455	72	2.000
21	1·170	48	1·500		
24	1·200	51	1·547		

Appendix.

TABLE OF THE SPECIFIC GRAVITIES OF SOLID AND LIQUID SUBSTANCES.

—◆◆—

Distilled Water at 60 = 1·000.

METALS.	Sp. Gr.	SATURATED SOLUTIONS OF SALTS at 42° *Fah.*	Sp. Gr.
Platinum.............. ...	21·5		
Gold	19·364	Of Lime	1·001
Mercury at 40°	5·612	— Arsenious acid	1·005
Ditto 'at 47°	13·545	— Corrosive Sublimate..	1·037
Lead	11·352	— Alum	1·033
Silver	10·510	— Sulphate of Soda	1·052
—— Sulphuret of	7·2	— Ditto of Potass	1·054
Bismuth	9·822	— Muriate of Soda	1·198
Copper	8·895	— Muriate of Ammonia .	1·072
Arsenic	8·310	— Carbonate of ditto ...	1·077
Iron	7·788	— Nitrate of Potass	1·095
— Sulphuret	4·518	— Tartrate of Potass and	
Tin..................	7·299	Soda.............	1·114
Zinc.............	6·861	—Subcarbonate of Potass	1·534
Antimony	6·712	— Sulphate of Copper ..	1·15
Sodium	·935	— Ditto of Iron	1·157
Potassium	·85	— Ditto of Magnesia....	1·218
		— Ditto of Zinc	1·306
INFLAMMABLES.			
Phosphorus	1·714	**EXTRACTS, GUMS, RESINS.**	
Diamond	3·521		
Charcoal.............	2·	Gum Arabic	1·515
Sulphur, melted	1·99	Aloes Spicat:	1·3796
		Ditto Hepatic ..	1·3586
ACIDS.		Ammoniacum..........	1·2071
		Assafœtida	1·3275
Sulphuric acid (of Com-		Benzoin	1·0924
merce)	1·855	Camphor..............	·9887
Ditto (real)	2·125	Catechu	1·4573
Nitric acid	1·5	Elemi................	1·0682
Muriatic acid	1·194	Euphorbium	1·1244
Acetic acid (Glacial) ...	1·063	Galbanum	1·2120
Distilled Vinegar	1·005	Gamboge..............	1·2216
Phosphoric acid.......	1·5575	Guaiacum	1·2289
Arsenious acid	3·7	Honey	1·45
		Labdanum	1·1862
ALKALIES.		Mastic	1·0742
		Myrrh	1·36
Potass (Hydrate)	1·708	Olibanum	1·1732
Soda .. (ditto)	1·336	Opium	1·3359
Ammonia (Gas)	·590	Opoponax	1·6226
Ditto (Liquefied)	·760	Resin of Jalap	1·2185
Lime.................	2·39	Rosin	1·0772
Baryta................	4·	Sagapenum............	1·2008
Magnesia	·346		

Extracts, Gums, Resins.	Sp. Gr.	SPIRITS.	Sp. Gr.
		Alcohol	·796
Scammony (Aleppo)....	1·2354	Proof Spirit	·930
Ditto (Smyrna)	1·2743	Rectified ditto	·835
Sugar (white)	1·6060	Brandy	·8371
Tragacanth...........	1·8161		
Turpentine	·991	ÆTHERS.	
Wax, Bees	·9648	Sulphuric	·700
——, White	·9686	Nitric	·9088
		Muriatic	·874
VOLATILE OILS.		Acetic	·866
Of Turpentine	·792		
— Juniper	·911	WOODS; BARKS.	
— Rosemary	·934	Cinchona.............	·7840
— Carraway	·940	Logwood.............	·9130
— Nutmegs	·948	Sassafras	·4820
— Mint..............	·975	Guaiacum	1·333
— Cummin	·975	Cork................	·24
— Pennyroyal	·978		
— Dill	·994	GASES.	
— Fennel............	·997	Hydrogen	·0688
— Cloves	1·034	Phosphurᵗ: ditto	·9022
— Cinnamon	1·035	Carburᵗ: ditto	·666
— Sassafras	1·094	Bi-carb: ditto	·970
		Ammonia.............	·590
		Steam	·622
FIXED OILS.		Carbonic oxide	·967
Of Olives	·9153	Nitrogen	·9691
— Almonds	·9170	Atmospheric air.......	1·000
— Poppies	·9238	Nitric oxide	1·05
— Linseed	·9403	Oxygen	1·1088
— Ben..............	·9119	Sulph: Hydrogen......	1·18
— Rapeseed	·9113	Muriatic acid	1·284
Tallow	·9419	Carbonic acid........	1·5236
Spermaceti	·9433	Sulphurous acid.......	2·23
		Chlorine	2·508

A TABLE OF THE REFRACTIVE POWERS OF DIFFERENT BODIES,

(Referred to at page 225.)

	Index of Refraction.		Index of Refraction.
A Vacuum	1·00000	Nitre	1·524
Air (mean)	1·00033	Canada Balsam	1·528
Water	1·336	Copal	1·535
Æther...............	1·358	Oil of Cloves	1·535
Alcohol	1·370	Ditto, adulterated	1·498
Oil of Lavender	1·467	Oil of Sassafras	1·536
— of Peppermint	1·468	Bees Wax	1·542
— of Olives	1·469	Phosphorus (Wollaston)..	1·579
— of Almonds	1·470	Ditto (Brewster)	2·224
— of Turpentine, (common)	1·476	Guaiacum	1·596
		Balsam of Tolu	1·600
Ditto Rectified	1·470	Arsenic	1·811
Camphor	1·487	Glass of Antimony	1·980
Oil of Nutmeg	1·497	Diamond	2·440
Balsam of Copaiba	1·507		

A TABLE
OF

ATOMIC WEIGHTS, AND CHEMICAL EQUIVALENTS.

TABLE I.

ACID, acetic	50	Antimony	44
crystallized (1 water)	59	chloride	80
arsenic	62	deutoxide	56
arsenious	54	peroxide	60
benzoic	120	protoxide	52
carbonic	22	sulphuret	60
chloric	76	potassa-tartrate	?
citric	58	Arseniate of ammonia	79
crystallized (2 water)	76	potash	110
ferrocyanic	?	soda	94
fluoric	17	Arsenic	38
gallic ?	63	acid	62
hydriodic	126	Arsenious acid	54
hydrocyanic	27	Azote	14
malic	70	Barium	70
muriatic	37	peroxide	86
nitric (dry)	54	sulphuret	86
liquid (sp. gr. 1·50 2 water)	72	Barytes	78
nitrous	46	carbonate	100
oxalic	36	chlorate	154
crystallized (4 water)	72	hydrate	87
perchloric	92	nitrate	132
phosphoric	28	muriate (crystal. 1 water)	124
phosphorous	20	oxalate	114
sulphuric (dry)	40	phosphate	106
liquid (sp. gr. 1·4838)	49	phosphite	98
sulphurous	32	sulphate	118
tartaric	67	tartrate	145
crystallized (1)	76	Benzoic acid	120
Alum (dry)	262	Bicarburetted hydrogen	7
crystallized (25 water)	487	Bismuth	72
Alumina	27	nitrate	134
sulphate	67	oxide	80
Aluminum	19	sulphate	120
Ammonia	17	sulphuret	88
acetate	67	Boracic acid	22?
bicarbonate (2 water)	79	acid crystallized (2 water)	40
carbonate	39	Borax (8 water)	158
sesquicarbonate (2 water)	118	Boron	6 ?
citrate	75	Calcium	20
fluoborate	39	chloride	56
hydriodate	143	oxide (lime)	28
iodate	182	phosphuret	32
muriate	54	sulphuret	36
nitrate	71	Calomel	236
oxalate	53	Camphoric acid	?
crystallized (1 water)	62	Carbon	6
phosphate	45	oxide	14
phosphite	37	phosphuret	18
sulphate	57	sulphuret	38
sulphite	49	Carbonic acid	22
tartrate	84	oxide	14

Appendix.

Appendix.

TABLE II.

Hydrogen	1	Zinc	34
Carbon	6	Chlorine	
Bicarburetted hydrogen	7	Protoxide of iron	
Oxygen		——————— manganese	
Silicium	8	——————— chromium	
Carburetted hydrogen		Peroxide of sodium	36
Water	9	Phosphuret of sodium	
Magnesium		Sulphuret of calcium	
Phosphorus	12	Oxalic acid (dry)	
Phosphuretted hydrogen	13	4 Water	
Azote		Muriatic acid	
Carbonic oxide	14	Phosphite of ammonia	37
Bihydroguret of phosphorus		Hydrate of lime	
Sulphur		Sulphuret of carbon	38
2 Oxygen	16	Arsenic	
Silica		Boracic acid (crystallized)	
Ammonia	17	Sulphuric acid	
Sulphuretted hydrogen		Potassium	
Alumium	18	Sulphuret of sodium	40
2 Water		Deutoxide of manganese	
Aluminum	19	Peroxide of iron	
Phosphorous acid		Phosphuret of iron	
Magnesia	20	5 Oxygen	
Calcium		Oxide of zinc	42
Carbonic acid	22	Carbonate of magnesia	
Nitrous oxide		Protoxide of chlorine	
Sodium		Peroxide of manganese	44
Phosphuret of magnesium	24	Protosulphuret of iron	
3 Oxygen		Antimony	
Alumina		Phosphate of ammonia	45
Cyanogen	26	5 Water	
Cobalt		Nitrous acid	46
Hydrocyanic acid		Phosphuret of zinc	
3 Water	27	Potash (dry)	
Sulphuret of magnesium		Phosphate of magnesia	48
Alumina		Chloride of magnesium	
Lime		6 Oxygen	
Phosphoric acid		Liquid sulphuric acid (1 water)	49
Phosphuret of sulphur	28	Carbonate of lime	
Iron		Acetic acid	50
Manganese		Sulphuret of zinc	
Hydrate of magnesia	29	Protoxide of antimony	52
Nitric oxide	30	Oxalate of ammonia	53
Sulphurous acid		Dry nitric acid	
Soda		Muriate of ammonia	
Phosphuret of calcium	32	Carbonate of soda	54
4 Oxygen		Arsenious acid	
		6 Water	

Sulphuret of potassium....... ⎫
Deutoxide of antimony........ ⎪
Chloride of calcium........... ⎬ 56
Phosphate of lime............. ⎪
7 Oxygen .... ⎭

Sulphate of ammonia.......... ⎫
Hydrate of potash .... ⎪
Muriate of magnesia .... ⎬ 57
Tin .... ⎪
Carbonate of manganese....... ⎭

Citric acid (dry).............. 58
Acetic acid crystallized 59

Chloride of sodium .... ⎫
Persulphuret of iron........... ⎪
Phosphate of soda .... ⎪
Sulphate of lime.............. ⎬ 60
Sulphate of magnesia (dry).... ⎪
Sulphuret of antimony........ ⎪
Peroxide of antimony .... ⎭

Arsenic acid .... ⎫
Chlorocyanic acid .... ⎬ 62
Oxalate of ammonia (1 water).. ⎭

Gallic acid................... ⎫
7 Water..................... ⎬ 63

Peroxide of potassium .... ⎫
Sulphite of soda .... ⎪
Protochloride of iron .... ⎪
Copper...................... ⎪
Chloride of lime............. ⎬ 64
8 Oxygen .... ⎪
Carbonate of zinc .... ⎪
Oxalate of lime .... ⎭

Protoxide of tin.............. ⎬ 66
Sulphate of alumina.......... ⎭

Sulphate of alumina .... ⎫
Tartaric acid (dry) .... ⎬ 67
Acetate of ammonia .... ⎪
Ferrocyanic acid .... ⎭

Sulphate of lime .... ⎬ 68
Oxalate of soda.............. ⎭

Carbonate of potash .... ⎫
Barium .... ⎬ 70
Acetate of magnesia .... ⎪
Malic acid .... ⎭

Nitrate of ammonia........... ⎬ 71
Tannin ? .... ⎭

Liquid nitric acid (2 water.... ⎫
Crystallized oxalic acid (4 water) ⎪
8 Water .... • ⎪
Sulphate of soda (dry)...... • ⎬ 72
Protoxide of copper.......... ⎪
Bismuth .... ⎪
9 Oxygen .... ⎭

Nitrate of magnesia .... ⎫
Sulphuret of tin............. ⎬ 74
Peroxide of tin .... ⎪
Sulphite of zinc............. ⎭

Citrate of ammonia............ 75

Chloric acid .... ⎫
Crystallized citric acid (2 water) ⎪
Crystallized tartaric acid (1 d°) ⎪
Chloride of potassium........ ⎬ 76
Phosphate of potash........... ⎪
Bi-carbonate of soda (dry) .... ⎪
Protosulphate of iron (dry).... ⎭

Baryta...................... ⎫
Acetate of lime .... ⎬ 78
Succinate of lime............. ⎭

Arseniate of ammonia........ ⎫
Bi-carbonate of ammonia ⎬ 79
  (2 water) .... ⎪
Oxalate of zinc .... ⎭

Sulphite of potash............ ⎫
Oxide of bismuth .... ⎪
10 Oxygen .... ⎬ 80
Peroxide of copper............ ⎪
Protosulphuret of copper...... ⎪
Phosphate of strontia .... ⎭

Chloride of antimony .... ⎬ 81
Sugar ? .... ⎭

Nitrate of lime.............. ⎫
Perchloride of iron ;....:.... ⎬ 82
Sulphate of zinc (dry) .... ⎪
Acetate of soda .... ⎭

Bi-phosphate of lime......... ⎫
Oxalate of potash .... ⎬ 84
Tartrate of ammonia.......... ⎭

Arsenite of soda............. ⎫
Sulphate of lime cryst. (2 water) ⎪
Peroxide of barium.......... ⎪
Nitrate of soda.............. ⎬ 86
Citrate of lime .... ⎪
Acetate of iron.............. ⎭

Hydrated baryta .... ⎬ 87
Tartrate of magnesia .... ⎭

Sulphate of potash .... ⎬ 88
11 Oxygen .... ⎭

Arseniate of lime,........... ⎫
10 Water .... ⎬ 90
Citrate of soda .... ⎭

Perchloric acid .... ⎫
Bi-carbonate of potash (dry) .. ⎬ 92
Acetate of zinc .... ⎭

Chlorate of ammonia .... ⎬ 93
Ammonia-phosphate of magnesia ⎭

Arseniate of soda.............. 94

Appendix.

Appendix.

Printed in the United States
By Bookmasters